Springer Finance

Springer Finance

Springer Finance is a programme of books addressing students, academics and practitioners working on increasingly technical approaches to the analysis of financial markets. It aims to cover a variety of topics, not only mathematical finance but foreign exchanges, term structure, risk management, portfolio theory, equity derivatives, and financial economics.

For further volumes:
http://www.springer.com/series/3674

Gilles Zumbach

Discrete Time Series, Processes, and Applications in Finance

Springer

Gilles Zumbach
Consulting in Financial Research
Saconnex d'Arve, Switzerland

ISBN 978-3-642-31741-5 ISBN 978-3-642-31742-2 (eBook)
DOI 10.1007/978-3-642-31742-2
Springer Heidelberg New York Dordrecht London

Library of Congress Control Number: 2012948091

Mathematics Subject Classification: 91B84, 91B70, 91G70, 62P20, 91G20, 91B30

JEL Classification: C22, G10, G17, G13, C23

Printed on acid-free paper

Springer is part of Springer Science+Business Media (www.springer.com)

To the men and women who have showed me various paths in life

– Yvette, without whom much would not have existed
– André, who showed me the exigence and difficulty of creating, and for sharing the pleasure and diversity of music
– Hervé, who taught me the rigor and beauty of mathematics and physics, and for his curiosity in science and literature
– Jean-Jacques, who introduced me to soaring above the Alps and for sharing unique moments gliding between glaciers and clouds. And for teaching me loud the difference between big bands west coast and east coast…
– Anne, for the many projects undertaken together

Contents

List of Figures

List of Tables

Chapter 1
Introduction

The main focus of this book is the construction of mathematical processes describing financial time series. An accurate description of the dynamics of the prices is central to many financial applications like portfolios optimizations, risk evaluations, or the valuation of contingent claims. For all these applications, forecasts are needed, and the key role of forecasts is a crucial point to acknowledge. For example, the optimization of a portfolio requires forecasts over the chosen investment period of the returns and covariance matrix in the considered universe. Similarly, the valuation of market risks and of derivatives is related to the possible future behaviors of the underlying(s), given the information available up to the present time. Random processes describing the time evolution of financial time series is a natural framework for producing the desired forecasts or possible time evolutions.

Louis Bachelier initiated this field in his celebrated Ph.D. thesis [12]. His original model uses a Gaussian random walk in order to describe the time evolution of the prices. This simple process puts in the central position the randomness, in sharp contrast with the deterministic description used at that time (see, e.g., [18] for a historical account of the evolution of ideas in finance). This paradigm shift took roughly 50 years to be acknowledged by the financial community and is now the central element used to build more complex processes. With the broad availability of financial data and of computers, with the increasing accountability and performance in the financial industry, more accurate processes are needed today.

The construction of processes in finance is similar to the description of the laws of nature by mathematical formulas as pursued in physics. Yet, the subject matter is fairly different, as finance is a man-made construction, whereas physics is "given" to us. As a consequence, some prejudices are very different in each field. In physics, there is a strong belief in esthetics: the universe must obey some simple laws, with rich symmetries. This guiding line has been used extensively in physics in order to develop and select theories with a large explanatory power. Behind the particular human constructions, there is a strong sense that the physical universe must be comprehensible and simple (to the extent that quantum mechanics and relativity are simple!). As finance is a human construction, such esthetics principle cannot be used as widely. Probably, the only strong guiding principles that can be followed

G. Zumbach, *Discrete Time Series, Processes, and Applications in Finance*,
Springer Finance, DOI 10.1007/978-3-642-31742-2_1,

broadly in finance are the following. First, the majority of the "homos economicus" behaves rationally (although what means "rationally" is debatable). Second, the human attitude toward profit and risk (or to take the seven sin words, toward greed and fear) consists in emphasizing more the potential profits and to obliterate the associated risks. These are weaker guiding lines, and it is not clear how to turn them into equations, but they can be considered the only guidelines that are broadly used in finance. On the other hand, finance draws its richness precisely from the diversity of the human behaviors, from the implications of the regulations, and from the major innovations in the industry (like life insurances or derivative contracts).

In finance, the broad principles that can be used are the following:

- Causality (what happens now is depending on the past but is not depending on future events)
- The limited rationality of the market participants
- Uncertainty (i.e., risk) must be rewarded (i.e., by an expected profit)
- The microscopic conservation of value in one transaction
- The time value of money (1$ available now is worth more than 1$ later)
- No arbitrage: two portfolios with the same stream of future cash flows must have the same value

For the mathematical description of financial time series, these broad principles give essentially no constraints, except for the causality. The consequence of the lack of constraints is that many different candidate models and equations can be written. In order to select among the competing processes, detailed statistical analyses have to be used both for the empirical time series and for the processes. This gives a much more delicate standpoint for many models used in finance as they cannot be validated or rejected by a formal mathematical construction.

Another important difference between physics and finance is the repeatability of the experiments. In physics, an experiment should be described with enough accuracy so that someone else can follow its description and obtain compatible results. Repeatability is crucial in order to reach a consensus about empirical facts. Even in statistical mechanics, which is inherently random, the large value of the Avogadro number (10^{23}) implies that quantities like pressure or energy are given by a single number despite their statistical origins. Finance has a much weaker status with respect to repeatability: the number of independent time series is not very large (of the order of 10^2 to 10^6 depending on the considered universe), they are of limited time span (a few decades), and they cannot be created or modified at the will of the observer. Moreover, the stationarity of finance at the scale of decades is debatable, whereas all empirical evidences point to stationary laws of physics at the cosmological time scale. With that respect, finance is somewhat similar to cosmology which considers a single universe that cannot be altered by the observer. Despite the more limited repeatability that can be achieved in finance, it is nevertheless important to extract "stylized facts" that characterize the majority of the time series. This is needed to move from a purely descriptive position to a synthetical description and to summarize efficiently the observed behaviors. This situation is somewhat similar to botanic or zoology, which moved from the descriptive classification of species to DNA sequencing and biochemistry. But clearly, the rigorous scientist is on softer grounds in finance.

Many things have been said and written about the respective contribution of the academics versus practitioners, and from people with a background in finance versus those with a background in physics (and the so-called "econophysics"). Beyond any judgment about the relevance and quality of the respective contributions, the approaches used by economists or physicists are quite different. In finance, the authors tend to postulate early on a model, which is then tested rigorously on empirical data using advanced tools from statistics and hypothesis testing. On the other hand, physicists tend to collect empirical facts with the least prejudices and then write a model that can describe the observed properties, using the mathematical structure as a guide line while being often sloppy on the statistical side. Both approaches are rooted in the history of their respective fields, and both have merits and drawbacks.

Roughly, this book follows the physicist's approach. Several chapters are devoted to collecting empirical statistics, trying to extract the leading empirical facts while not using any models or processes. An important point in the empirical analysis is the multiscale statistics and their capacity to extract information over many time scales. Another important aspect of the empirical investigation consists in the ranking of observed properties, in order to select the dominant features. Following the empirical analysis, processes are introduced, with various levels of complexity and accuracy. The goal is to reproduce qualitatively in mathematical models as many empirical features as possible. The focus point is on the mathematical structure, as the parameters in a model can be modified to accommodate quantitatively one or another particular time series. Yet, we suffer from the embarassement of the riches, namely that too many processes can be written at a given level of accuracy. Therefore, parts of this book are also devoted to the selection of the best processes, by eliminating the constructions that cannot capture some observed properties. At the end of the selection, a few processes are left, presenting different compromises between accuracy and complexity.

Several leitmotives appear in this book, in the empirical analyses and in the mathematical processes:

- An analysis at multiple time scales, ranging from minutes to months
- The important dynamics of the volatility, the so called "heteroscedasticity"
- The universal occurrence of fat-tail distributions and the slow convergence toward a normal distribution
- The flexibility of the processes with a finite time increment, to be contrasted with the difficulties with the continuous time limit
- The complex multivariate structures
- The success of ARCH processes in describing empirical data

The description of financial time series by various models followed an explosive grow in the last two to three decades. This book aims to synthesize the present status of the field, but it also represents a subjective snapshot of the current situation, as viewed by the author. Necessarily, many topics and contributions are left out as it is beyond the scope of this book to survey extensively the literature. For this reason, the author would like to apologize to all the authors and papers not mentioned. Despite many important progresses, many questions are still left open, and further improvements can be expected in several areas. These open directions which deserve further investigations are pointed out.

The focus of this book is on "mesoscopic" time scales, ranging from one hour to one year. At shorter time ranges, the microscopic mechanism of the transactions takes place, with a new phenomenology due to the bid and ask prices, the finite tick size, the order queues, and the various possible market orders. Above a time interval ranging from ten minutes to a few hours, most of these effects can be neglected, and only prevailing prices are used. In the other direction, for time intervals above one year, an economic description is appropriate, with a focus on the relations between major economic indicators, regulations, and political decisions. The boundaries between the microscopic, mesoscopic, and macroscopic is a research area where most remains to be done. With a focus on the intermediate time range, a phenomenological description by random processes of the financial time series is appropriate. The range from a few hours to one year is of major concern for many financial and economic decisions, for example, for portfolio allocations, risk evaluations, or option pricing. Because of its practical implications, the "mesoscopic" time range is the main subject matter for this book.

Members of the academic community will find most of this book at a fairly basic level with respect to its mathematical content. Indeed, many existing good books present more advanced and rigorous theoretical results. Yet, the interest of this presentation lies elsewhere. The goal is to offer an original and comprehensive statistical analysis of empirical time series, and to apply the same tools to a wide range of theoretical processes. The statistical analysis focuses on the multiscale properties of the time series. This different approach allows one to extract much more information and ultimately to decide which mathematical structures can or cannot capture the generic empirical properties. As a result, some of the basic assumptions widely used in finance and taken for granted are revisited. Then, many of the frequently used processes are found to be inadequate. Critical comparisons between competing models allow one to narrow down the field and to select a small set of ARCH processes of increasing accuracy and complexity. Essentially, important results that are different from the more traditional financial literature are obtained thanks to the focus on the statistical properties and to the broad coverage of the process structures. Some applications of the selected models are presented, like for risk management and option pricing.

A large set of statistics have been computed for Chaps. 3 to 12 of this book, both for empirical time series and for theoretical processes. This large amount of information is best conveyed in the form of graphs. The emphasis is mostly on qualitative features that can be apprehended easily from graphical shapes, and less on peculiar numerical values (as they can be altered by modifying the process parameters). The most relevant graphs are summarized in the *mug shots* (introduced in Chap. 4), which show a common set of information about empirical data and processes. Much more statistics and figures have being made during this research, which cannot be displayed in a book. The companion web site www.finanscopics.com presents the complete set of graphs, with the possibility to view all the graphs for a given time series or processes, or to extract one selected graph across time series, or to compare side by side the pairs of graphs for two selected time series. The site has been build by the author as a tool to explore conveniently the wealth of information generated by the systematic exploration of the empirical data and of the various processes. I invite the interested reader to pursue his or her exploration on this web site.

Acknowledgements This book was written over many years, incorporating the author experience in many areas and with many colleagues. At *Olsen & Associates* in Zurich, high-frequency data was the main topic, and the author wishes to thank the research group for sharing this exiting adventure: Giuseppe Ballochi, Fulvio Corsi, Michel Dacorogna, Hoss Haukson, Richard Olsen, Ulrich Müller, Olivier Pictet, Rejeb Sadruddin, and Adrien Trapletti. Many people visited the "Mühle Tiefenbrunnen", leading to fruitful collaborations and research papers, and among them I wish to thank particularly Ramo Gençay and Tim Bollerslev. Paul Lynch and Oliver Masutti deserve a special mention for spending beautiful Summer days taking pictures of volatility landscapes instead of pictures of the Swiss Alps.

At the *Risk Metrics Group* in Geneva, I worked on improving the existing risk methodologies in several directions, and I would like to thank my colleagues for sharing their expertise as well as some fun travels and beers: Fabien Couderc, Christopher Finger, Tarik Garidi, Cathy O'Neil, Luis O'Shea, and Daniel Straumann. A special thank go to Romain Cosandey for the project done together on large covariance matrices.

Swissquote Bank near Geneva is another interesting place where finance meets algorithms and the Internet. I would like to thank the members of the quantitative group: Abdi Bennani, Floren Galien, Serge Kassibrakis, Luis Martinez, David Morton de Lachapelle, and Eugenio Rodriguez. Caroline Weber deserves a special mention for bringing her enthusiasm in studying new directions in capturing the complexity of real stocks and option pricing.

In this global world, there are also a number of persons I keep meeting at conferences and workshops. The lively discussions we had been a great source of inspiration for many papers, and I would like to thank particularly Lisa Borland, Robert Engle, and Marc Potters. A special thanks go to Jean-Philippe Bouchaud for its long term friendship and for advocating after a seminar many years ago that I should write up this stuff. Ultimately, this paper grows until it becomes the present book.

I should particularly thank my wife Anne for her love over the years, for sharing the good times and the difficult times, and for her support and help on this book. Finally, my sons Yves and Hervé deserve a special note for bringing their bits of salt, their joy and curiosity, and for keeping me distracted from my computer screen.

Chapter 2
Notation, Naming, and General Definitions

2.1 Time, Time Interval, and Time Scale

A time is noted with a t, a time interval with δt, Δt, ΔT and as τ in process parameters. A clear distinction should be made between a time (or an epoch), like in "today at one o'clock", and a time interval, like in "one hour ago". A time interval is measured essentially by the beats of a pendulum. Some convenient time intervals used as standard unit are $1h$, $1d$, and $1y$, corresponding respectively to one hour, one day, and one year. A time interval is a dimensionfull quantity (with units, like second, hour, day, or year) and is not a pure number. The ratio of two time intervals gives a real number. For example, for a time interval δt of one week, the ratio $\delta t/1h = 168$ is a real number corresponding to the number of hours in one week. Such ratios of time intervals appear in exponential moving average (EMA), or in process definitions, in the form $\delta t/\tau$, with τ a characteristic time interval of the process. In finance, it is convenient to "annualize" the volatilities. This is done with the ratio $\delta t/1y$, where δt denotes a characteristic time of the measured volatility. Although time is familiar to everyone, it is a fairly abstract concept corresponding to the time interval elapsed since a reference date, often taken as Christ's birth. The algebra of times and time intervals are different. With obvious notations, correct expressions are $\delta t_3 = \delta t_1 + \delta t_2$, $t_2 = t_1 + \delta t$, $\delta t = t_1 - t_2$, $x = \delta t_1/\delta t_2$ with $x \in \mathbb{R}$. Meaningless expressions are $t_1 + t_2$, $\exp(-t)$, t_1/t_2, $\sqrt{\delta t}$, $\exp(-1/\tau)$.

The reader should be aware of the confusion existing in most of the literature at daily frequency between times, time intervals, integers, and real numbers. This occurs because daily time series are often viewed as a vector, indexed by an integer. Implicitly, all the time intervals are measured in business day, therefore reducing them to integers. This confusion is reinforced by most statistical packages, as they do not have specific classes to model times and time intervals, which are therefore represented as numbers. We have been careful to avoid such confusion as much as possible.

Physical events can be "indexed" by the beats of any pendulum, because time is the "same" regardless of the domain (Newtonian mechanics, quantum dynamics, heat equation, etc.). But finance is another world, created by the human activity, and

G. Zumbach, *Discrete Time Series, Processes, and Applications in Finance*,
Springer Finance, DOI 10.1007/978-3-642-31742-2_2,

subject to days and nights, working days and week-ends. Because of this fundamental difference, the most appropriate way to measure the flow of the financial events is not necessarily the physical time given by a pendulum, but it is the typical human activity that indexes appropriately the financial events. More generally, time and time intervals are measured on a given *time scale*. Usually, this is the physical time scale, as assumed in the discussion in the previous two paragraph. In finance, other time scales can be more appropriate, like a business time scale, or a transaction time scale [42]. A simple daily business time scale contains 5 days per week (Monday through Friday) and removes the week-end (Saturday and Sunday). For daily data and for a given market, this simple recipe gives an efficient way to obtain a steady flow of financial events. Yet, extending the simple daily business time scale at a world scope raises many issues related to time zones, Holidays and the definition of the week-end. With high-frequency data, the simple daily business time scale needs to be extended in the intra-day regime. We use the words "time scale" for a scale to measure times and time intervals, and the words "time interval" or "time horizon" for time differences.

In order to deseasonalize the empirical high-frequency data, we use the dynamic business time scale developed in [32]. Similarly to the daily business time scale, the dynamic business time scale contracts periods of low activity (night, week-end) and expands periods of high activity. The activity pattern during the week is related to the measured volatility, averaged on a moving sample of 6 months. Holidays and day light saving time are taken into account. The normalization is chosen such that the business time scale flows on average at the same pace that the physical time scale. Therefore, a long time interval on the business time scale corresponds to the same physical time interval, but this is not true for short time interval. Moreover, for short time intervals, the corresponding physical time interval depends on the time series and on the time in the week. Take, for example, the foreign exchange market, which is open essentially 24 hours per day, 5 days per week, and closed on the week-end. The business time scale has to be regularized on the week-end, by giving it a total "mass" of $\delta\theta_{\text{w-e}}$ for Saturday and Sunday ($\delta\theta_{\text{w-e}}$ is measured in day in the computation below). Moreover, let us assume for the sake of the argument that the activity is constant during the working week. With this simplified time scale, a 1 minute physical time interval corresponds to $(7 - \delta\theta_{\text{w-e}})/5 \cdot 1'$ during the market opening hours (i.e., between $1'$ and $1'24''$ depending on $\delta\theta_{\text{w-e}}$) and to $\delta\theta_{\text{w-e}}/2 \cdot 1'$ during the week-end (i.e., between 0 and $1'$). This simple computation shows the expansion during the opening hour, and the contraction during the week-end. In the other direction, a 3 minutes business time interval corresponds to $5/(7 - \delta\theta_{\text{w-e}}) \cdot 3'$ during the market opening hours (i.e., between $2'8''$ and $3'$, depending on $\delta\theta_{\text{w-e}}$) and to $2/\delta\theta_{\text{w-e}} \cdot 3'$ during the week-end. In practice, the transformation of physical to business time interval is depending on the mean activity at this time in the week. In more technical term, the Jacobian of the transformation from physical to business time is given by the mean activity pattern, where the activity is measured by a short-term volatility (see [32] for more details).

The homogeneous time series used for the empirical analysis are computed from high-frequency filtered tick-by-tick data. First, the incoherent effect is filtered from the high-frequency data: a very short exponential moving average is taken on the prices in order to attenuate the tick-by-tick incoherent price formation noise (see Sect. 10.7 or [46]). Then, the high-frequency time series are sampled using a linear interpolation every 3 minutes in dynamic business time scale. The author is grateful to Olsen & Associates, in Zürich, Switzerland, to provide the high-frequency data.

2.2 Time Series

A time series is denoted by a single letter, like r for the price change (or return) time series. The value of a time series at a given time is denoted with parenthesis, like $r(t)$ for the return at a given time t. The parameter(s) of a time series are denoted between square bracket $[\cdot]$, like $r[\delta t]$ for the return time series with the returns computed over a given time interval δt. The return at a given time t and for a given time interval δt is denoted $r[\delta t](t)$.

2.2.1 *Historical, Centered and Realized Quantities*

The price time series gives the value of a given security or process at a given time. It is an instantaneous value. All the other time series are derived from the prices time series and involve some time interval(s), like for the returns or volatilities. These quantities are therefore not instantaneous and measure some properties of the prices inside the given time interval. An important issue is how these time intervals are related to t, namely if they are located before, around or after t.

- A quantity computed using data up to t is called *historical*. For example,

  $$r_{\text{historical}}[dt](t) = p(t) - p(t - dt)$$

  is the (usual) historical return (beware that our definition of the return is different from this example, see the next section).
- A quantity computed using data in the future of t is called *realized*. For example,

  $$r_{\text{realized}}[dt](t) = p(t + dt) - p(t)$$

  is the realized return. This is a standard convention in time series forecasting: a forecast is constructed using historical values, and the forecasts are compared to the realized values.
- Some quantities are computed around t and are called *centered*. They use information in the past and future of t. For example, the centered price change can be defined as

  $$r_0[dt](t) = p(t + dt/2) - p(t - dt/2)$$

and measure the price change occurring around t over a time horizon dt. In the empirical analysis, the centered volatility change is studied in relation with historical quantities, as this gives a dynamic measure of the volatility behaviors.

In formulas and axes, historical, centered, and realized quantities are denoted with the indices h, 0, and r, respectively.

2.2.2 Returns

The historical return measured at a time scale δt_r is given by the logarithmic price difference

$$r_h[\delta t_r](t) = \frac{\ln(p(t)) - \ln(p(t - \delta t_r))}{\sqrt{\delta t_r / \delta T_{\text{ref}}}} \tag{2.1}$$

with δt_r the time interval used to measure the return, and $\ln(p)$ the logarithm of the prices. This is a historical return definition, using information only in the past of t. Simple time translation can be used on the right-hand side to define the centered and realized returns.

The denominator is used to annualize the return using a random walk scaling, and we take by convention $\delta T_{\text{ref}} = 1$ year. In this way, the returns can be compared directly across different time horizons δt_r. For example, $E[r^2[\delta t_r]] \simeq 5\ \%$ to $50\ \%$ regardless of δt_r for most of the empirical time series. Similarly, the probability density $p(r[\delta t_r])$ can be directly compared. When studying scaling properties, this return definition already eliminates the random walk scaling. Hence, for an iid random walk, the annualization leads to the scaling relation $E[|r[\delta t_r]|^k] \simeq$ constant (whereas a simple price difference leads to $E[|\delta \ln(p)[\delta t]|^k] \simeq \delta t^{k/2}$). This systematic annualization of the returns proves to be the convenient definition for the systematic study of the time series properties across time horizons. It works very well because this definition removes the leading diffusion property of a random walk. This allows one to search for differences, therefore revealing the sub-leading properties we are interested in.

Definition (2.1) is widely used and therefore no often criticized. This definition is fine when applied to empirical time series but shows significant convergence issues when used for a process. The logarithmic versus relative return definitions are discussed in detail in Chap. 6. Both return definitions are natural when applied to a price time series, as the returns should be independent of the monetary unit. For other time series like interest rates, spreads, or volatilities, it is much less obvious to justify a particular definition for the returns. For example, in [154], the returns of interest rates are computed as differences of $x = \log(1 + R/R_0)$ with $R_0 = 4\ \%$ and R the rate at a fixed time to maturity. The justification for this transformation is that this mapping essentially decouples the interest rate R from its volatility, allowing one to write convenient process equations.

2.2.3 Volatilities

The volatility is a fairly intuitive concept, but its precise definition is rather subtle. The core idea of the volatility is to measure the magnitude of the price fluctuations during a specified time interval. For the empirical studies, a "hard" definition is used with all the returns having the same weight in a given time window. Processes and high-frequency data lead to other definitions that will be introduced later in this book.

The historical volatility σ_h at time t measures the price fluctuation at the time horizon δt_r, in a window of length δt_σ in the past of t:

$$\sigma_h^2[\delta t_\sigma, \delta t_r](t) = \frac{1}{n} \sum_{t-\delta t_\sigma+\delta t_r \leq t' \leq t} r^2[\delta t_r](t') \tag{2.2}$$

$$n = \sum_{t-\delta t_\sigma+\delta t_r \leq t' \leq t} 1.$$

Because the returns are already annualized, the volatility is also annualized.

Usually, the price time series is sample regularly in some time scale, giving a uniform weight in the chosen time scale. The canonical example is the business time scale where the data are sampled regularly according to the physical time, except on the week-end and holidays. Our continuous time scale works essentially in the same way, but gives a weight (i.e., an integration measure) corresponding to the mean activity. A weight given by the mean tick density is implicit in the summation in (2.2) (in the continuum, the weight corresponds to the integration measure on the time axis). Different time scales will give different measure for the volatility. For example, using the physical time scale retains all the seasonalities present in the empirical data.

The volatility definition involves two parameters δt_r and δt_σ. The most important parameter is δt_σ, which fixes the time interval over which the volatility is estimated. In practice, for a given δt_σ, the volatility has a weak dependency on δt_r. Most of the figures in this book are computed with a fixed ratio $\delta t_\sigma / \delta t_r = 24$, namely by aggregating simultaneously the return and volatility. This choice corresponds roughly to our intuition that short-term intra-day traders use tick-by-tick data, while medium term traders use daily data to take decisions. In order to explore the δt_r dependency, the parameter δt_σ is kept fixed while δt_r is changed in Sect. 3.8.

Definition (2.2) involves a summation over a finite interval δt_σ, with the same weight for all the squared returns. This emphasizes the idea of a volatility measured over a specified time interval. This definition is convenient for the empirical study as it allows one to define similarly the realized and centered volatilities, as well as the volatility increment in the next section. With processes appears the idea of an instantaneous volatility measured over a time interval δt_r corresponding to the time increment of the process, but such volatility definitions are depending on the analytical form for the process as well as on its parameters. A simple example of such definition is the "RiskMetrics" exponential moving average, corresponding to an I-GARCH(1) process (see Chap. 15). Such definitions are explored later in this book, when a detailed enough understanding of the heteroscedasticity allows us to focus on a few relevant formulas.

2.2.4 Volatility Increments

The historical volatility increment is defined by

$$\delta\sigma_h[\delta t_\sigma](t) = \sigma_h[\delta t_\sigma](t) - \sigma_h[\delta t_\sigma](t - \delta t_\sigma). \tag{2.3}$$

This quantity measures the volatility changes occurring at a time scale δt_σ. Indeed, for a stochastic variable evolving on a typical time scale δt_σ, this is the meaningful definition to evaluate the "process increments" and is more appropriate than a continuum derivative. Another definition of the historical volatility increment is to compute relative changes

$$\delta\sigma_{\ln,h}[\delta t_\sigma](t) = \ln\left(\frac{\sigma_h[\delta t_\sigma](t)}{\sigma_h[\delta t_\sigma](t - \delta t_\sigma)}\right). \tag{2.4}$$

The correlations computed with both definitions (2.3) and (2.4) are almost identical. The correlation figures displayed in this book correspond to definition (2.4) for the logarithmic increment.

We will use also the "centered" volatility increment defined by

$$\delta\sigma_0[\delta t_\sigma](t) = \sigma_r[\delta t_\sigma](t) - \sigma_h[\delta t_\sigma](t) \tag{2.5}$$

and similarly for the logarithmic centered volatility increment. This centered volatility increment measures the change of volatility occurring around t and uses the information in the past and the future of t.

2.3 Average, Expectation

Statistics on random variables need to be computed. Depending on the context, different notations are used to avoid confusion. In a theoretical setting, where the distributions are known by hypothesis, the expectation of x is denoted by $E[x]$. As we are dealing mostly with time series, x can be dependent of the time t and of the past events. The set of previous events up to the time t is denoted by $\Omega(t)$ (i.e., $\Omega(t)$ is the filtration induced by the realizations of the random variables up to and including t). In this setting, the random variables up to t are known as they are included in $\Omega(t)$, and the expectation for later time $t' > t$ is denoted by $E_t[x(t')] = E[x(t') \mid \Omega(t)]$. Essentially, the value of this expectation is a function of the random process between t and t', and of the realizations of the random variables included in $\Omega(t)$.

When dealing with empirical data, much less mathematical structures are present. The expectations are computed with respect to the single realization of the time series, and denoted by $\langle x \rangle$. Although absent in the notation, such expectations are dependent on the available sample and in particular on its start and end times. This dependency is implicitly assumed to be weak, so that the sample statistics is a representative of an "infinite" sample computation. This is not necessarily the case, and a good example of a simple statistics with a strong dependency on the sample boundary is the computation of the mean return. Another point worth emphasizing is that this computation assumes that the data are stationary, namely have a well-defined distribution that does not change with time. A simple example of a random variable that does not have this desirable property is the price itself: all processes for the prices are based on diffusion for which the distribution is not stationary. Less trivially, the sample expectations are computed mainly for the returns, volatilities, and innovations which have—we believe—stationary distributions. Yet, this is not such a trivial statement for long times.

With many time series available (say in a given universe for stocks), statistics can be computed across the universe direction. Such studies are called cross-sectional, say, for example, the returns on a given day can be studied for all the stocks traded on a given exchange. Consequently, cross-sectional statistics can be introduced. In this book, the emphasis is on the time direction, and only a few cross-sectional studies are presented.

2.4 Scaling, Annualization, and Reference Time Interval

The dominant characteristic of financial time series is the diffusive nature of the prices. Consequently, the returns measured as a simple logarithmic or relative price difference has a scaling behavior given by $r[\delta t] \simeq \sqrt{\delta t}$, and similarly for the volatility. This diffusive scaling dominates by far any other properties but is not very interesting. Instead, deviations from the base scaling are much more informative. For this reason, our basic definitions scale away the random walk behavior in order to focus on the differences from a pure random walk. When scaling away the observation time horizon, a reference time interval needs to be chosen. The usual convention is to annualize the return and volatility as done with definitions (2.1) and (2.2). This is the convention used mainly in this book, as it corresponds to the order of magnitudes that most practitioners have in minds. Clearly, other reference time horizons can be chosen, and natural choices are 1 day and the time increment δt of a homogeneous time series. Changing from one reference to another is done through a simple square root ratio.

With process equations, the situation is a bit more complex because the returns should be computed at the time horizon of the process increment δt. The two main choices are illustrated here for a basic constant volatility diffusion:

$$r[\delta t] = \sigma|_{1\text{y}} \sqrt{\frac{\delta t}{1\text{y}}} \epsilon \tag{2.6a}$$

$$= \sigma|_{\delta t} \epsilon. \tag{2.6b}$$

The reference time horizon for the volatility is denoted explicitly after the vertical bar. With the form (2.6a), the volatility parameter is annualized, and the next term scales back the volatility to the process time increment. The advantage of this form is that the volatility has the "usual" values, and the scaling appears explicitly in the equation. With the form (2.6b), the volatility is expressed at the scale δt. The advantage of this form is that the equation is simpler, in particular for complex analytical computations. Clearly, both forms are equivalent, and it is a matter of taste, convention, and context which one is preferred. In this book, the process equations are given in the second form, but process parameters are given as annualized values.

2.5 EMA, MA, and Operators on Time Series

Financial time series are dominated by noise, and a common operation is to extract a local average. This is conveniently done by an Exponential Moving Average (EMA). For a generic time series z, with event z_i occurring at time t_i, the EMA$[\tau; z]$ of z is defined by the iterative equation

$$\begin{aligned} \mu_i &= \exp(-\delta t_i/\tau) \\ \text{EMA}_i &= \mu_i \, \text{EMA}_{i-1} + (1-\mu_i) z_i \end{aligned} \tag{2.7}$$

where $\delta t_i = t_i - t_{i-1}$ is the time interval between the last two consecutive events. The decay of the exponential average is characterized by the time interval τ. The resulting time series is denoted by EMA$[\tau; z]$. In this form, an EMA can be computed for an inhomogeneous time series, namely with irregular time points. Many time series are known on a regular grid with a fixed δt, an in this case μ is also fixed. For homogeneous time series, the relation between τ and μ is

$$\tau = -\frac{\delta t}{\log(\mu)} \simeq \frac{\delta t}{1-\mu}. \tag{2.8}$$

For example, the common value $\mu = 0.94$ applied to daily data corresponds to $\tau \simeq 16$ business day. Beware that while τ has a dimension (of a time interval), μ is a number depending on δt, and one should be careful that δt is often only given implicitly. The domains for the parameters are $\tau > 0$ and $0 < \mu < 1$. This EMA operator is used in many part of finance, as in the empirical analysis of time series or in the definition of processes. The reasons behind its broad application are that the numerical computations are very efficient, that only a minimal amount of computer memory is required, and that the recent events have a larger weights compared to the distant past.

The recursive definition of the EMA can be unwinded in order to obtain explicit formula. Because the EMA definition involves multiplications by μ, the resulting weights are decaying as an exponential. For an inhomogeneous time series, the value of the EMA at time t_i is

$$\mathrm{EMA}_i = \sum_{j \le i} e^{-(t_i - t_j)/\tau} (1 - \mu_j) z_j \tag{2.9}$$

and for an homogeneous time series the value is

$$\mathrm{EMA}_i = (1 - \mu) \sum_{k \ge 0} \mu^k z_{i-k} = (1 - \mu) \sum_{j \le i} e^{-(t_i - t_j)/\tau} z_j. \tag{2.10}$$

These formulas take the form of a convolution where the weight applied to $z(t)$ decays exponentially with a characteristic time τ. With the lag operator L defined by $(Lz)_i = z_{i-1}$, the EMA operator can be rewritten as

$$\mathrm{EMA}_i = (1 - \mu) \sum_{k \ge 0} (\mu L)^k z_i = (1 - \mu)(1 - \mu L)^{-1} z_i. \tag{2.11}$$

The EMA operator is very convenient, but the kernel shape is fixed. In order to build operators with other kernel shapes, the observation is that an EMA is also a time series, and therefore an EMA can be applied on it. And applied again. A recursive definition of an iterated EMA$[\tau, n]$ of order n is

$$\mathrm{EMA}[\tau, n; z] = \mathrm{EMA}\big[\tau; \mathrm{EMA}[\tau, n - 1; z]\big] \tag{2.12}$$

with EMA$[\tau, 1; z] =$ EMA$[\tau; z]$. The explicit formula takes the form of a convolution with a kernel

$$\mathrm{ema}[\tau, n](\delta t) = \frac{1}{(n-1)!} \left(\frac{\delta t}{\tau}\right)^{n-1} \frac{e^{-\delta t/\tau}}{\tau}. \tag{2.13}$$

Linear combinations of iterated EMA allow one to build more operators in order to extract various information from a time series. Such a construction is pursued systematically in [170].

In the direction of computing moving averages with other kernels, the Moving Average (MA) operator has a more rectangular kernel. The MA operator is defined as

$$\mathrm{MA}[\tau, m] = \frac{1}{m} \sum_{j=1}^{m} \mathrm{EMA}\big[\tau', j\big] \tag{2.14}$$

$$\tau' = \frac{2}{m+1} \tau.$$

The characteristic time τ' is set so that the memory length of the MA operator is τ (the memory length is defined as the first moment of the kernel, see [170]). The parameters m control the shape of the kernel, and $m = 32$ gives an almost rectangular kernel.

Similarly, differences of EMAs allow one to build differential-like operators. The general idea is to evaluate the changes in a time series at a selected time horizon τ. The simplest form is the frequently used $z - \text{EMA}[\tau; z]$, but this formula is quite sensitive to the noise. The following combination is a convenient and smooth differential operator which evaluates the changes in the argument at the scale τ:

$$\Delta[\tau] = \gamma\big(\text{EMA}[\alpha\tau, 1] + \text{EMA}[\alpha\tau, 2] - 2\text{EMA}[\alpha\beta\tau, 4]\big) \tag{2.15}$$

with $\gamma = 1.22208$, $\beta = 0.65$, and $\alpha^{-1} = \gamma(8\beta - 3)$. The value of γ is fixed so that the integral of the kernel from the origin to the first zero is one. The value of α is fixed by the condition that the differential of a constant time series is zero. The value of β is chosen in order to get a short tail.

2.6 Computation of the Histograms and Pdf

The probability distributions corresponding to a given time series need to be computed by a binning algorithm. The first step is to select the bin boundaries, or equivalently the sampling grid. The best results are obtained when the bin boundaries are chosen so that roughly an equal number of points fall into each bins (i.e., the cumulative density function (cdf) is divided equally by the sampling points). In the previous section, the volatilities and returns are defined so that the leading scaling dependency is removed. The resulting distribution are then fairly stationary, allowing one to chose a priori a (nonuniform) sampling grid for each variable.

On the selected sampling grid, a histogram is estimated by aggregating the empirical data using a linear interpolation scheme: a fraction is added to the left and right grid points according to the respective distance from them. When all the data have been inserted into the histogram, the pdf is computed by normalizing the aggregated values by the distance between the sampling points. This procedure allows one to obtain the pdf on the nonuniform sampling grid, with a fairly uniform sampling error on each grid point. Although more complex, this procedure allows one to have a more accurate view of the distribution, in particular in the tails. Essentially, this is similar to a kernel estimation of the pdf, with a "tent shape" kernel, and with a width inversely proportional to the pdf.

Chapter 3
Stylized Facts

3.1 Introduction

The most important stylized fact about financial time series is the random walk nature of the prices. The basic probabilistic nature of finance was recognized by [12] in a celebrated Ph.D. thesis. Following Bachelier's work, the basic model for financial time series is a Gaussian random walk with independent increments. Its importance is due to its simplicity, good analytical tractability, and existence of a convenient continuum limit. More than a century after Bachelier's work, this model is the basis of large parts of finance, say, for example, in option pricing with the Black–Scholes model.

With the availability of long time series and powerful computers, the Gaussian random walk model can be probed in details. Clear differences appear in several aspects of their properties, revealed by many statistics used to characterize in details financial time series. The most important differences are the multiscale dynamical properties of the volatility and the fat tail distributions of the returns and innovations. This chapter provides for a detailed analysis of the empirical time series, using a broad coverage of statistics applied on the returns and volatilities. Our goal is to extract as much information as possible from empirical data, in order to build processes that describe accurately financial time series.

A basic property of random walks is the square root diffusion of the prices with time, often expressed as the variance of the returns being proportional to the corresponding time interval. Essentially, this property follows from the assumption of independence for the price increments (but the increment distribution do not need to be Gaussian). Nowadays, the diffusive characteristic of the prices is well established, and we want to search for deviations from an iid random walk. For this purpose, our basic set-up is such that the basic random walk is already scaled away by the $1/\sqrt{\delta t_r}$ time interval normalization in the return and volatility definitions. In this way, returns and volatilities should have properties independent from their characteristic time intervals, and many statistics would be zero if the time series are indeed (Gaussian) random walks. Then, the deviations from zero signal departures from the Bachelier model.

G. Zumbach, *Discrete Time Series, Processes, and Applications in Finance*,
Springer Finance, DOI 10.1007/978-3-642-31742-2_3,

If the daily returns are independent but not Gaussian random variables, the returns at longer time intervals are expected to converge fairly quickly to iid Gaussian random variables. The reason is the central limit theorem (CLT) and the fact that one return at a long time interval is the sum of many returns at smaller time intervals. The conditions for the CLT to hold are fairly mild, essentially that the dependency between the random variables decays fast enough, namely at least exponentially fast [66]. Because the correlations between returns must be very small (otherwise forecasts for the price changes can be build and traded), it is reasonable to expect convergence to a Gaussian distribution. Moreover, the convergence must occur at the pace given by the expansion around the Gaussian fixed point distribution, essentially as inverse power of the aggregation factor. The catch in the present line of though is the difference between liner correlation and general dependency, and the fact that the dependencies in the volatility decays slowly. As we will see, the textbook CLT fails for financial data. Several statistics below are meant to measure the pace of the convergence toward the CLT fixed point.

All the figures for this chapter are produced using the foreign exchange rate CHF/USD. The raw data are high-frequency tick data, which has been sampled regularly at a 3 minutes interval in business time following [32] in order to remove the seasonality for the volatility. The available sample starts on 1.1.1988 and ends on 31.7.2001. The statistics for the figures are evaluated over the period 1.1.1990 to 1.7.2001, with the extra data at the beginning and end of the available sample used to compute the historical and realized values of the returns and volatilities. The "mug shots" for other foreign exchange rates given in Chap. 4 are based on the same procedure and span.

3.2 Probability Density Function

3.2.1 Pdf for the Return

Several views of the probability density for the price changes at different time horizons are given in Figs. 3.1, 3.2 and 3.3. The linear scale shows the overall shape of the pdf, while the logarithmic scale emphasizes the tails. This set of graphs shows several features of the return pdf.

- The variances of the pdf are nearly stationary due to our definition of the return that discounts already the random walk scaling. Eliminating at first the "trivial" scaling will prove itself very useful in the subsequent computations, because the most interesting signals are due to the *deviations* from the simple random walk scaling.
- On the lin–log scales in Fig. 3.2, the overall pdf has a characteristic "tent shape" at short time intervals, similar to $\exp(-|r|/\sigma)$ (with empirical data at the smallest time interval, the peak at $r = 0$ is due to gaps or to stale market conditions). This shape persists over a large range of time interval δt. This shows that in a large domain of returns, the pdf is different from a Gaussian, in particular at the daily scale.

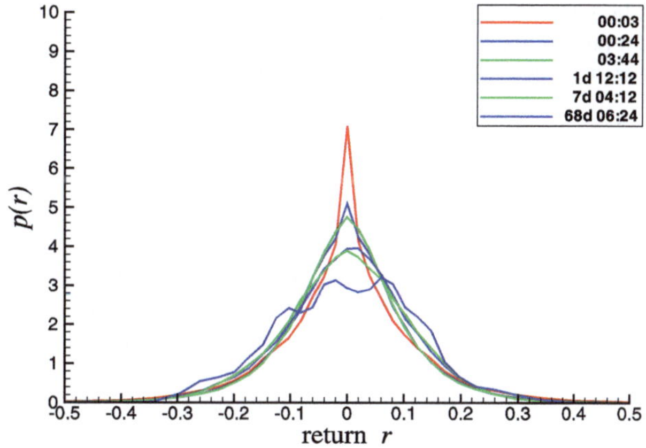

Fig. 3.1 Probability density for the returns, in linear–linear scales

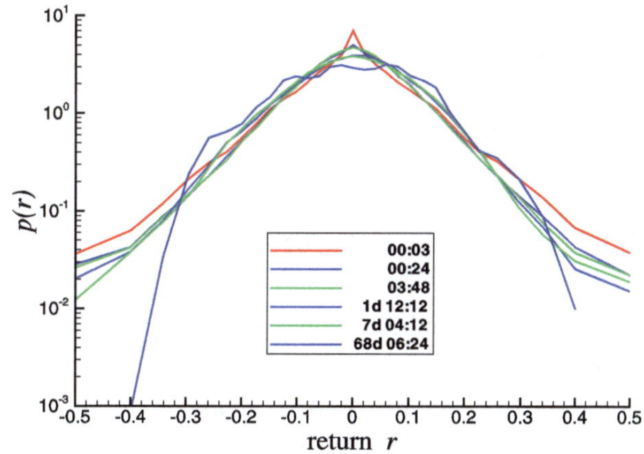

Fig. 3.2 Probability density for the returns, in linear–logarithmic scales

- With the increasing time horizons, the pdf converges toward a Gaussian distribution due to the returns aggregation. Yet, this convergence is very slow, and even at the longest time horizon, the pdf is still a poor approximation of a simple parabola in the lin–log scale (the longest time horizons also suffer from poor statistics due to the small number of independent returns in the 11 years sample). This slow aggregation toward the Gaussian distribution will be examined quantitatively in the next two subsections.
- The shape for the tail of the pdf is best shown on the log–log scale. A power law tail distribution, as, for example, a Student or a Pareto distribution, exhibits a straight line shape. Clearly, the empirical behavior is not very close to a power law. Yet, an accurate study of the tail behavior must be done based on the cdf computed with all events (and not from a histogram), and with a proper statistical estimator for the tail exponent like the Hill estimator [81]. A large literature already exists on this topic [48, 51]. The current consensus on tail exponents is that the exponents are between 3 to 5, but anyway larger that 2 (thus excluding the

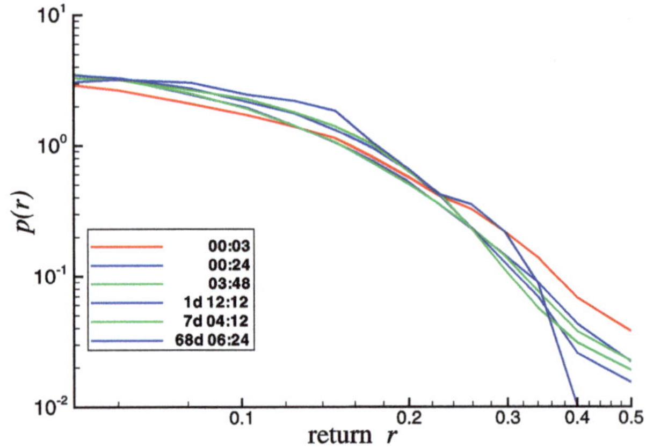

Fig. 3.3 Probability density for the positive returns, in log–log scales

Lévy stable distributions proposed by Mandelbrot [95]). A theoretical argument exists to show that the exponent must be 3 [72], but the premises of the argument are debatable. An overall characterization of the return pdf is investigated further in Chap. 13.

3.2.2 Pdf for the Volatility

Figures 3.4, 3.5 and 3.6 show the probability density for the volatility at different time intervals. The most noticeable features are similar to the ones of the return pdf.

- The pdf are nearly stationary due to our discounting of the random walk scaling.
- The overall shape is clearly unimodal. This is a simple argument against regime switching processes for the volatility (see Chap. 9). In a model with a few distinct states, we can expect that the pdf will show maxima at the corresponding values. This is not the case, and therefore many states must be used to describe the empirical pdf. But then, the concept of a state process becomes less relevant, and a continuum of volatility is a more parsimonious description of the volatility dynamics.
- On the lin–log scales, at short time horizons, the overall pdf has a characteristic shape, with a long quasi-straight decline on the large volatility side. This shape persists over a large range of time horizon δt. On the other hand, the shape around the maxima is close to a parabola.
- For independent Gaussian returns, the volatility must have a χ distribution (the variance should have a χ^2 distribution). This is clearly not the case as illustrated on Fig. 3.7, showing that the returns are not independent. In the set of usual distributions, the log-normal pdf provides for a good overall approximation for the volatility distribution but does not capture correctly the upper tail.
- The shape for the upper tail of the pdf must be similar to the tail of the return pdf. This occurs because the same extreme returns dominates both pdfs.

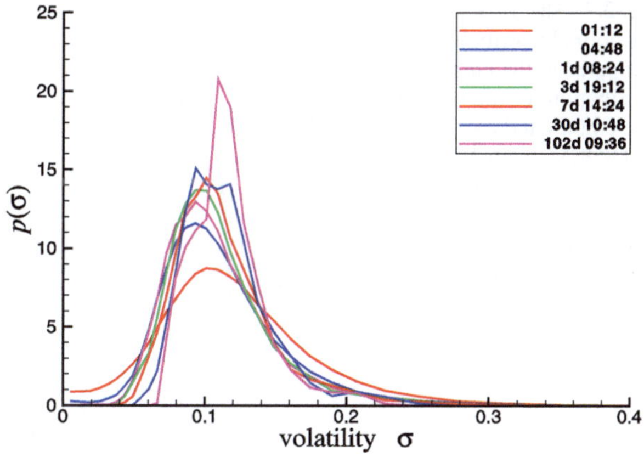

Fig. 3.4 Probability density for the volatility, in linear–linear scales

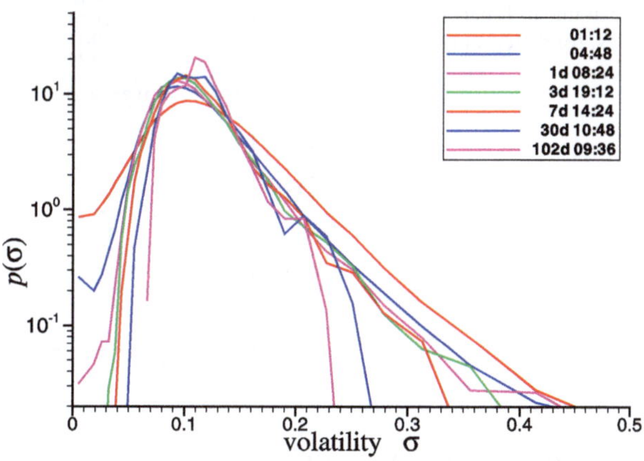

Fig. 3.5 Probability density for the volatility, in linear–logarithmic scales

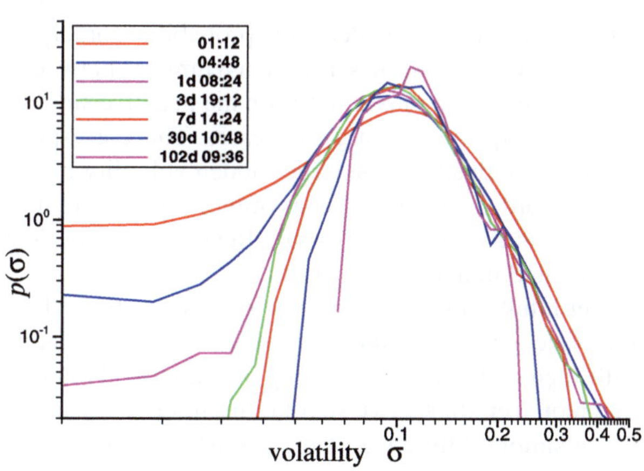

Fig. 3.6 Probability density for the volatility, in log–log scales

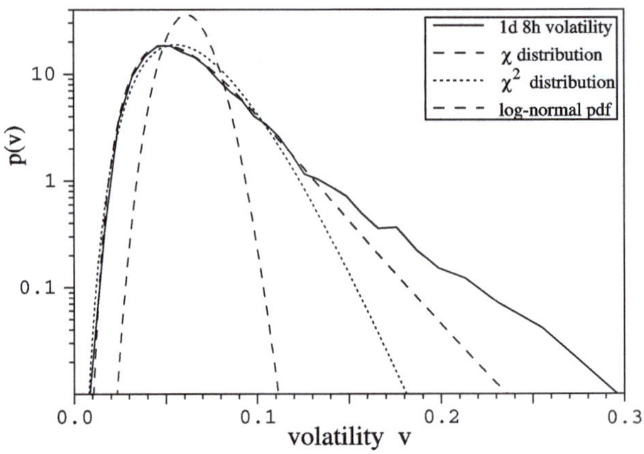

Fig. 3.7 The χ, χ^2 and log-normal distribution compared to the empirical distribution for the volatility. The empirical data is for USD/DEM, from 1.1.1987 to 1.1.1998, and with $\delta t_\sigma = 1$ d 8 h. The distributions are rescaled to have the same mean as the empirical volatility

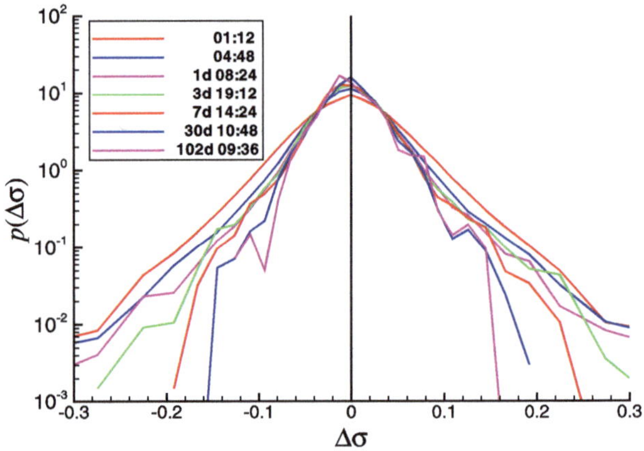

Fig. 3.8 Probability density for the volatility increment $\Delta\sigma$, in linear–logarithmic scales, for different time horizons

3.2.3 *Pdf for the Volatility Increment*

The probability density for the volatility increment $\Delta\sigma$ is shown on Fig. 3.8. The pdf has a simple shape, with decaying probability for large volatility changes. A more attentive observation shows a slight asymmetry between positive and negative volatility increments: the maximum is on the negative side, while large volatility increases are more frequent than large decreases. This corresponds to our idea of the volatility dynamics, with sometime large sudden volatility increase followed by a slow relaxation toward the mean. This asymmetry is quantitatively small but very important as it shows a non-time reversal effect in the empirical data. This question will be investigated in depth in Chap. 11.

Our intuition for this asymmetry is based typically on a GARCH(1, 1) volatility time trace, where the empirical returns are used as input in a GARCH volatility filter. With such computation, a strong asymmetric dynamics emerges from the volatility time plot. Yet, the GARCH volatility filter is already asymmetric. Our computation uses a simple difference of rectangular kernels and therefore weights equally past and future. But then, the computed asymmetry in the dynamic of the empirical data is small. This shows that our image of the volatility dynamics as "large increases followed by a relaxation toward the mean" is qualitatively correct but quantitatively overrated.

3.3 Scaling for the Moments: The Width of the Pdf

3.3.1 Introduction

The goal of this section is to measure in a quantitative way the width of the pdf for increasing time horizons. The width is measured using moments μ_k with different exponents k defined by

$$\mu_k[\delta t] = E\big[\big|r[\delta t]\big|^k\big]^{1/k}. \tag{3.1}$$

The exponents k range from 0.5 to 4 (larger values for k are meaningless due to a tail exponents between 3 to 5). As the definition of the returns already includes a factor $1/\sqrt{\delta t_r}$, the "trivial" random walk scaling is already discounted. With this convenient normalization, these moments as function of δt capture the *deviation* from a random walk. For small (large) k, the moment μ_k measures the width with an emphasize on the center (the tail) of the pdf.

The dependency of the widths $\mu_k[\delta t]$ with respect to δt is interesting. The scales for the moment graphs are log–log, in order to search for power laws. Each curve is multiplied by a constant such that the moment at $\delta t = 1$ day equal to 1. This causes all the curves to cross at 1 day, but this is only a convenient convention.

3.3.2 Scaling for the Return

The scaling for the absolute value of the returns $|r[\delta t]|$ versus the time horizon δt is displayed in Fig. 3.9. Empirically, the moment for $k = 2$ is nearly flat. This is a consequence of the absence of correlation for the return, and in practice, the correlations are small enough to be neglected. For all the exponents k, in roughly 2 decades from 0.1 to 10 days, the moments are nearly linear (in log–log scales), indicating a power law behavior. The scaling for $k < 2$ ($k > 2$) show a clear positive (negative) slope. Overall, this behavior is consistent with the published literature [68, 101]. Yet, notice also the different behaviors at very short and very long time intervals, as well as the deviations from a straight line in the "scaling regime". These deviations from a simple straight line indicate secondary order effects, on top of a leading scaling.

A power law scaling of the form

$$\mu_k[\delta t] = E\big[\big|r[\delta t]\big|^k\big]^{1/k} = \left(\frac{\delta t}{dt_0}\right)^{\zeta'(k)} \tag{3.2}$$

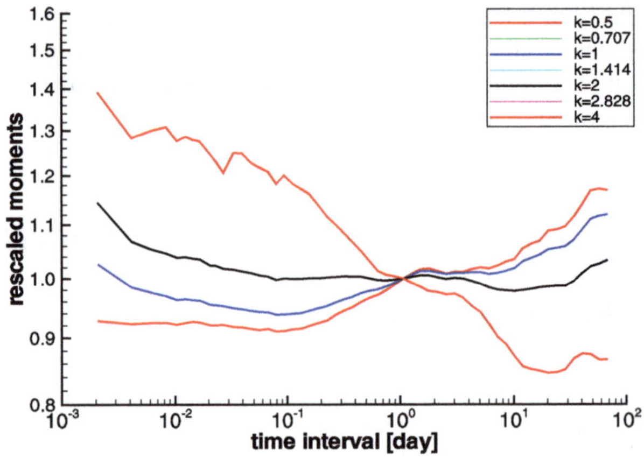

Fig. 3.9 The k-moment of the return at time horizon δt, in log–log scales

constitutes a definition of a multifractal process. The usual exponent $\zeta(k)$ found in the literature is defined for unanualized return (i.e., without the $1/\sqrt{\delta t_r}$ factor in the return definition) and without the $1/k$ exponent after the expectation. These different definitions lead to the simple relation between the exponents $\zeta'(k) = \zeta(k)/k - 1/2$. A simple Gaussian random walk has $\zeta(k) = k/2$, or $\zeta' = 0$. This shows that $\zeta'(k)$ measures the deviation from a simple random walk. Reference [102] gives a good review of the many variations around the scaling exponent concept, while [91, 148] are critical about simple empirical "multifractal" analysis.

Several articles start from the above empirical relation and construct processes so that the scaling relation is exact up to an "integral time". The construction of [13] involves a volatility cascade from long time horizon to short time horizon, followed by a limit to the continuum. In the construction of [96], it is the process time scale which is constructed as a cascade, followed by the appropriate continuum limit. These constructions allow one to obtain nontrivial processes for which, in some case, the above scaling law can be computed exactly from the process equations. This relation can then be tested against empirical data [68, 69].

Closely related to the multifractal properties is the local singularity spectrum of the price $p(t)$ (for a continuous time). One can pass from one to another by a Legendre transform [43, 44, 91]. Then, the nonzero $\zeta'(k)$ function is interpreted as a continuous distribution for the local singularities of the price process, whereas a simple random walk with $\zeta'(k) = 0$ has everywhere square root singularities. With this point of view, the nontrivial multifractal properties are related to the local behavior of a continuous price process. In the above theoretical price construction, the cascade followed by a continuum limit is needed to construct a multifractal measure on a line segment. In turn, this nontrivial measure induces interesting multiscaling properties in the process. We feel at odd with these constructions because the empirical price process is never continuous. At very short time, using high-frequency data, one can measure effects related to the price formation process. Particularly strong in the foreign exchange market is the incoherent effect [46]: the market participants need to publish quotes and to trade in order to exchange information and build an

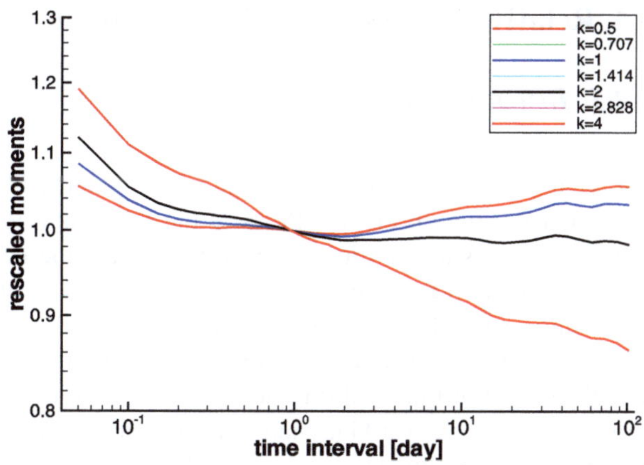

Fig. 3.10 The k-moment of the volatility at time horizon δt, in log–log scales. Notice that the moments are not centered, namely the mean is not subtracted (contrary say to a standard deviation)

ever elusive consensus price. This process takes time of the order of minutes. As a result, the price has an inherent fuzziness (similar to some degree with the quantum uncertainty of elementary particles). Another effect is induced by the conventional accuracy with which prices are quoted, say of 1 BIPS, and by order queues for securities exchanged through electronic systems. The defined accuracy causes the price to move by discrete increments. All these very short term effects, related to the details of the price formation mechanism, to the exchange of informations and to the transactions, make the actual price process rather different from an abstract continuum random walk. Therefore, an explanation of the nontrivial multiscaling in terms of the derivability of such limiting continuum price is somewhat counter-intuitive. Moreover, the construction of a volatility cascade from long to short time horizons, similar to a turbulent cascade of eddies with decreasing sizes, does not bring an intuitive explanation in terms of the behavior of the market participants.

As we will see in Chap. 7, the ARCH like processes have nontrivial multifractal scaling properties. In particular, processes with multiple time scales have the correct empirical behavior. If the market components are included in the process, even the deviation from the simple multifractal can be observed. This agreement between ARCH process and empirical data shows clearly that the singularity spectrum is not needed to explain the multifractal behavior.

3.3.3 Scaling for the Volatility

The same moment computation is done for the volatility at different time horizons. As we do not use (integer) centered moments and as the volatility is positive, this quantity is mostly influenced by the large volatilities. With empirical data, the computed properties on Fig. 3.10 are very similar to the return moments, with power law behavior between 0.1 to 10 days.

3.4 Relative Excess Kurtosis: The Shape of the Pdf

3.4.1 Introduction

The goal of this section is to measure in a quantitative way the *shape of the pdf* for increasing time horizons. Because the return at a long time horizon is the sum of returns at a smaller time horizon, we expect the central limit theorem to take place. For example, a daily price change is "made" of hundreds or thousands of elementary price changes. Using the central limit theorem, the distribution of daily price changes is expected to be close to a Gaussian. This argument is used to validate the model of financial time series by a Gaussian random walk. Moreover, with an expansion around a Gaussian distribution for iid random variables, known as an Edgeworth expansion (see, e.g., [66]), it can be shown that the deviations from the limiting Gaussian law vanish as a power law with increasing aggregation. As we now know, this aggregation argument is incorrect, even though the returns are essentially uncorrelated.

The returns are indeed not independent, due to the volatility correlation. Moreover, the volatility correlations show clusters at all time horizons, as measured by the slow decay of the return magnitude or of the volatility. This slow decay of the volatility correlation makes the convergence to the Gaussian much slower, appearing essentially as the logarithm of the aggregation factor. To understand intuitively the mechanism, consider for now a process with one exponential characteristic time, say a GARCH(1, 1) process. For short time lags, a regime occurs with a strong dependency between the magnitude of the return and a slow convergence to the central limit law. For time lags larger than the characteristic time of the GARCH(1, 1) process, the returns can be considered as independent, and the central limit theorem can take place, with a power law convergence. The cross-over between these two regimes take place at a time lag comparable to the exponential correlation time of the GARCH process (see Sect. 7.1). For empirical data (and for some of the processes), several characteristic times distributed like a geometric series lead to a cascade of cross-overs and therefore to the slow convergence toward the Gaussian fixed point of the central limit theorem with a pace given by the logarithm of the aggregation factor.

As discussed in the previous section, the width of the probability distribution is measured using moments with different exponents

$$\mu_k[\delta t] = E\left[\left|r[\delta t]\right|^k\right]^{1/k} \tag{3.3}$$

where the exponents k range from 0.5 to 4. A measure of the shape of the probability distribution can be obtained by constructing dimensionless ratio of moments. Moreover, the goal of the relative excess kurtosis is to measure the deviation of the return pdf from a Gaussian distribution, similarly to the kurtosis. The usual kurtosis is the ratio of the fourth moment and the second moment square. The excess kurtosis is the difference between this ratio and the same ratio for a Gaussian distribution (i.e., 3). The excess kurtosis measures the importance of the tails (the fourth moment of the empirical pdf), in dimensionless unit (the ratio with the second moment square), compared to this ratio for a Gaussian distribution. In order to build a more robust measure of the distribution shape, the kurtosis definition is generalized to

$$\kappa_k[\delta t] = \frac{\langle |r[\delta t]|^{2k} \rangle}{\langle |r[\delta t]|^k \rangle^2} \Big/ \frac{\langle |r[\delta t]|^{2k} \rangle_{\text{Gauss}}}{\langle |r[\delta t]|^k \rangle^2_{\text{Gauss}}} - 1. \tag{3.4}$$

For all k, the relative excess kurtosis is constructed to converge to zero when the empirical pdf converges to a Gaussian distribution. For large k (small k), it measures the convergences of the tail (the center of the pdf) toward a Gaussian. For an iid process, with increasing time interval δt, the relative excess kurtosis converges as a power law toward zero, as dictated by aggregation.

3.4.2 Relative Excess Kurtosis for the Return

The relative excess kurtosis for the returns is shown in Fig. 3.11 for increasing time interval δt. The horizontal axis for the time intervals is logarithmic. Therefore, for an iid process, we should observe an exponential shape convergence toward zero (as a function of $\ln(dt)$). The salient feature of the empirical data is the very slow convergence to a Gaussian pdf, appearing essentially as a linear decay as function of $\ln(dt)$. In more details, the decays is better described by a succession of regimes, with alternating slower and faster convergences. This empirical description would be in agreement with the above explanation of the slow convergence toward the Gaussian as caused by a succession of cross-overs due to a cascade of characteristic time intervals.

3.4.3 Relative Excess Kurtosis for the Volatility

The same statistical measure can be computed for the volatility. Yet, as the moments are not centered, this definition of κ_k is not appropriate for positive random variables that do not converge toward a Gaussian distribution.

3.4.4 Relative Excess Kurtosis for the Volatility Increment

For an iid random walk, the volatility increment is a sum of random variables, essentially $\delta\sigma_h[\delta t_\sigma] \simeq \sum \pm\epsilon^2$. The central limit theorem is therefore applicable, and the distribution should converge toward a Gaussian distribution. For empirical time series, the heteroscedasticity creates a decreasing dependency between subsequent volatilities, and the convergence toward a Gaussian distribution becomes more interesting. The volatility increment $\delta\sigma_h[\delta t_\sigma]$ has essentially a zero mean, and therefore

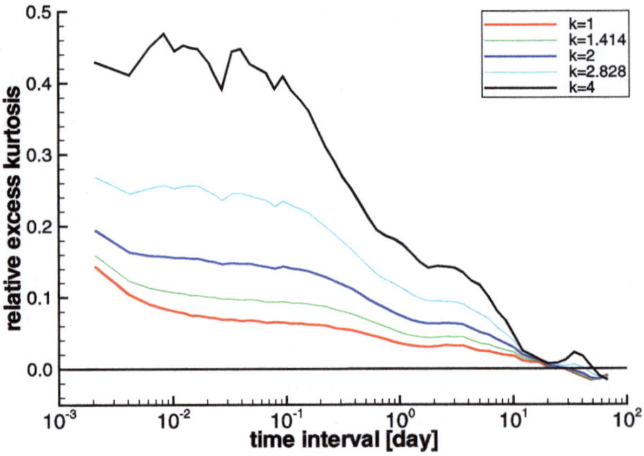

Fig. 3.11 The relative excess kurtosis $\kappa_k[\delta t]$ of the returns at various time intervals δt. The exponents k range from 1 to 4

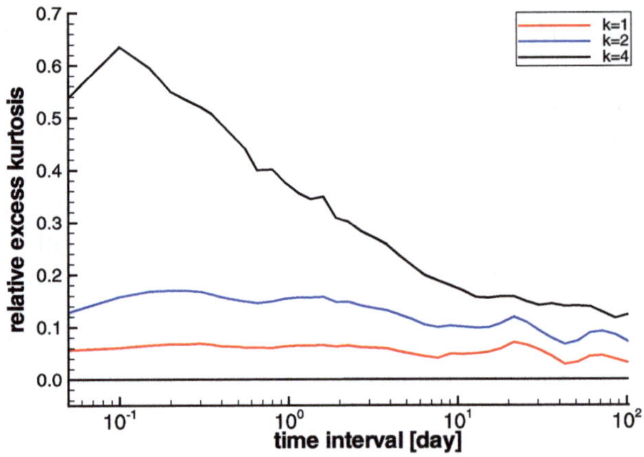

Fig. 3.12 The relative excess kurtosis $\kappa_k[\delta t]$ of the volatility increment at various time intervals δt

the above definition for the relative excess kurtosis is a meaningful measure for the shape of its distribution, relevant for the dynamics of the volatility. The relative excess kurtosis is shown on Fig. 3.12 for the CHF/USD. After a slow decay for time interval up to a month, the empirical relative excess kurtosis seems to stay at a finite value for long time intervals. The comparison with other empirical time series shows a large variability, but the excess kurtosis is always positive for medium to long time horizons. This is yet another measure of the slow convergence toward a simple random walk.

3.5 Lagged Correlations

3.5.1 Introduction

The lagged correlation measures the linear dependency between subsequent values in a time series as the time difference increases. Let us emphasize the difference between no correlation and no dependency: the correlation measures only the linear dependency between two random variables. Indeed, the financial time series show no correlation for the returns but positive correlation for the squared returns, giving a good example of the difference between the concept of dependence and linear correlation. The absence of correlation for the return can be explained by a statistical arbitrage argument: any large enough correlation for the return can be used for trading as it gives a directional information. This correlation can be exploited until it becomes too small to be tradeable. In essence, this argument says that the correlations for the returns have to be very small. Yet, no such trading argument can be built for measure of volatility, and therefore the possible nonzero values for these correlations.

Several measures of volatility can be built. With the constraint that they are homogeneous of degree one with the returns, two natural definitions are $|r|$ and $\sqrt{\sum r^2}$. Essentially, all volatilities measure the magnitude of the price changes inside a given period δt_σ, and all are even in the returns. On the other hand, the quality of the measured volatilities depend on the definition. With $\sigma = \sqrt{\sum r^2}$, the estimated volatility becomes better for an increasing number of returns in the given time interval δt_σ. This trade off and a robust definition of the correlation is explored systematically in Chap. 12.

For a time series $x(t)$, the lagged correlation is defined as

$$\rho(\tau) = \frac{\sum_t x(t)x(t+\tau)}{\sum_t x^2(t)}. \tag{3.5}$$

The interesting quantity to analyze is the lagged correlation for different measures of volatility, as done in the following sections. All these measures of volatility depend on one time horizon, say, for example, $|r[\delta t_\sigma = \delta t_r]|$. The graphs below present the lagged correlations as functions of the lag τ, for three characteristic time horizons δt_σ, and for both measures of volatility. The computations are done for all the volatility measures on a 3 minutes grid for the time lag. For lags τ shorter than δt_σ, this results in an overlap of the two random variables. Therefore the correlation decays from the value 1 for $\tau = 0$ to lower values as the overlap decreases.

3.5.2 Lagged Correlations for the Absolute Return

With the volatility measured as the absolute value of the return, the lagged correlation is given on Figs. 3.13 and 3.14. The salient feature of the empirical time series is the slow decay of the lagged correlations. This slow decay shows that financial time series are heteroscedastic, with an interesting dependency over several time horizons. Beyond this raw observation, a simple characterization of the shape of such curves is a delicate business. The log–log plot in Fig. 3.14 shows that the lagged correlation can be described by a power law, from a few minutes to one month. Yet, the decay seems faster for lags larger than 1 to 2 months, albeit the small statistics

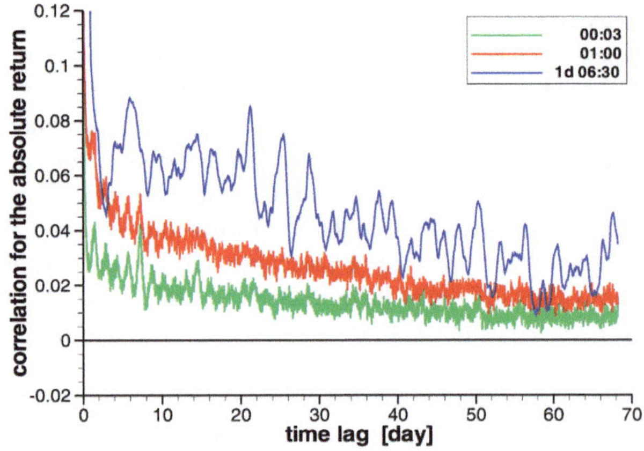

Fig. 3.13 The lagged correlations of the absolute value of the returns, in linear–linear scales

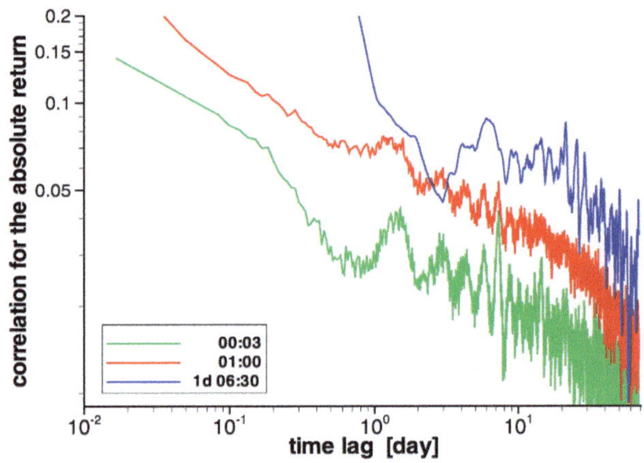

Fig. 3.14 As above, but in log–log scale

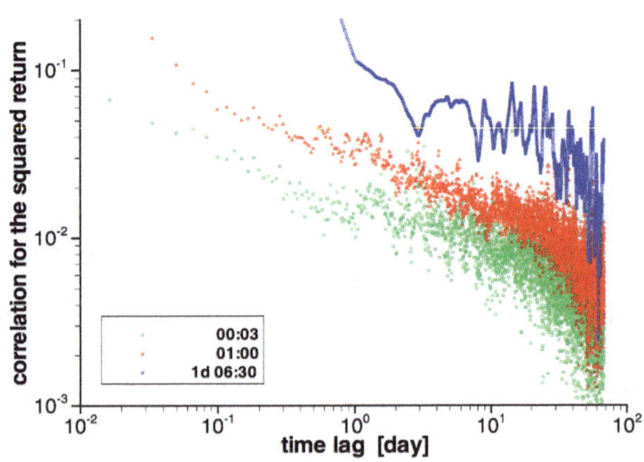

Fig. 3.15 The lagged correlations for the squared return r^2, in log–log scale

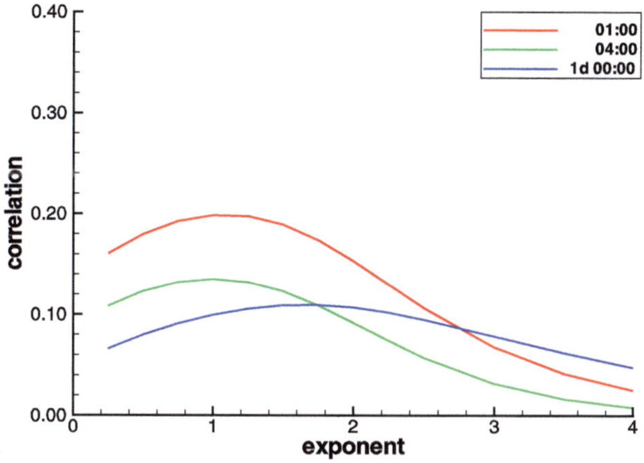

Fig. 3.16 The lagged correlation at lag $\tau = \delta t_r$, for $|r[\delta t_r]^\alpha$. The exponent α is given on the horizontal axis, the correlation on the vertical axis

for such large lags make difficult to be more conclusive. A broad characterization of the lagged correlation is the subject of Chap. 12.

For the 3 minutes and the 1 hour returns (green and red curves), a seasonality remainder is clearly visible with peaks a days and weeks multiples. This shows that either our deseasonalization algorithm is not perfect or that there is a "heat wave" effect [62] in the empirical data. More statistical studies would be needed in order to decide for the best explanation for the remaining peaks. Yet, it is very likely that the peak at 1 week is due to the weekly seasonality discounted incompletely by our deseasonalization algorithm.

On Fig. 3.15, the lagged correlations for the squared returns show essentially the same overall behavior, but with a smaller level of correlations. This indicates that larger moves have smaller dependency on previous events. This quantitative difference can be measured best by the lagged correlation of $|r[\delta t_r]|^\alpha$ at $\tau = \delta t_r$, namely with the absolute value of the return raised to the exponent α. For small (large) exponent α, this measures the dependency between the small (large) moves. The results of this computation is reported on Fig. 3.16 for exponent α between 0.2 to 4. The curves show that the dependency is largest for exponents between 1 and 2. This indicates that medium size moves have the largest dependency, while small and large moves have a smaller correlations.

3.5.3 Lagged Correlations for the Volatility

The lagged correlations for the volatility are similar to the one for the absolute return, but with a higher levels and less fluctuations. They are shown on Figs. 3.17 and 3.18. Because the volatility is a better fluctuation estimator than the absolute return, the correlations have a higher level. The slow decay of the lagged correlations is again clearly visible, but giving a simple analytic characterization of the decay requires some care. The lin–log plot shows a very straight behavior, pointing to a

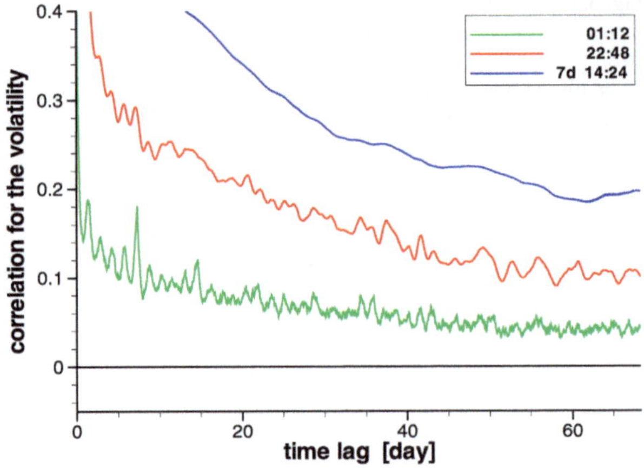

Fig. 3.17 The lagged correlations of the volatility for three time horizons, in linear–linear scales

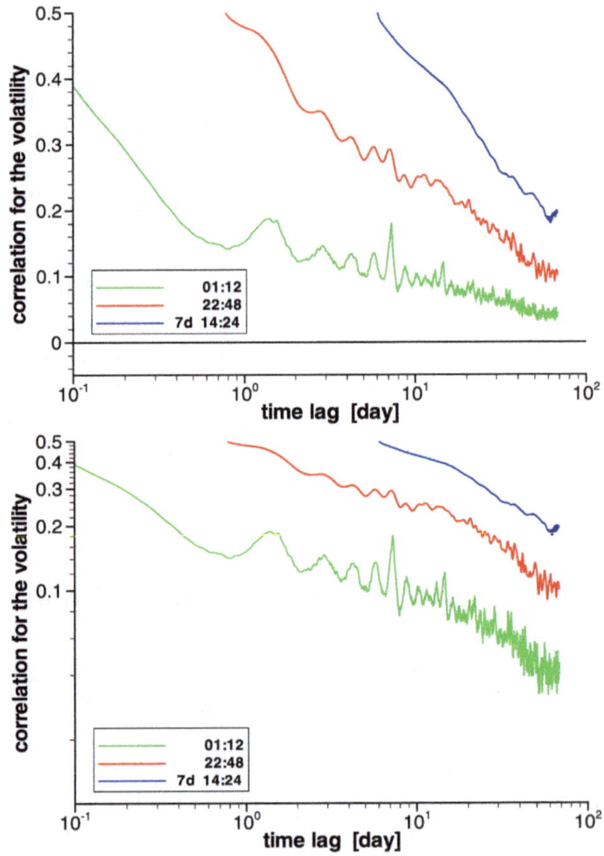

Fig. 3.18 As above, but in lin–log and log–log scales

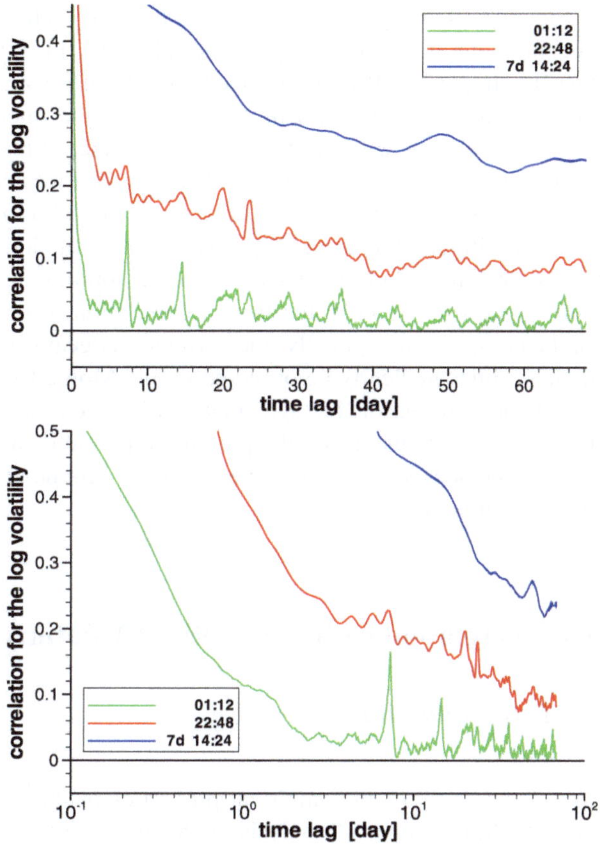

Fig. 3.19 The lagged correlations of the logarithm of the volatility for three time horizons, in lin–lin and lin–log scales

logarithmic correlation decay

$$\rho(\tau) \simeq 1 - \ln(\tau)/\ln(\tau_0). \tag{3.6}$$

The log–log plot shows that the lagged correlation can also be well described by a power law

$$\rho(\tau) \simeq (\tau_0/\tau)^{\nu}. \tag{3.7}$$

Over the analyzed time intervals, the first form seems slightly better, and a similar conclusion is reached for other time series. Yet, a definitive statement on the analytical shape would require a study over a large set of time series and with longer time spans. This is the subject of Chap. 12.

Finally, we give the lagged correlation for the logarithm of the volatility for three time horizons. The rationale for studying the logarithm of the volatility is that this transformation makes the volatility distribution closer to a Gaussian, and most analytical convergence results are established for Gaussian variables. As can be seen on Fig. 3.19, the overall level of correlations is lower, but the slow decay is again clearly visible.

Figure 3.19 in log–lin scale is used to infer the "integral time", which is an important parameter in multifractal processes [13, 112, 113]. Essentially, the integral time gives the upper cutoff in multifractal models above which they become essentially random walks. The integral time is given by the intersection of the extrapolated lagged correlation with the horizontal axis, and corresponds to the parameters τ_0 in (3.6). The extrapolation from the 22 h volatility (red curve) would give an integral time of the order of 1 to 2 years, in agreement with more rigorous procedures used in the multifractal literature [13]. Clearly, extrapolation is a difficult procedure, and the use of longer daily time series is more appropriate for this exercise. For some multifractal model [113], the lagged correlation of the logarithm of the volatility can be computed analytically: the correlation decays linearly with the logarithm of the lag. This can clearly be observed on the Monte Carlo simulations for a similar model and is in reasonable agreement with the empirical data. Yet, it is difficult to draw strong conclusions on the present empirical data, as the linear regime covers at most one decade (and a possible change of regime at 1 to 2 months is right in the middle of this range).

3.6 Correlation with the Realized Volatility

3.6.1 Introduction

So far, the study of the empirical properties of the financial data was essentially along traditional lines of analysis, with studies of the pdf, moments and their scalings, and lagged correlations. The lagged correlations for different measures of volatility are important as they give information on the long time dependencies. Yet, the lagged correlations are not optimal, in the sense that they are not extracting the most information from a time series. A more powerful way of getting the time dependency for a range of time horizons is to compute the correlations between historical and realized quantities. For example, given a historical quantity $x_h[\delta t](t)$ and a realized quantity $y_r[\delta t'](t)$, the correlation $\rho[\delta t, \delta t']$ between the two time series measures the linear influence of x_h on y_r. This correlation summarizes at best the influence of the immediate past on the immediate future. In fact, a market participant acting at a given time t has the knowledge of $x_h[\delta t](t)$ for all time horizon δt to make his decisions, and he care much less about the past values of x_h. Therefore, these correlations measure at best the influence on the future values y_r of the information set which is composed of all possible $x_h[\delta t]$ at t and before. For the scientist, the task is to pick the quantities x_h and y_r that best reveal the properties of the time series. As the dominant feature of the financial time series is the heteroscedasticity, the obvious choice is different combinations of the volatility σ and volatility increment $\Delta\sigma$. This section investigates mainly the correlation with the realized volatility, and the next section the correlation with the volatility increment.

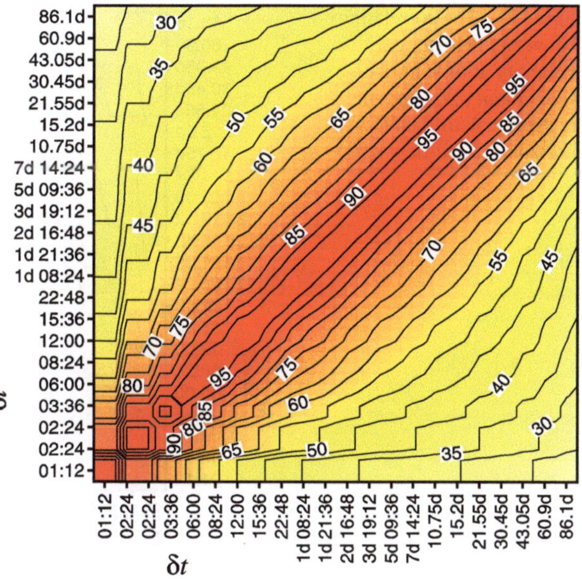

Fig. 3.20 The correlations between the volatility at time horizon δt_σ (*horizontal axis*) and itself at time horizon $\delta t'_\sigma$ (*vertical axis*)

3.6.2 Autocorrelation for the Volatility

Figure 3.20 shows the correlations between the historical volatility at different time horizons. The salient feature on the empirical time series is the slow decay of the correlation with the increasing difference between δt_σ and $\delta t'_\sigma$. For uncorrelated returns (i.e., a simple random walk with constant volatility), this volatility should decay as $\min[\delta t_\sigma, \delta t'_\sigma]/\max[\delta t_\sigma, \delta t'_\sigma]$. Taking $\delta t = 1$ week and $\delta t' = 8$ hour leads to a 5 % correlation, whereas a 50 % correlation is observed on the empirical data. The volatility memory leads to this high correlation across time horizons.

3.6.3 Correlations Between the Historical and Realized Volatilities

The correlation between the historical volatility (horizontal axis) and the realized volatility (vertical axis) is shown in Fig. 3.21. This figure shows the dependency structure between past and future volatilities. The salient features are the following.

- An overall asymmetry across the diagonal. This asymmetry is a measure of the breaking of the time reversal invariance. Essentially, the asymmetry shows that historical volatilities at all time horizons influence the realized volatility at short time horizon, whereas historical volatility at short time horizon does not influence the realized volatility at long time horizon. Another way of expressing the asym-

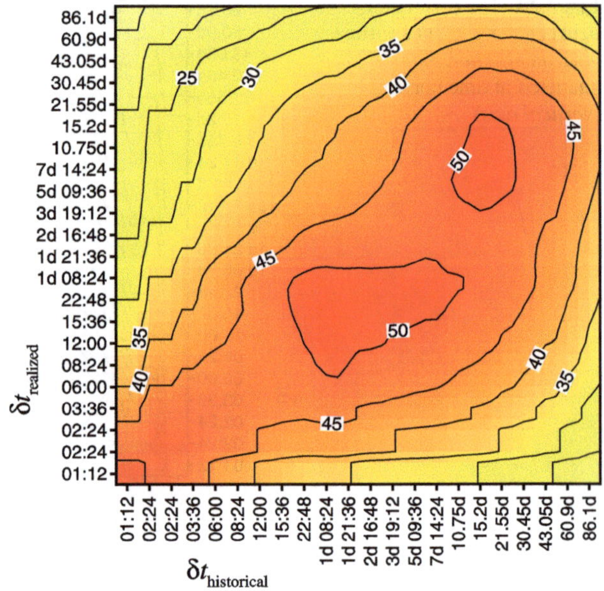

Fig. 3.21 The correlations between the historical volatility at time horizon δt_σ (*horizontal axis*) and the realized volatility at time horizon $\delta t'_\sigma$ (*vertical axis*)

metry is that a realized volatility at a given time horizon is mostly influenced by historical volatility at longer time horizons.

- A set of local maxima, located at time horizons corresponding to the natural human cycles: intra-day, day, week, and month. These maxima are observed essentially at the same locations for all empirical time series. This gives a direct measure of the heterogeneity of the market participants, with different groups of agents acting at different time horizons.

  The relationship between the observed market heterogeneity and the efficient market hypothesis is interesting. In particular, this rules out a strong form of the efficient market hypothesis where all the new informations are incorporated immediately into the current prices through trading by all market participants. There exist clearly groups of market agents acting on medium to long time horizons, of the order of one week and one month.

For data generated by a process, when the historical/realized volatility correlation is reasonably close to the empirical graph, many other volatility properties are also close to the corresponding empirical properties. This is the case for the lagged correlations and most of the volatility/volatility increment correlation figures. The opposite is not true, as, for example, several processes can reproduce a long memory for the volatility lagged correlation but cannot capture the information unveiled in the historical/realized volatility correlation. Therefore, this figure gives a very strong visual criterion to select processes that duplicate the empirical data. If a single plot were chosen to summarize the overall properties of a financial time series, the historical/realized volatility correlation would give a very selective signature.

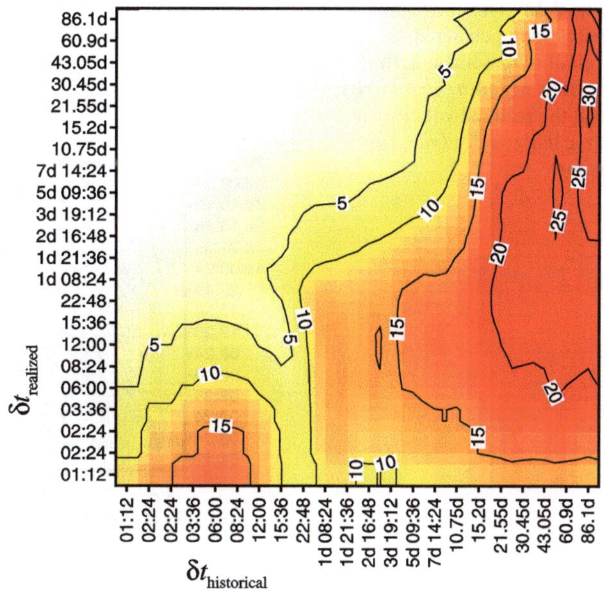

Fig. 3.22 The correlation between the historical volatility increment at time horizon δt_σ *(horizontal axis)* and the realized volatility at time horizon $\delta t'_\sigma$ *(vertical axis)*

3.6.4 Correlations Between the Realized Volatility and the Historical Volatility Increment

Figure 3.22 displays the correlation between the historical volatility increments and the realized volatilities. This correlation measures how past changes in the volatility influence the realized volatility. The intuitive idea behind this correlation is that increase (decrease) of the volatility can lead market participants to trade (keep their positions) and therefore to increase (decrease) the realized volatility. This intuitive argument leads to positive correlations, in agreement with the empirical findings.

Beyond this overall argument, the time structure of this correlation is interesting: the empirical correlation shows a very strong asymmetry around the diagonal. In particular, volatility changes at short time horizons have no influence on realized volatility at longer time horizon. At a finer level, the same intra-day, day, week, and month characteristic times can be observed.

The implications for process equations of this figure is interesting. The question is if explicit terms involving the volatility derivative must be included in process equations in order to reproduce the empirical findings, or if the structure of a multi-scale GARCH-like process, for example, is enough to obtain these correlations. The Monte Carlo simulations of the Market component model show that this figure is reproduced nicely without including explicitly specific terms. Therefore, these correlations can be considered to a large extent as by-products of the volatility structure: it is obtained through the mean reversion and multiple time horizons memory of the volatility.

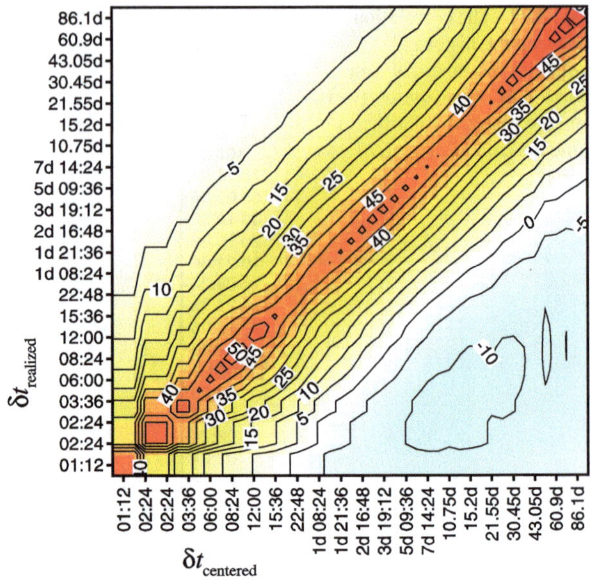

Fig. 3.23 The correlation between the centered volatility increment at time horizon δt_σ (*horizontal axis*) and the realized volatility at time horizon $\delta t'_\sigma$ (*vertical axis*)

3.6.5 Correlations of the Realized Volatilities with the Centered Volatility Increment

The correlation between the centered volatility increment and the realized volatility is given on Fig. 3.23. This correlation measures the dependency between the dynamics of the volatility (i.e., its derivative) and the subsequent volatility. Notice that it is a noncausal correlation, because the centered volatility increment is computed using information in the future of t. The figure shows essentially positive correlations, as the volatility is usually larger (smaller) after an increase (decrease).

3.7 Correlation for the Volatility Increment

The previous section shows correlations for the volatilities. In order to investigate the dynamic of the volatility, this section focuses on the volatility increments. The goal is to investigate the possible correlation structures present in the volatility dynamics.

Figure 3.24 shows correlations for the volatility increments at different time horizons. With empirical data, this correlation decays rapidly with increasing time difference, showing the quite weak correlation across time horizons in the volatility dynamics. Beyond the decay away from the diagonal, the correlation is mostly structureless. A comparison of this figure with the one obtained for a constant volatility random walk shows that the correlations decay more slowly away from the diagonal in the empirical data. This is an effect of the volatility dynamics that is correctly reproduced by most multiscale processes.

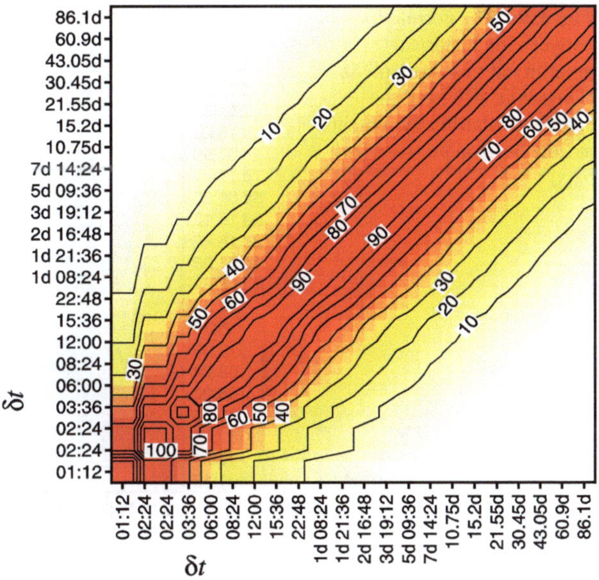

Fig. 3.24 The correlation between the centered volatility increment at time horizon δt_σ *(horizontal axis)* and itself at time horizon $\delta t'_\sigma$ *(vertical axis)*

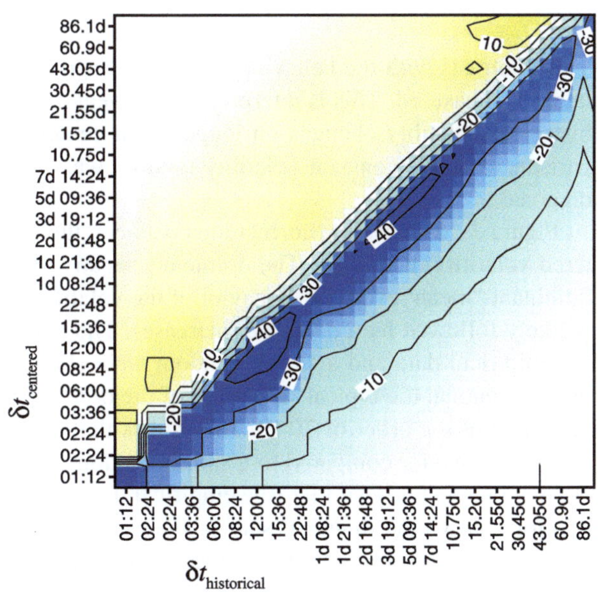

Fig. 3.25 The correlation between the historical volatility increment at time horizon δt_σ *(horizontal axis)* and the centered volatility increment at time horizon $\delta t'_\sigma$ *(vertical axis)*

The dynamical behavior of the volatility changes is investigated in Fig. 3.25 using the correlation between historical and centered volatility increments. For identical time horizons, the negative correlations are due to the "mean reverting" behavior of the volatility, namely a volatility increase is likely followed by a decrease (and vice-versa). This negative correlation extend below the diagonal, namely a historical volatility change influences the centered volatility change at shorter time horizon.

Fig. 3.26 The correlation between the historical volatility at time horizon δt_σ (*horizontal axis*) and the centered volatility increment at time horizon $\delta t'_\sigma$ (*vertical axis*)

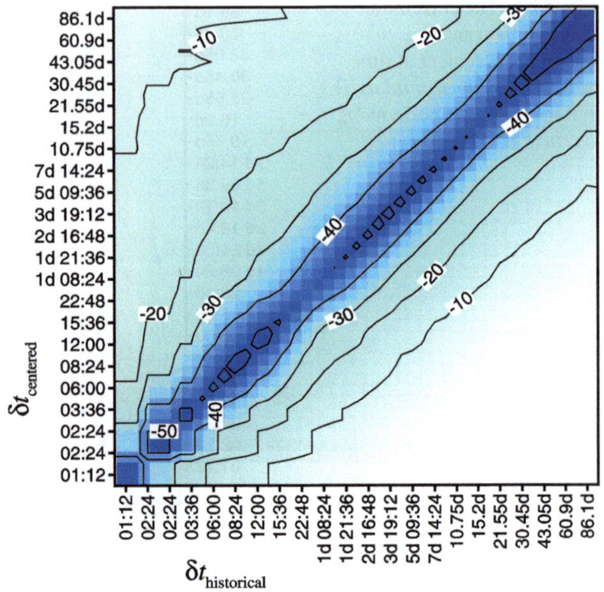

This contrasts with the behavior above the diagonal, where a slight positive correlation is measured. This is interpreted as a weak positive influence from short term historical volatility changes on longer term centered volatility changes. By comparison, a simple constant volatility random walk has a zero correlation above the diagonal.

Figure 3.26 shows the correlations between the historical volatility and the centered volatility increment. The dominant negative values give another view of the dominant "mean reverting" behavior of the volatility, namely a high (low) volatility is likely followed by a decrease (increase). The comparison of this figure between the empirical data and a simple random walk shows interesting differences. First, on the diagonal the typical values are of the order of 50 %, whereas a random walk has value of the order of 70 %. This shows that empirical time series have a more dynamic behavior compared to a simple random walk. Second, the behavior below the diagonal is fairly similar (the empirical data has a slower decay). Third, the correlations above the diagonal stay in the range of -10 % to -20 %, whereas the correlations are essentially 0 for a random walk. This shows the influences of the volatility at short time horizons on the subsequent volatility change at longer time horizons, even though the time interval difference could be large.

Overall, these figures for the correlations of the volatility increment are dominated by the "mean reverting" behavior of the volatility, as shown by the mostly negative correlations. Beyond this dominant feature, there appear structures related to the multiscale long memory dynamic of the volatility, but overall the differences with a simple constant volatility random walk are small. Moreover, this set of figures is well reproduced my most multiscale ARCH processes. This shows that the explicit inclusion of volatility increment terms in the process equation are not needed in order to reproduce the empirical correlations.

3.8 Volatility Graining

The volatility is essentially an estimator for the standard deviation of the return on a given time interval δt_σ. A simple quadratic definition for the volatility σ is given in Sect. 2.2.3:

$$\sigma_h^2[\delta t_\sigma, \delta t_r](t) = \frac{1}{n} \sum_{t-\delta t_\sigma+\delta t_r<t'\leq t} r^2[\delta t_r](t') \tag{3.8}$$

The integer n is the number of returns included in the sum. The summation boundaries are such that only the prices in the interval $[t - \delta t_\sigma, t]$ are used. This definition makes clear that the volatility depends on *two* time horizons:

- the time interval δt_r of the return $r[\delta t_r]$,
- the time interval δt_σ over which the returns square are summed.

The most important parameter is δt_σ: as long as there are enough independent returns in the sum for a given δt_σ, i.e., $\delta t_\sigma/\delta t_r$ large enough, σ measures the price fluctuations and has only a weak dependency on δt_r.

In this section, we want to explore the role of the volatility "graining" given by the ratio $\delta t_\sigma/\delta t_r$, by fixing δt_σ and changing δt_r. In most of the literature, the volatility depends explicitly only on the parameter δt_σ, whereas δt_r is fixed either implicitly or explicitly. With daily data, the return time interval is taken to be one day: $\delta t_r = 1$ day almost always, regardless of δt_σ. As in general the ratio $\delta t_\sigma/\delta t_r$ is not large, this is an optimal choice. With high-frequency data, a common choice is to fix the ratio $\delta t_\sigma/\delta t_r$. For example, in most of this book, we took $\delta t_\sigma/\delta t_r = 24$ for the empirical study. Yet, it should be clear that different choices for the two parameters are simply different definitions of the volatility, measuring the price fluctuations with different estimators. When the number of independent returns entering in the estimator is too small, a poor estimator is obtained. A typical example of a very poor estimator is the estimate of daily volatility with daily time series, as only one return is available to compute the volatility.

The importance of the volatility granularity is measure by the correlation between historical and realized volatility, with the same δt_σ but different δt_r:

$$\rho_{\mathrm{gr}}(n, n') = \rho\big(\sigma_h\big[\delta t_\sigma, 2^n \delta t\big], \sigma_r\big[\delta t_\sigma, 2^{n'} \delta t\big]\big) \qquad 1 \leq n, n' \leq n_{\max} \tag{3.9}$$

and $\delta t_\sigma = 2^{n_{\max}} \delta t$. For the empirical figures, the value $\delta t_\sigma = 2^{10} \delta t = 2$ day 03:12 was selected. The horizontal (vertical) axis gives the value for n (n'), and the correlation is given by the color. Large (small) value for n corresponds to a coarse (fine) grained definition of the volatility. Notice that with a coarse definition of the volatility (large n value on the axes), the number of returns used to compute the volatility is small, and hence the statistical volatility estimator has a large variance. This effect leads to the lower correlations on the upper border and on the right border. The efficiency of the volatility estimator is combined with the "graining" effect to produce the observed graph. On the empirical figure, notice that the maximum for the correlation is not at the finest granularity, namely for the best estimator. This indicates that coarse grained volatility contains more information.

Dacorogna et al. [50] (see also [48]) discovered an important effect related to the granularity of the volatility. They summarize the finding in the sentence: "coarse volatility predicts better fine volatility than the other way around." The effect was found by studying lagged correlation between two volatilities with the same time horizon δt_σ but different granularity δt_r. In [50], the authors went further by using this asymmetry to build the HARCH model. The HARCH model uses several volatility components with time horizons taken as a geometric series and with a fixed ratio $\delta t_\sigma / \delta t_r$. The "intensity" coefficients for all volatility components are taken as parameters of the model, and a log-likelihood optimization shows larger coefficients at the natural human time horizons (few hours, 1 day, 1 week, ...). This paper is important as it is the first to show indirectly the presence of different market components. However, the model includes different ingredients, in particular multiple time horizons and the volatility coarse graining. A subsequent study [152] shows that the success of the model is due to the multiple volatility components, but a log-likelihood estimator does not differentiate models with different granularity. In our systematic study of the ARCH processes in Chap. 7, we include several models in two versions differing by the volatility granularity: a "microscopic" one with $\delta t_r = \delta t$ (i.e., the volatility is a measure at the finest granularity) and an "aggregated" one with $\delta t_\sigma / \delta t_r$ fixed. The comparison between the empirical and simulated figures for the volatility graining shows that only the aggregated models are able to reproduce the observed asymmetry.

Figure 3.27 is slightly different compared to the study of [50] as all possible granularities are studied (by using the correlation between historical and realized volatility), but only the first lag at δt_σ is computed. Then, the coarse/fine volatility asymmetry corresponds in the figure to the dissymmetry around a diagonal axis. Figure 3.28 shows the asymmetry for the lagged correlation between fine and coarse grained volatility. More precisely, this figure measures the asymmetry in the correlations between historical fine grained and realized coarse grained versus historical coarse grained and realized fine grained volatilities. The positive correlation (yellow–red) in the lower right corner indicates that historical coarse grained volatility predicts better realized fine grained volatility than the other way around, in agreement with [50].

Both graphs measuring the granularity effect proved very useful when selecting models for the price process. For example, the GARCH(1, 1) process has a completely symmetric volatility granularity figure, and similarly for all the "microscopic" ARCH-like processes. This shows that the correct amount of granularity has to be put into the measures of volatility entering into a process. This is in agreement with our expectation about the market participants: peoples working at long (short) term horizons use low (high) frequency data. On the graphs corresponding

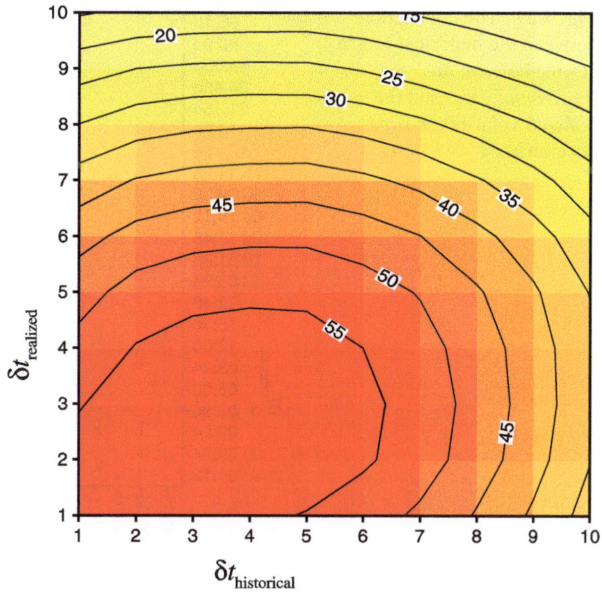

Fig. 3.27 The lagged correlation between fine and coarse grained volatility. The *axes* gives the logarithm (base 2) of the return time horizons δt_r used to compute the volatility

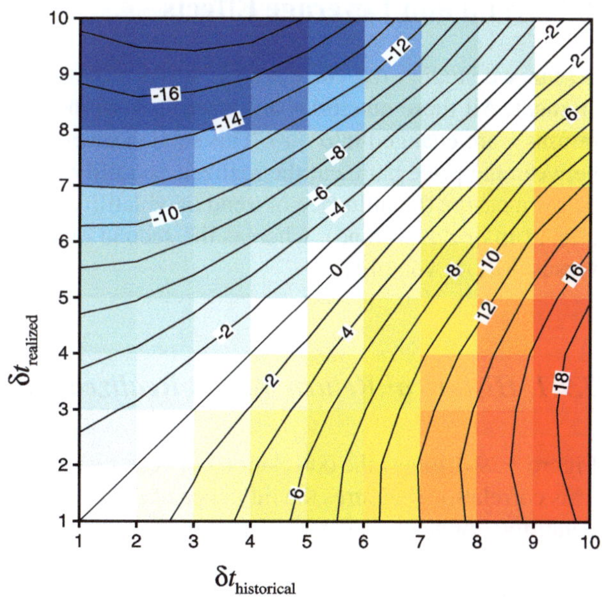

Fig. 3.28 The lagged correlation asymmetry between fine and coarse grained volatility

to empirical data, the observed asymmetry is related to time reversal, as the exchange of the axes is equivalent to permute the historical and realized volatilities. This asymmetry is studied in detail in Chap. 11, both for empirical time series and for processes.

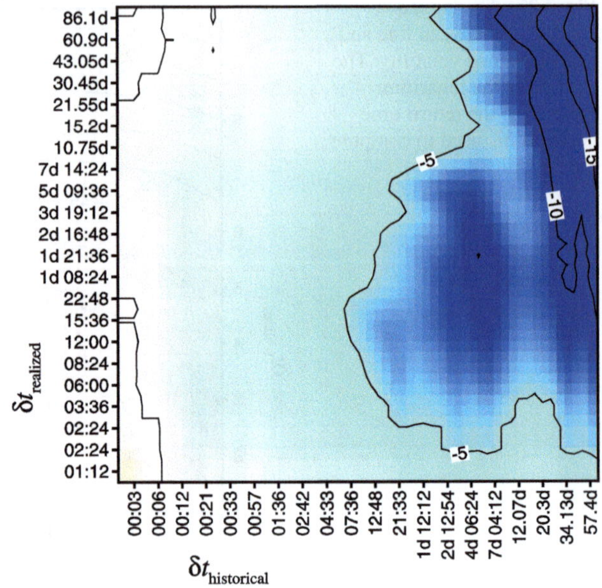

Fig. 3.29 The correlation between the historical return at time horizon δt_r (*horizontal axis*) and the realized volatility at time horizon δt_σ (*vertical axis*)

3.9 Trend and Leverage Effects

So far, we studied the correlations between the volatility and/or the volatility increment. All these quantities are insensitive to the sign of the return (i.e., if all the returns r are changed into $-r$). This section investigates possible direct connection between the price path and the realized volatility. In a broad sense, the correlations below measure the influence of trend on volatility. The first one seeks for directional effect (up or down trend), whereas the second one measures the influence of trend versus drift on volatility.

3.9.1 Historical Return Versus Realized Volatility Correlation

Figure 3.29 displays the correlation between historical return and realized volatility. This correlation measures the influence of raising or falling prices on the subsequent volatility. For stocks and stock indexes, downward moves are considered as bad (as most of the market participants have long positions in these securities). A strong down move can trigger a sale and therefore increases the subsequent volatility, whereas an up move entices securities holders to keep their positions, reducing the subsequent volatility. This argument leads to a negative correlation between historical return and realized volatility. For stock index, using daily data, this effect has been demonstrated by [31, 39].

This argument does not work for foreign exchange, where the two currencies play a symmetrical role. If the interchange of the two currencies is an exact symmetry, all the return-volatility correlations should be zero, because they are odd and even quantities under the currencies exchange (exchanging currencies is equivalent to take the inverse of the price, which is mapped into minus the return after the log transformation). Yet, this symmetry is only plausible, and there is no fundamental reason for its existence. Indeed, for countries with very dissymmetric economies, the two currencies can well have a dissymmetric role. Similarly, the USD plays a unique central role in the world economy during the 20th century.

The return-volatility correlation for CHF/USD is found to be overall negative, with a small magnitude but a consistent pattern across time horizons. For a long memory ARCH process which has exactly zero return-volatility correlation by symmetry, a Monte Carlo simulation over a similar time span shows much smaller correlations than the empirical one. Moreover, other currency pairs show similar magnitude for the return-volatility correlations. These statistics cast a doubt on the exact currencies exchange symmetry argument. This topic should deserve a full study in itself, possibly with long daily time series and rigorous statistical bounds on the estimated correlations.

3.9.2 Trend/Drift Versus Realized Volatility Correlation

We can expect that large market swings have an impact on the subsequent volatility. The idea is that large moves, up or down, can induce market participants to modify their positions, whereas in "drifting" conditions, with prices evolving in a small range, participants keep their positions. The problem is to construct an indicator sensitive to a "trending" prices versus a "drifting" market. A simple indicator with such property was discovered in [171] using genetic programming (GP) to investigate systematically volatility forecasts. In this work, the usual GP is enhanced by a typing system, and the types are used to enforce an even/odd symmetry in the tree values. For an objective function measuring the volatility forecast accuracy, an even parity of the GP tree is enforced. Interestingly enough, the GP fairly quickly discovers models equivalent to GARCH$(1, 1)$. The next systematic enhancement corresponds to a trend versus drift indicator. The fact that one similar indicator beyond GARCH was discovered systematically by the GP algorithm shows its quantitative importance in our understanding of the volatility dynamics.

The trend/drift indicator is given by $r[\delta t_r](t) \cdot r[\delta t_r](t - \delta t_r)$, namely by a product of two nonoverlapping returns. It is even in the returns, with a positive value indicating a trend when both returns have the same sign and a negative value indicating a drift when both returns have opposite sign. Using the lag operator L, we denote this term by $r \cdot L[r]$, with $(r \cdot L[r])(t) = r[\delta t_r](t) \cdot L[\delta t_r; r[\delta t_r]](t) = r[\delta t_r](t) \cdot r[\delta t_r](t - \delta t_r)$. In [157], the inclusion of such term in ARCH processes is systematically investigated. The effect of trend/drift on the subsequent volatility is measured by the correlation between $rL[r]$ (a historical quantities) and the realized volatility σ_r. The study in [157] covers the richer class of indicators given by $r[\delta t_r](t) \cdot r[\delta t_r'](t - \delta t_r)$. As the realized volatility $\sigma_r[\delta t_\sigma]$ has its own time horizon δt_σ, the correlation is depending on the three parameters δt_r, $\delta t_r'$, and δt_σ. The

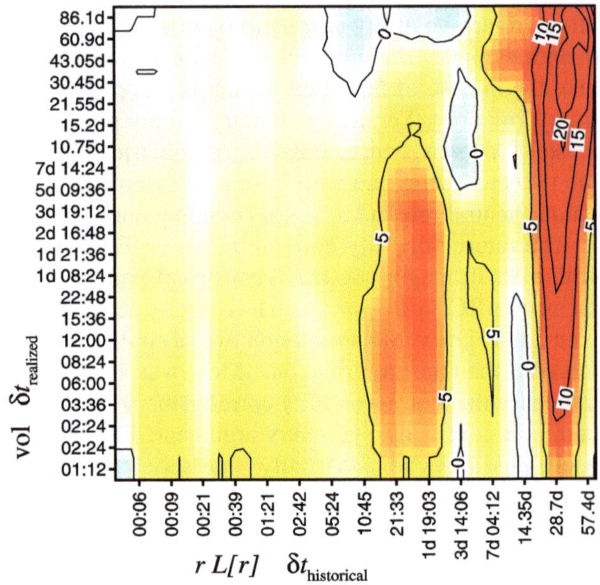

Fig. 3.30 The correlation between the realized volatility $\sigma_r[\delta t_\sigma]$ at time horizon δt_σ (*vertical axis*) and the product of the historical return and lagged historical return $r[\delta t_r]L[r[\delta t_r]]$ at time horizon δt_r (*horizontal axis*)

empirical analysis shows that the largest correlations are obtained for $\delta t_r = \delta t'_r$, and we have kept this condition to reduce the dimensionality of the problem.

The correlation between the product of (non overlapping) historical returns $rL[r]$ and the realized volatility at time horizon δt_σ is shown on Fig. 3.30. This correlation measures the influence of historical trend versus drift for the prices on the subsequent volatility. According to the argument above, we expect that price swings induce trades, hence volatility. This argument leads to a positive correlation between $rL[r]$ and the realized volatility, as is indeed observed on empirical data. The empirical pattern is again nonuniform but shows maxima at time intervals corresponding to the natural human activity cycles of day, week, and month.

The implication of such dependencies for process building and selection is interesting. Clearly, the effect is absent in all "simple" processes, and specific terms should be added. In a stochastic volatility framework, the trend/drift effect cannot be inserted in a natural way as the volatility is a process independent of the returns. By contrast in ARCH processes, it is easy to add terms of the form $rL[r]$ at different time horizons in the processes. Such path is explored in Sect. 7.13.

3.10 Time Reversal Invariance

The core question of time reversal invariance (TRI) is if a time series of prices originating from the financial market is time reversed, can we "see" the difference? As an example, Fig. 11.1 shows four time series, and one had been reversed, namely is presented backward. Can the reader spot which one? Clearly, at the level of prices or of the returns, it is indeed very difficult to notice a difference because financial time series are dominated by a randomness that hides possible structures and orders. Therefore, an appropriate formulation of the question is whether statistics can show the presence or absence of time reversal invariance.

The answer to the above question is yes, namely some statistics are sensitive to TRI, and they show clearly that empirical data are not TRI. These statistics are based on the probability distribution of the volatility increment (Sect. 3.2.3), on the historical/realized volatility correlations (Sect. 3.6.3), and on the correlation between fine grained and coarse grained volatilities (Sect. 3.8). Essentially, changing the direction of time is changing the role of the historical and realized volatilities. For TRI time series, Figs. 3.21 and 3.27 are symmetric under a reflection by a 45 degree line. The clear and consistent level of asymmetry in these figures points to a clear asymmetry under time reversion. The same criterion applied to processes is very selective with respect to choosing a structure that can duplicate the empirical facts. Because the comparison of TRI between empirical data and processes is very instructive, and indeed selective with respect to the structure of the processes, we prefer to postpone this discussion until Chap. 11. For the time being, we will add to the list of empirical facts that the empirical financial time series are not TRI, namely an appropriate statistics can detect which series has been reversed in Fig. 11.1.

Chapter 4
Empirical Mug Shots

In the previous chapter, empirical facts have been collected one by one. The goal was to extract as many features as possible, and to discuss the various computations, parameters, and results. After this fairly exhaustive collection, we want to summarize the key findings in one page that characterizes best empirical time series and processes used to model them. We call this summary a "mug shot".[1] The mug shot for CHF/USD is given in Fig. 4.1, with the following subplots.

- 1st row, left: The scaling of the moment of the probability distribution of the returns.
- 1st row, right: The lagged correlation for the absolute value of the return, in a linear–linear scale. This graph characterizes the heteroscedasticity.
- 2nd row, left: The relative excess kurtosis of the probability distribution of the returns. This graph characterizes the convergence toward a Gaussian distribution.
- 2nd row, right: The lagged correlation for the volatility, in a linear–logarithmic scale. This graph gives a precise characterization of the volatility clustering.
- 3st row, left: The correlation between historical and realized volatilities, with the respective time horizons given on both axes. This graph characterizes the dependency structure between time horizons.
- 3st row, right: The asymmetry of the correlation between historical and realized volatilities. This graph characterizes the asymmetry with respect to time reversal invariance.
- 4st row, left: The correlation between the change in volatility (centered) and the realized volatilities, with the respective time horizons given on both axes. This graph characterizes the dynamic of the volatility evolution
- 4st row, right: The correlation between historical and realized volatilities, at the same time horizon but with different granularity. This graph characterizes the time horizon's return structure of the volatility, and its asymmetry gives another measure of time reversal invariance.

[1] In police slang, a mug shot is the pair of pictures of someone face, one front, one profile.

G. Zumbach, *Discrete Time Series, Processes, and Applications in Finance*,
Springer Finance, DOI 10.1007/978-3-642-31742-2_4,

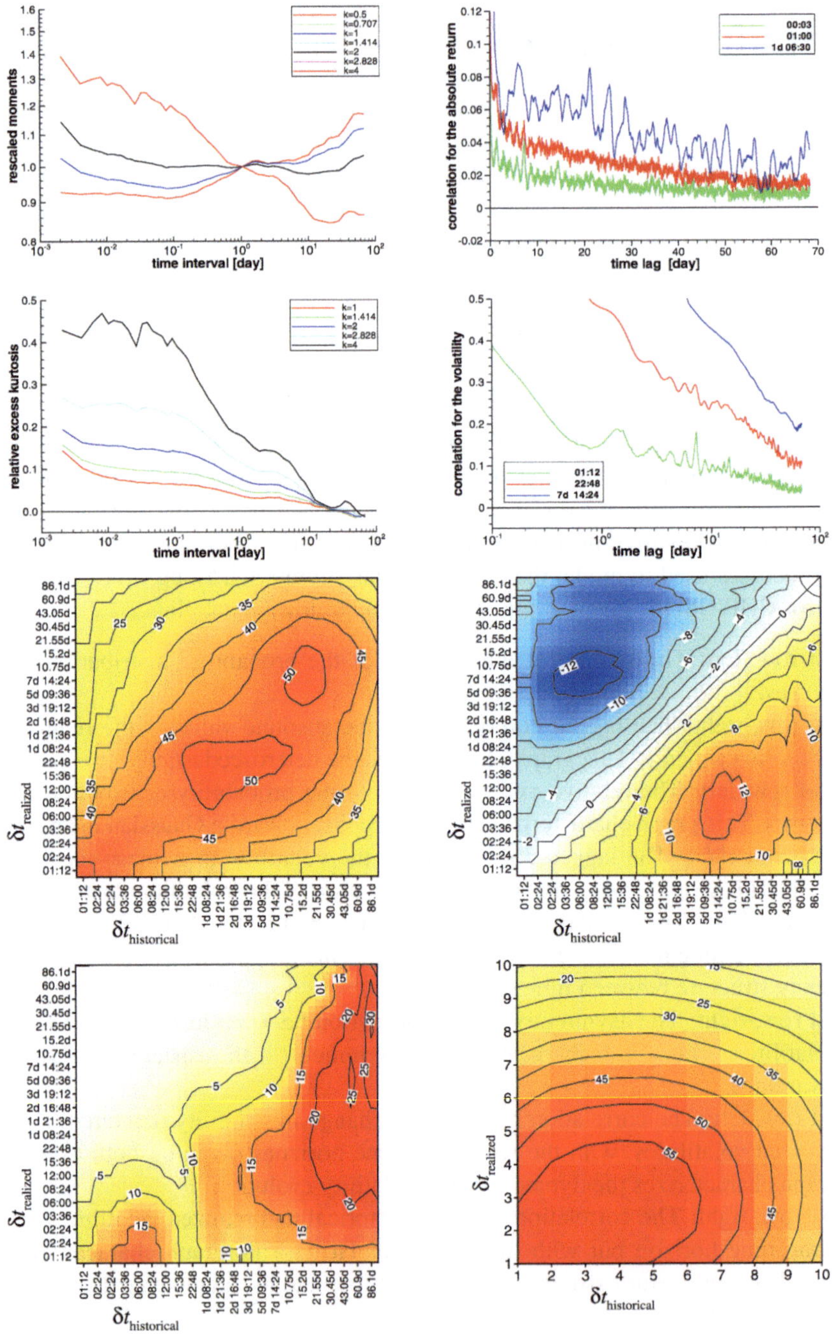

Fig. 4.1 The mug shot for the exchange rate CHF/USD

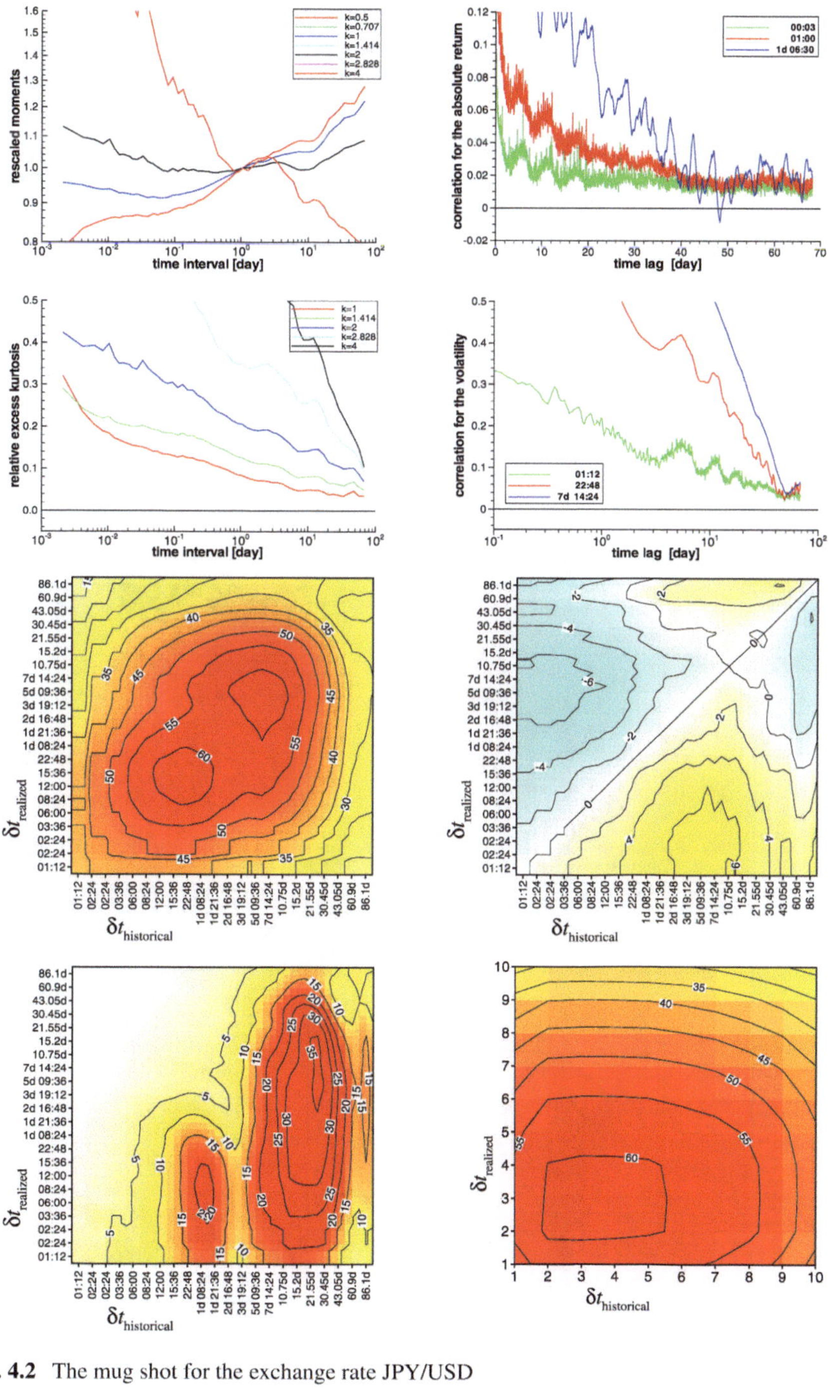

Fig. 4.2 The mug shot for the exchange rate JPY/USD

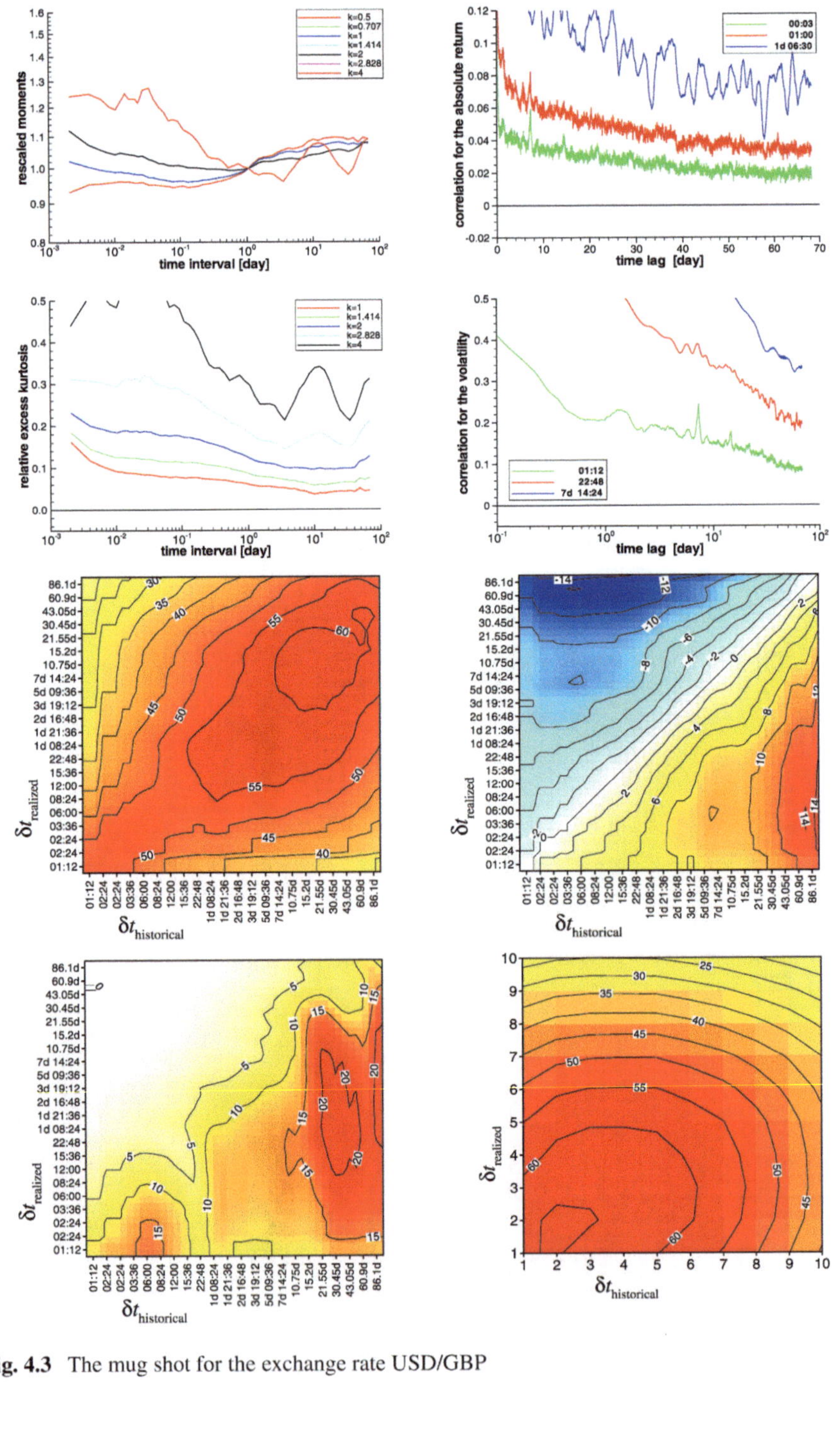

Fig. 4.3 The mug shot for the exchange rate USD/GBP

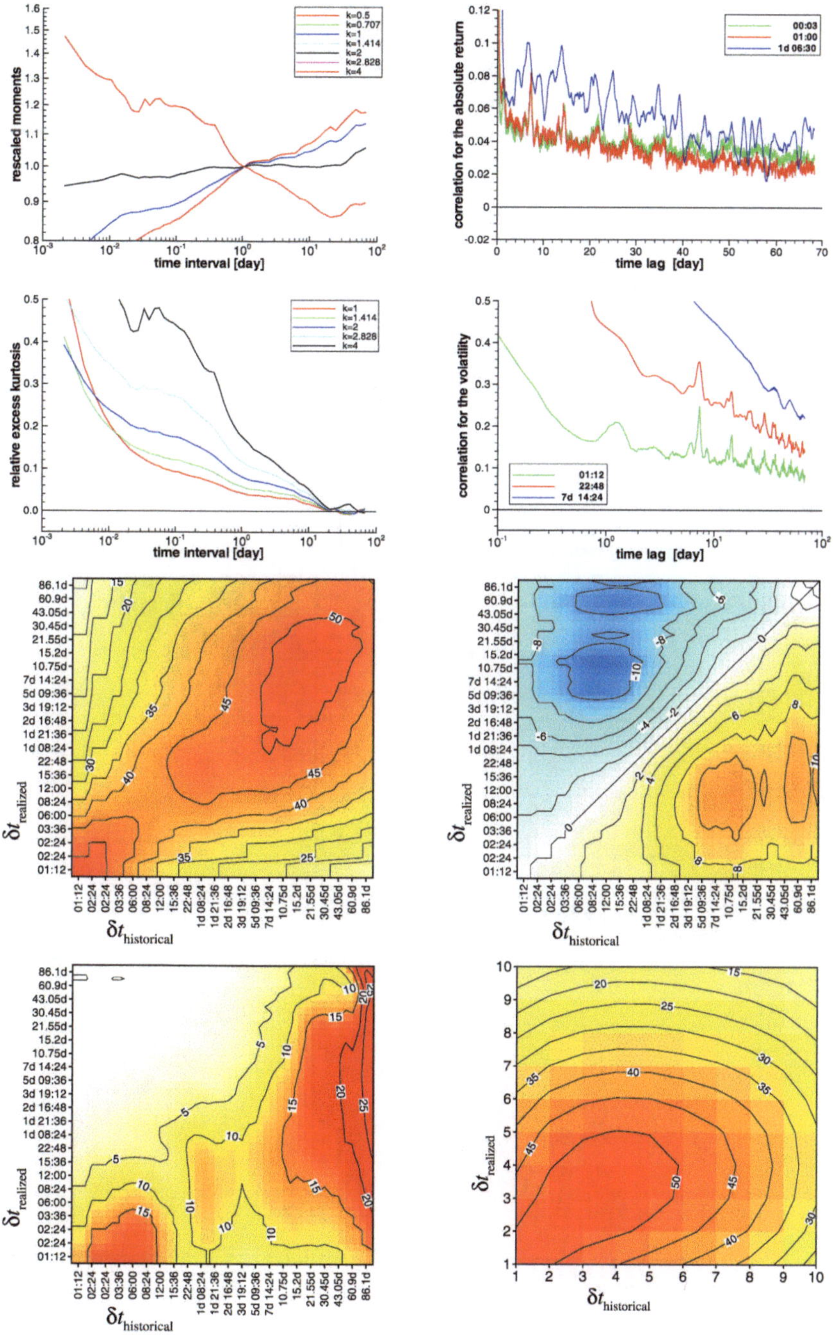

Fig. 4.4 The mug shot for the exchange rate DKK/USD

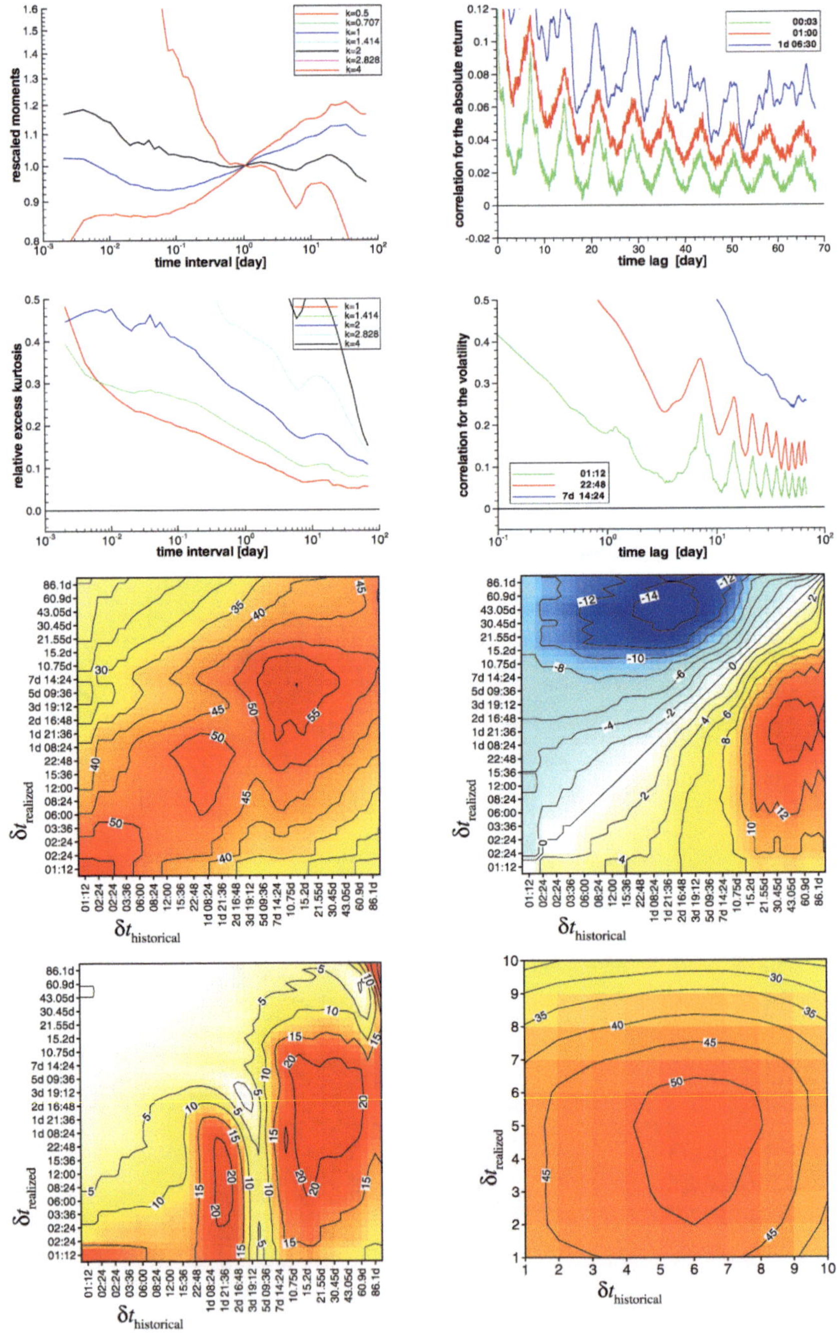

Fig. 4.5 The mug shot for the gold expressed in dollar XAU/USD

As will be clear when studying processes, both graphs on the third line are key in reproducing the distributional properties and the clustering properties, namely when the time structure of a process is correct, the graphs on the first and second rows are also correct.

The mug shots given in Figs. 4.2, 4.3, 4.4, and 4.5 correspond respectively to JPY/USD, USD/GBP, DKK/USD, and XAU/USD. The most important overall feature is the strong similarity of the empirical time series with respect to their key statistical properties. At least in the space of the foreign exchange rates, all financial time series exhibit the same most important scaling and dependencies. Essentially, this set of mug shots will be used as a guide line for the study of the theoretical processes, and in particular the CHF/USD is used to set the parameters of the processes in the forthcoming chapters.

The graphs included in the mug shot have been selected because they convey the most salient characteristics of the empirical data and processes. Much more graphs have been made in order to explore thoroughly the properties of the time series and to understand the relations between the mathematical structure of a process and its statistical properties. As mentioned in the introduction (page 5), all the graphs are available on the site www.finanscopics.com. The web site includes more processes and sets of parameters than given here, and will be extended over time. With this public tool, the reader can easily study and compare the various properties of the empirical time series and processes.

Chapter 5
Process Overview

So far, the focus was on extracting as much information as possible from the financial time series, with a special concern on generic features. This overview of the "stylized facts" sets a benchmark against which various processes can be gauged. The next goal is to construct mathematical processes that can reproduce some, possibly all, stylized facts. This is of primary concern for many areas in finance, for example, portfolio construction, derivative pricing, or risk control.

This chapter gives a first "grand tour" of the existing processes used to model financial time series. The goal is to set the notations and basic ideas, as well as to present the *structures* for the main models used in finance. Then, subsequent chapters delve into more details and extensions of the basic processes. The challenge then is to be realistic, namely to reproduce as many stylized facts as possible. This will require more complex structures, but possibly some empirical features cannot be reproduced at all inside a given structure. The "mug shots" give a convenient summary of the main properties, either of the empirical data or of the processes.

5.1 Why Using a Finite Time Increment for the Processes?

The process equations are given in this book in a discrete form, with a finite time increment δt. There are several reasons for this choice.

- The empirical data are known with a finite time increment, mostly of one day, possibly at a shorter time increment with tick-by-tick data. The processes should reproduce the empirical time series at this time scale and longer, and the behavior of the data and processes at shorter time horizon than δt is irrelevant, as unaccessible empirically.
- A continuous-time process needs to be discretized in order to perform Monte Carlo simulations, to compute a forecast, or to evaluate a figure of merits like a log-likelihood. The discretization of a continuous-time stochastic equation is a nontrivial business. Because in general stochastic equations involve terms of

G. Zumbach, *Discrete Time Series, Processes, and Applications in Finance*, Springer Finance, DOI 10.1007/978-3-642-31742-2_5,

order $\sqrt{\delta t}$ and δt, inequivalent discretizations can be constructed. A good example is the Heston (continuous-)time process which can be discretized in different ways, with different properties with respect to positivity [73].

- The continuous-time limit of some discretized process may not exist, or be quite different to a "naive" limit. An example of a process with a nonobvious limit is the GARCH(1, 1) process [45, 115]. At a deeper level, the description of financial time series at the tick-by-tick time horizons involves a new phenomenology, related to order queue, limit orders and market orders, and to the exchange of information [46, 144, 162]. This new phenomenology is important at time horizons ranging from a few seconds to a few hours. Only at longer time scales, the details of the price formation mechanism are small enough to be safely neglected. This very short term effects are not included, and are not needed, at a mesoscopic scales ranging from a few hours to months. Therefore, the continuous limit of a mesoscopic model is anyway missing some part of the price formation mechanism.
- The comparison of the process properties with empirical data shows that for *all* processes, fat-tailed innovations should be used. The most common choice is to use Student innovations with a number of degrees of freedom in the range from 3 to 7, albeit other distributions are possible. The Student distribution is not a stable distribution, namely the sum of independent Student variables is not distributed according to a Student distribution. For this reason, it is not possible to construct a process where the increments over several time intervals are distributed according to a Student distribution. The Student distribution has a weaker property, namely is infinitely divisible. An infinitely divisible distribution $P(y)$ is such that for any integer n, a probability distribution $p_n(x)$ can be found so that the sum $y = \sum x_i$ of n independent variables $x_i \sim p_n(x)$ is distributed according to the distribution P. This condition is needed to construct Lévy processes (see, e.g., [44]). Therefore, a Lévy process can be constructed so that the innovations over one selected time interval δt are distributed according to a Student distribution. But the independence condition for the increments leads to a central limit type convergence toward a normal distribution for longer time intervals.

In the continuum formulation, the Itô calculus is based on Wiener processes. The Itô calculus is very important to price derivatives and to formulate rigorously the associated replication strategy. In order to price contingent claims in a Black–Scholes scheme, the process equations can be fairly general, but the source(s) of randomness has to be infinitely divisible. The scheme can be generalized to accommodate random volatility, processes with jumps or a broader class of distributions (but still with infinite divisibility) [44]. Yet, stochastic calculus cannot accommodate dependency involving the process history like in a GARCH model.

In a discrete-time formulation, there is no rigorous Itô calculus, and this creates a problem with respect to option pricing as an important computational tool is now missing. Yet, if the time increment is sufficiently small, a "physicist" approach can be used in algebraic manipulation by using the simple rule that $E[dz^2] = dt$ and neglecting higher-order terms. Therefore, the key issue is not the continuum limit *per se*, but that many interesting process equations do not admit a straightforward continuum extension. Yet, it would be wrong to select a particular process structure because of its mathematical tractability in an idealized continuum limit. It is to the mathematics and our models to adapt to the empirical properties of the financial time series, as clearly the financial world will not change just to conform to an elegant mathematical idealization. The formulation of the process equations directly in a discretized framework avoid these difficulties. Recent progresses for option pricing has been made in order to accommodate a broader class of volatility processes like GARCH [40] and fat-tailed innovations [116]. These formulations are based on a discrete-time increment for the underlying processes and are presented in Chap. 16.

5.2 The Definition of the Returns

The vast majority of the time series models in finance are derived from a random walk for the logarithm of the price

$$\log\bigl(p(t+\delta t)\bigr) = \log\bigl(p(t)\bigr) + r(t+\delta t). \tag{5.1}$$

The change of variable $x = \log(p)$ makes this equation slightly simpler. The (logarithmic) return is then

$$r(t+\delta t) = \mu(t) - \frac{1}{2}\sigma_{\text{eff}}^2(t) + \sigma_{\text{eff}}(t)\epsilon(t+\delta t) \tag{5.2}$$

with μ fixing the drift and σ_{eff} the effective volatility. The Itô term $-1/2\sigma_{\text{eff}}^2$ is such that the mean expected price is μ. In general, the innovations ϵ are i.i.d. random variables with

$$E[\epsilon(t)] = 0 \tag{5.3}$$

$$E\bigl[\epsilon^2(t)\bigr] = 1 \tag{5.4}$$

but the distribution is otherwise not specified. Both conditions are needed so that the drift μ and volatility σ_{eff} can be identified.

This is not the only way to write the basic random walk equation for a finite discretization step δt. A slightly different discretization is given by

$$p(t+\delta t) = p(t)\bigl(1 + r(t+\delta t)\bigr) \tag{5.5}$$

with the (relative) return

$$r(t+\delta t) = \mu(t) + \sigma_{\text{eff}}(t)\epsilon(t+\delta t). \tag{5.6}$$

Up to higher-order terms in δt, both discretizations are equivalent, and therefore they share the same continuum limit. Yet, they have different properties for fat-tailed innovations, and in particular the discretization in term of the logarithmic return leads to diverging expectations. This issue is explored in more details in Chap. 6.

The Gaussian random walk corresponds to no drift $\mu = 0$, a constant volatility $\sigma_{\text{eff}}(t) = \sigma_\infty$, and to a normal distribution for the innovations $p(\epsilon) = N(0,1)$. An example of returns for a Gaussian random walk is given in Fig. 5.4 in the top panel. This random walk model was written by Bachelier in 1900 [12], but for the price (i.e., with p instead of $x = \ln(p)$ in Eq. (5.1)). It is only in the 1960s that it becomes clear that the model should be written for the logarithm of the price. This change of variable can be deduced from a change of numeraire argument, namely the change $p \to \alpha p$ for any positive α corresponds to a change of the currency unit, and this change should not affect the economy. The logarithm transformation changes the scaling by α into an additive transformation, and this leaves the logarithmic price differences $r(t) = x(t) - x(t - \delta t)$ invariant for any α. This argument can be used for any contract which basically trades the numeraire, like stocks, stock indexes, or FX rates. For other time series, like interest rates, implied volatilities or spreads, the change of numeraire argument cannot be used (even if denominated in a currency unit, a bond price is trading the underlying interest rates). For these other time series, the change of variable that transforms the time series into a random walk is open as the change of numeraire argument cannot be used. For example, with interest rates, [154] uses the condition that the volatility $\sigma = \sigma(x)$ computed with the transformed price should be uncorrelated with the price, namely $\rho(p, \sigma) \simeq 0$.

5.3 The Most Important Stylized Facts

As a daily return is the sum of many intra-day transactions, the central limit theorem justifies to model the returns with a Gaussian distribution, at least at first glance. In the other direction, as a Gaussian random variable can be decomposed as a sum of random variables distributed according to a Gaussian law, a continuum limit can be constructed where the infinitesimal random variables are Gaussian. This construction is such that, at any time scale, the corresponding random variable is Gaussian. This construction can be made rigorous, leading to Wiener process. Yet, as explained in the later sections, the relationship between return distribution and the central limit theorem is quite complex because the returns are not independent. Independence is one of the prerequisites for the central limit theorem to hold, and as the dependency between returns is subtle, the return distribution is converging to a Gaussian but at a much slower pace.

The basic Gaussian random walk is very successful at giving a first good description of the behavior of the prices, and today a large part of finance is relying on this model. However, a simple visual comparison of returns time series shows that this model is deficient on two important aspects.

1. The volatility is constant, namely the time series is homoscedastic.
2. The return distribution is Gaussian instead of the observed fat-tailed distribution.

Both points can be seen directly on plots of the returns as function of time, like in Figs. 5.1, 5.2, and 5.3 for empirical time series and on Fig. 5.4 for some popular

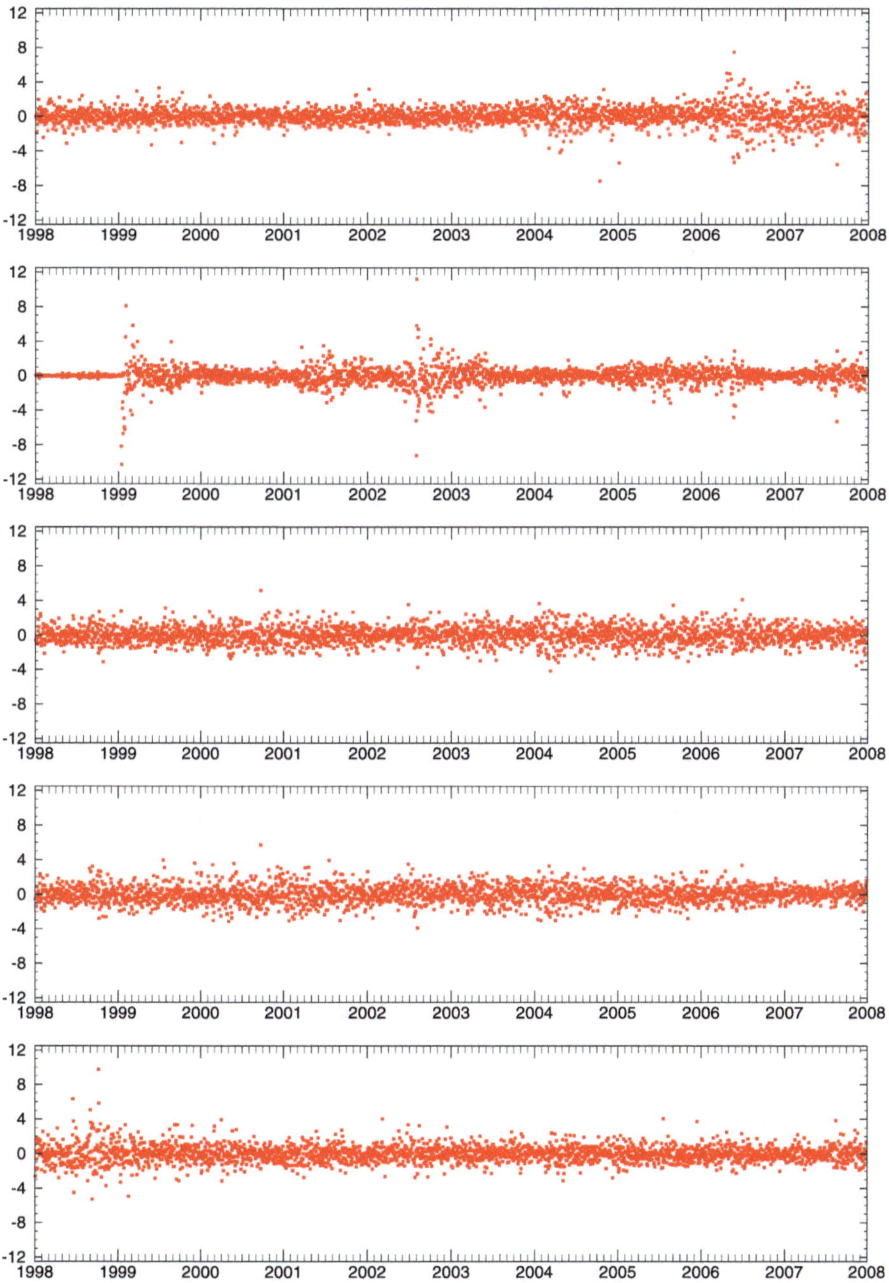

Fig. 5.1 The standardized return for one commodity and some FX against the USD (from top to bottom): 3 month future on copper, BRL/USD (Brazil), GBP/USD (GB), EUR/USD (Euroland), and JPY/USD (Japan)

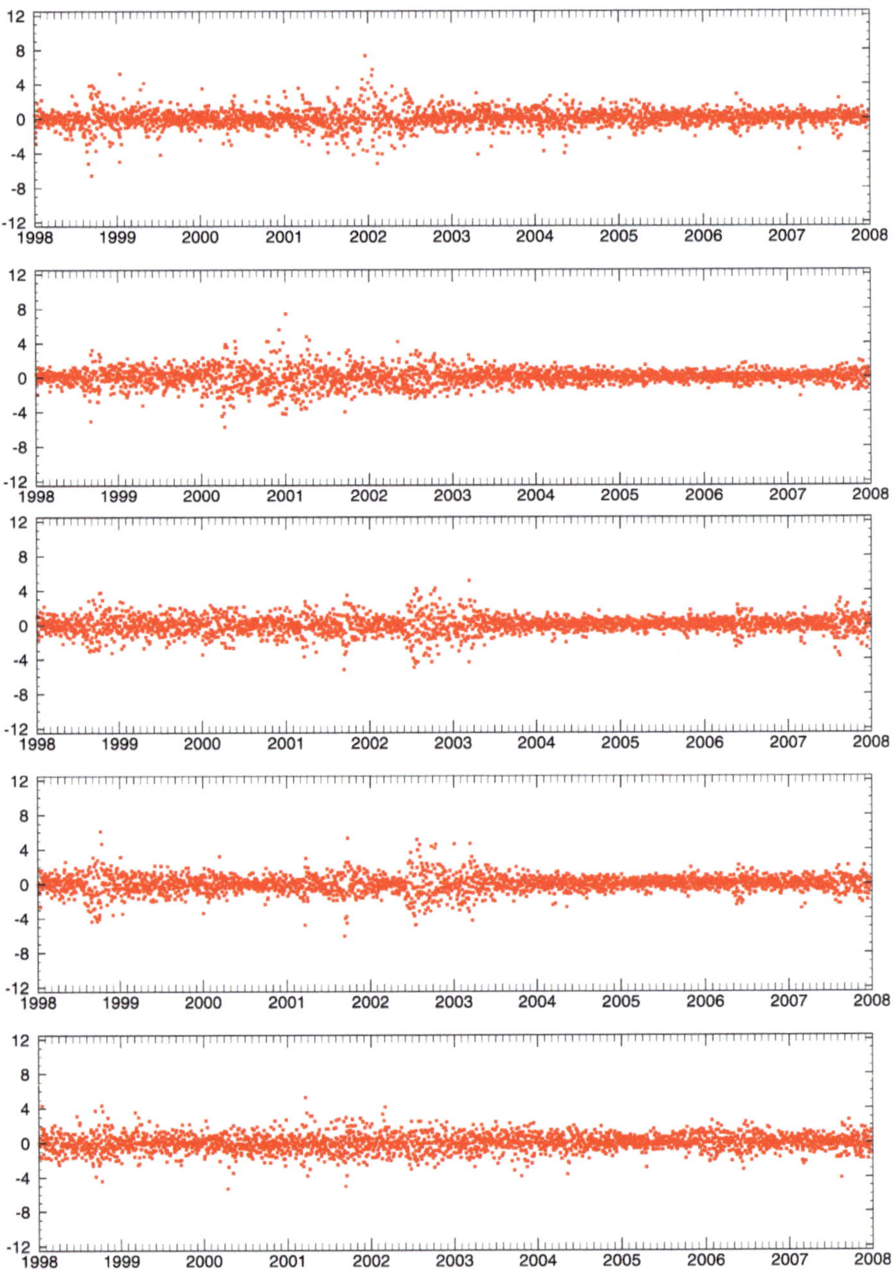

Fig. 5.2 The standardized return for several stock indices (from top to bottom): Merval (Argentina), Nasdaq (USA), FTSE-100 (GB), SMI (Switzerland), and Nikkei-225 (Japan)

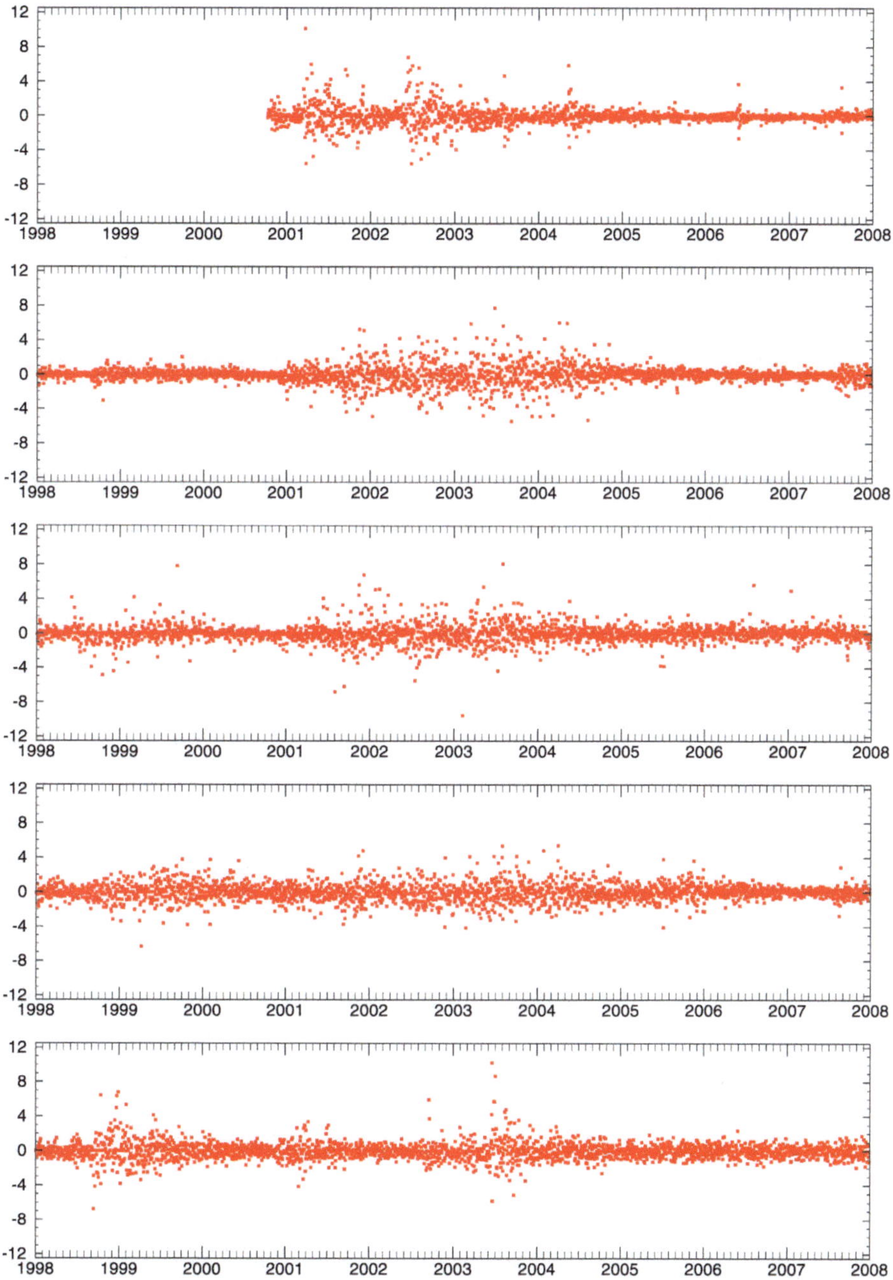

Fig. 5.3 The standardized return for several interest rates (from top to bottom): Brazil at 1 year, US at 1 year (interest rate swap), UK at 1 year (interest rate), Euro at 1 year (Government debt), and Japan at 10 years (Government debt; this period had very (very) low rate in Japan for IR up to 1 year)

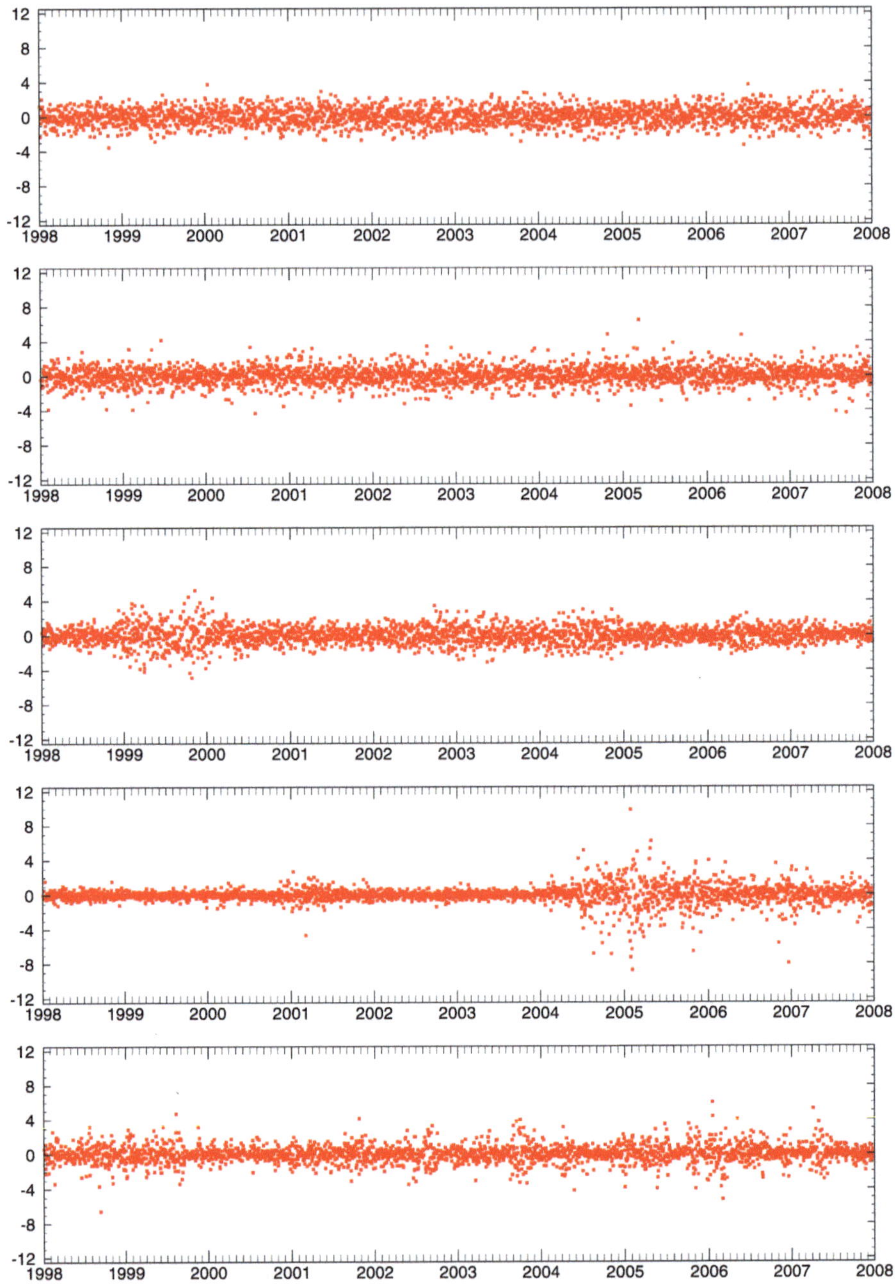

Fig. 5.4 The standardized return for a few processes (from top to bottom): constant volatility with normal innovations, constant volatility with Student innovations, GARCH(1, 1) with normal innovations, GARCH(1, 1) with Student innovation, and Heston process with normal innovations. Student innovations have 5 degrees of freedom, and the characteristic time of the GARCH and Heston processes is of one month (21 business day)

theoretical processes. For all plots, the returns are normalized using the empirical standard deviation computed on the visualized sample. In this way, the returns are displayed directly in terms of the sample empirical standard deviation. The empirical time series have been chosen to display many geographic areas and many instruments types over the same 5 year period, but have not been selected specifically otherwise. When visually comparing a Gaussian constant volatility random walk (Fig. 5.4, top panel) with the empirical time series, the deficiencies can be directly seen without any statistical analysis. Let us discuss both points in turn.

The magnitude for the returns is not constant, but there are periods of low volatility and periods of large volatility. The figure is at a time scale of half a decade, and the cluster of volatility with a characteristic length between months and years are clearly visible. If only one year of data is used for the figure, a similar picture emerges, but with a characteristic length for the clusters between weeks and months. This intuitively shows that there is not a single characteristic time for the volatility clusters, but that many time scales are needed to describe their dynamic. The multiple time scales structure is quantified by the lagged correlations for the volatilities, which show a slow decay (and not an exponential decay). This important empirical stylized fact was discussed in Chap. 3 and is illustrated by two graphs in the mug shots.

The range for the y-axes $(-12, +12)$ has been chosen so that the largest standardized daily return appears on the graphs. This range is very wide, and many daily returns are beyond a 6 sigma level. On the other hand, the Gaussian random walk does not display any point above a 5 sigma level. This simple comparison shows clearly that the returns have a fat-tailed distribution.

Both features need to be quantified by a serious statistical analysis of empirical data, yet the dominant characteristics of any financial time series can be seen with naked eyes on most graphs of the time series of returns. The next question is how to incorporate them in a process used to describe the evolution of prices. The easiest part is the fat tail, as it is enough to use fat-tailed innovations. More complex mechanism can be used, as, for example, in a GARCH process, but in practice all realistic models need to include fat-tailed innovations. The more complex part is the heteroscedasticity: to capture this feature, some memory from the past volatility needs to be included in the price process. Essentially three paths can be taken that are introduced in the next three paragraphs. A full discussion of each process classes are presented in the subsequent chapters, with Monte Carlo simulations of the key processes and the corresponding mug shots.

5.4 ARCH Processes

In a (G)ARCH process, the volatility is a function of the past return magnitudes, and this dependency on the past brings in the volatility clustering. The simplest model is to have the squared volatility (i.e., the variance) to be a weighted sum of past squared returns. For most ARCH processes, this dependency is induced by exponential moving averages of the squared returns. This structure provides for a efficient mean to construct estimators of the past volatilities at various horizons, but the dependency of the past returns can be more complex. Ultimately in a GARCH process, the volatility is a function of the past (squared) returns. The key point is that the volatility can be removed completely from the equations and the magnitude of the next return is a function of the past returns only. In this family of processes, the volatility is a parsimonious way to construct the equations and to interpret them, but is nothing more than a convenient intermediate variable that could be eliminated from the equations.

5.5 Stochastic Volatility Processes

In a stochastic volatility process, the variance has a higher status as it becomes an integral part of the process with its own source of randomness. The heteroscedasticity is introduced by a dependency of the volatility on its past, typically by a moving average. As the volatility has to be mean reverting, a natural candidate is an Ornstein–Uhlenbeck process. Then, the return appears as a slave process, with the magnitude given by the volatility process. In particular, there is no feed-back from returns to the volatility. Many variations can be constructed around this theme, while preserving the core feature of a volatility depending on its past, but not on the returns.

The process for the volatility has to preserve the positivity of the volatility. One simple way to enforce this restriction is to write a process for the logarithm of the volatility. Another way is to modify the amplitude of the random term by a power of the volatility, so that when the volatility decreases, the random term decreases, leaving the mean reverting term to "pull up" the volatility. Both families will be explored.

A closely related family of processes is based on the idea of subordinating the return to a random time. The time is viewed as progressing randomly, for example, increasing according to the events on the market. The process for the time can depend on its own past, in order to introduce heteroscedasticity. Then, the return follows a simple random walk with constant volatility but is indexed by the random time. This setup is in fact similar to a stochastic volatility process, with the volatility acting like the "speed of time". This equivalence is explained in Sect. 8.1.1.

5.6 Regime-Switching Processes

In a regime-switching process, a new hidden stochastic variable is introduced. This variable define the "state of the word". The state of the world follows its own process, typically given by a Markov chain over a discrete space with only a few states, for example, with two states taken as "quiet" and "crisis". The volatility takes values depending only of the state of the world, and the return follows a simple random walk with the given state dependent volatility. In this family, the heteroscedasticity is introduced by the Markov chain for the state of the world, while the volatility and return are simple slave processes, without feed-back on the state of the world.

5.7 The Plan for the Forthcoming Chapters

The next four chapters discuss in details the possible processes inside the different structures. First, the discretization of price process is investigated, using either the logarithmic returns or the relative returns. This part investigates whether (5.1) or (5.5) should be used. Then, the heteroscedasticity is added to the price random walk in the next three chapters which investigate in details respectively the ARCH, stochastic volatility, and regime switching structures. This exploration gives many mug shots which can be compared to the empirical ones. The systematic process investigation allows us to select the relevant mathematical structures to be used in our mathematical description of the price dynamics. Some graphs in the mug shots acquire a more important status, as they allow one to decide clearly in favor of some process structures. In particular, the time reversal invariance, the precise form for the heteroscedasticity, and the leverage effect for stocks are very important and deserve further chapters in their own. Finally, the distribution of the innovations can be analyzed. With a clear view on the process structure, the applications to market risk evaluations and to option pricing are presented. This completes the discussion of the univariate case, and the subsequent chapters move to the analysis of multivariate time series.

[illegible]

6.7 The Plan for the Forthcoming Chapters

[illegible]

Chapter 6
Logarithmic Versus Relative Random Walks

6.1 Motivations

As mentioned in Chap. 5, a random walk can be discretized in two forms which are in principle very close. The usual way is to use a logarithmic random walk

$$\log\bigl(p(t+\delta t)\bigr) = \log\bigl(p(t)\bigr) + r(t+\delta t) \tag{6.1}$$

but a geometric random walk is another discretization

$$p(t+\delta t) = p(t)\bigl(1 + r(t+\delta t)\bigr). \tag{6.2}$$

In this chapter, the properties of both discrete processes are investigated for constant volatility return $r(t) = \sigma\epsilon(t)$ (the Itô term should be added for a logarithmic random walk, but this is irrelevant for most of the discussion). For normal innovations, the continuum limit for both processes are identical. Yet, the logarithmic and geometric processes are different for a finite time increment, in particular some of their mathematical properties.

The core of the problem when using fat-tailed innovations is fairly simple. For a logarithmic random walk, the random price $p(t+\delta t)$ given $p(t)$ is given by

$$p(t+\delta t) = p(t)e^{-1/2\sigma^2+\sigma\epsilon}. \tag{6.3}$$

The computation of the expected price involves the integral

$$E[\exp(\sigma\epsilon)] \tag{6.4}$$

where the expectation is over the distribution $p(\epsilon)$ for the innovations. For this integral to converge, $p(\epsilon)$ has to decay faster than an exponential. This is fine for normal innovations, but this integral diverges for any fat-tailed distribution (regardless of σ!). Yet, all theoretical processes need to include fat-tailed innovations in order to have realistic distribution for the returns. This is a substantial road-block

G. Zumbach, *Discrete Time Series, Processes, and Applications in Finance*,
Springer Finance, DOI 10.1007/978-3-642-31742-2_6,

for logarithmic processes as this restricts the innovations to have a normal distribution (or small modifications of it) in order to have well-defined statistics for the simplest quantities.

The unpleasant features of the processes are created by using a random process increment for the *logarithm* of the prices. When computing the (random) prices, the exponential mapping is then used, with unwanted consequences for large volatility and/or fat-tailed innovations. The textbook process used to describe financial time series, like stocks or foreign exchange rates, is a random walk for the logarithm of the prices, with constant volatility and normal innovations. This setup allows us to have convenient analytical tractability. The most important analytical computations are that the aggregated logarithmic returns also have a normal distribution, that the prices have a log-normal distribution, and that the infinite divisibility of the normal distribution allows one to construct a proper continuum limit (the Wiener process). These computations are at the core of the Itô calculus and of derivative pricing following Black and Scholes. Yet, the above mathematical argument shows that it is not possible to include fat-tailed distributions for the innovations, and therefore all the processes built using logarithmic random walks miss an important feature of the empirical data. Clearly, something has to be abandoned to build more realistic processes.

The equations for a geometric process eliminate the log/exp mappings, and this simple modification makes the expectations have a finite value. The geometric process is mathematically well behaved for fat-tailed innovations (within some limits). For example, the price expectation for a geometric discretization involves the integral

$$E[\sigma\epsilon] \tag{6.5}$$

which is finite for realistic tail exponents (see Chap. 13). As we will see, the move from logarithmic processes to geometric processes leads to abandoning the convenient analytical properties mentioned in the previous paragraph. Yet, realistic processes have to include fat-tailed distributions (which are not stable under aggregation for $\alpha > 2$) and to include heteroscedasticity using ARCH processes (for which there is no continuum limit) as Chap. 11 shows. Therefore, we have already given up on analytical tractability, and using a geometric discretization is not worsening our situation.

This chapter explores the similarities and differences of the logarithmic versus the geometric random walks. The setting is the simplest, namely a constant volatility. The cases of univariate and of multivariate processes are investigated for constant volatility, while the inclusion of a dynamics in the volatility is deferred until Chap. 7.

6.2 The Definitions of the Return

Consider a time series of price $p(t)$. The return over an interval δt can be defined in three different ways, each with its advantages and drawbacks.

The *difference return* is defined by

$$r_{\text{diff}}(t) = p(t) - p(t - \delta t). \tag{6.6}$$

This is a convenient definition for time series as the return over a period $n\delta t$ is given by a simple sum of the one period returns over the corresponding time interval. This definition gives a return in monetary units and is valid for positive and negative prices (in general, the prices of financial contracts are not positive, say, for example, for a long/short portfolios or for swaps). Because this definition is valid for all prices, it must be used to compute the relevant financial statistics at a given time t, like standard deviations or risk measures. It is however not a good definition for empirical investigations or for process specifications, as the drift and volatility would be proportional to the price. For long-term statistics on time series, only the logarithmic or relative return definitions are appropriate.

The *logarithmic return* is defined by

$$r_{\log}(t) = \log\bigl(p(t)/p(t - \delta t)\bigr) \tag{6.7}$$

with the condition that $p > 0$. Similarly to the difference return, this definition has a simple additive time aggregation property. For the price $p(t)$, the time-aggregation formula is

$$p(t + \Delta T) = p(t) \exp\left(\sum_{i=1}^{n} r_{\log}(t + i\delta t)\right) \tag{6.8}$$

with $\Delta T = n\delta t$. This aggregation formula leads to simple analytical computation.

The *relative return* is defined by

$$r_{\text{rel}} = \frac{p(t) - p(t - \delta t)}{p(t - \delta t)} = \frac{p(t)}{p(t - \delta t)} - 1 \qquad r_{\text{rel}} > -1 \tag{6.9}$$

with the condition that $p > 0$. This relative return definition can be expressed for $p(t)$ and iterated in order to obtain the aggregation formula

$$p(t + \Delta T) = p(t) \prod_{i=1}^{n} \bigl[1 + r_{\text{rel}}(t + i\delta t)\bigr] \tag{6.10}$$

which can be compared to (6.8) for the logarithmic return definition. Consequently, the relative return definition leads to products of random variables, and most of the analytical tractability is lost.

The exact change of variable between log and relative returns are

$$\begin{aligned} r_{\text{rel}} &= \exp(r_{\log}) - 1 \\ r_{\log} &= \log(1 + r_{\text{rel}}). \end{aligned} \tag{6.11}$$

A Taylor series expansion around zero shows that $r_{\text{rel}} \simeq r_{\log}$ for small returns. At the daily scale, the precise definition of the returns is mostly a matter of convenience, and because of its analytical tractability, the logarithmic return definition is usually preferred. Because both definitions are nearly identical for small returns, their statistical properties are very similar. In particular, fat-tailed distributions of empirical daily returns are observed with both definitions.

6.3 Logarithmic Process: One Asset, Constant Volatility

This section presents a brief summary of the properties of a constant volatility process defined using logarithmic returns, so as to compare with processes defined using relative returns. The *constant volatility logarithmic random walk process* is given by

$$\log\bigl(p(t)\bigr) = \log\bigl(p(t-\delta t)\bigr) + r_{\log}(t) \tag{6.12}$$
$$r_{\log}(t) = \mu - \frac{\sigma^2}{2} + \sigma\epsilon(t)$$

where $\mu = \mu[\delta t]$ and $\sigma = \sigma[\delta t]$ are respectively the drift and volatility at scale of the time increment δt. The random variable $\epsilon(t)$ has a (stationary) distribution with the constraints $E[\epsilon] = 0$ and $E[\epsilon^2] = 1$, and is independent for different times. The term $-\sigma^2/2$ in the drift is the Itô term needed to make the price process a martingale.

Let ΔT denote n time periods, so $\Delta T = n\delta t$. Assume that the random variable ϵ is *normally* distributed. Starting from a given present price $p(t)$, the (random) difference return after n time steps, using the logarithmic random walk model, is given by

$$r_{\text{diff}}[\Delta T] = p(t)\left\{\exp\left(n\bigl(\mu - \sigma^2/2\bigr) + \sigma\sum_{i=1}^{n}\epsilon_i\right) - 1\right\}. \tag{6.13}$$

We have the following properties for $p(t+\Delta T)$ and $r_{\text{diff}}[\Delta T]$:

$$\begin{aligned} E\bigl[p(t+\Delta T)\bigr] &= p(t)e^{n\mu} \\ E\bigl[r_{\text{diff}}[\Delta T]\bigr] &= p(t)\bigl\{e^{n\mu} - 1\bigr\} \\ \text{stdDev}\bigl(r_{\text{diff}}[\Delta T]\bigr) &= E\bigl[\bigl(r_{\text{diff}}[\Delta T] - E\bigl[r_{\text{diff}}[\Delta T]\bigr]\bigr)^2\bigr]^{1/2} \\ &= E\bigl[p(t+\Delta T)\bigr]\bigl\{e^{n\sigma^2} - 1\bigr\}^{1/2}. \end{aligned} \tag{6.14}$$

The above computations follow from straightforward calculations using $E[e^{\alpha\epsilon}] = \exp(\alpha^2/2)$ for Gaussian random variables. When $n\sigma^2 \ll 1$, Taylor expansions lead to

$$\text{stdDev}\bigl(r_{\text{diff}}[\Delta T]\bigr) \simeq E\bigl[p(t+\Delta T)\bigr]\sqrt{n}\sigma \tag{6.15}$$
$$\text{Skew}\bigl(r_{\text{diff}}[\Delta T]\bigr) \simeq 3\sqrt{n}\sigma \tag{6.16}$$

and the standard deviation grows as the square root of time. When $n\sigma^2 \gg 1$, we have instead

$$\mathrm{stdDev}\big(r_{\mathrm{diff}}[\Delta T]\big) \simeq E\big[p(t+\Delta T)\big]e^{\frac{1}{2}n\sigma^2} \tag{6.17}$$

namely the volatility for the return grows exponentially fast with ΔT. This (surprising) point is discussed in Sect. 6.5 for the logarithmic and geometric processes.

The previous computations are done for normal innovations. Let us emphasize that the situation is dramatically different when fat-tailed innovations are used. With the above logarithmic random walk, assuming that the random variable ϵ is fat-tailed, the price expectation $E[p(t+\delta t)]$ involves the integral $E[e^{\sigma\epsilon}]$, which diverges for any fat-tailed distribution, regardless of $\sigma > 0$, and already at one step.

Is there some way to salvage the situation when using log-returns and fat-tails? For $\sqrt{n}\sigma \ll 1$, the expected difference return could be defined by introducing a cut-off in the distribution for ϵ, in order to make the pdf decay faster (or even have compact support) for large enough ϵ. This would make the integral finite, although the answer would depend weakly on the cut-off. Indeed in a Monte Carlo simulation, which has a finite size and where the cut-off is implicitly given by the size of the simulation, this is effectively what happens. Unfortunately, this approach fails when $\sqrt{n}\sigma \geq 1$ since the result depends strongly on the cut-off. Similarly in a Monte Carlo simulation, the statistical estimates depend directly on the number of simulations. Therefore, there is no way to define for all values of $\sqrt{n}\sigma$ a regularized pdf for the innovations, depending on a cut-off, so that the end results are finite and with a weak dependency on the cut-off.

Figure 6.1 displays the mug shot for a constant volatility process with Student innovations. The simulations length is equivalent to 200 years, with a process increment of 3 minutes. The fast convergence to a Gaussian random walk is clearly observed on the moment graphs, otherwise the process is structureless. This figure gives also a fair measure of the Monte Carlo noise present in the mug shots for the other processes.

6.4 Geometric Process: One Asset, Constant Volatility

Better behaved processes with respect to fat-tailed innovations are obtained when the relative returns are used for the increments. The *geometric random walk process* is defined by

$$\begin{aligned} p(t+\delta t) &= p(t)\big(1 + r_{\mathrm{rel}}(t+\delta t)\big) \\ r_{\mathrm{rel}}(t+\delta t) &= \mu + \sigma\epsilon(t+\delta t) \end{aligned} \tag{6.18}$$

and is a natural discretization of a continuous geometric random walk. Let us emphasize that both processes (6.12) and (6.18) have the same continuum limit. At a daily scale $\delta t = 1$ day, the wealth of empirical data might allow one to discriminate empirically which one of definitions (6.12) and (6.18) describes best financial time

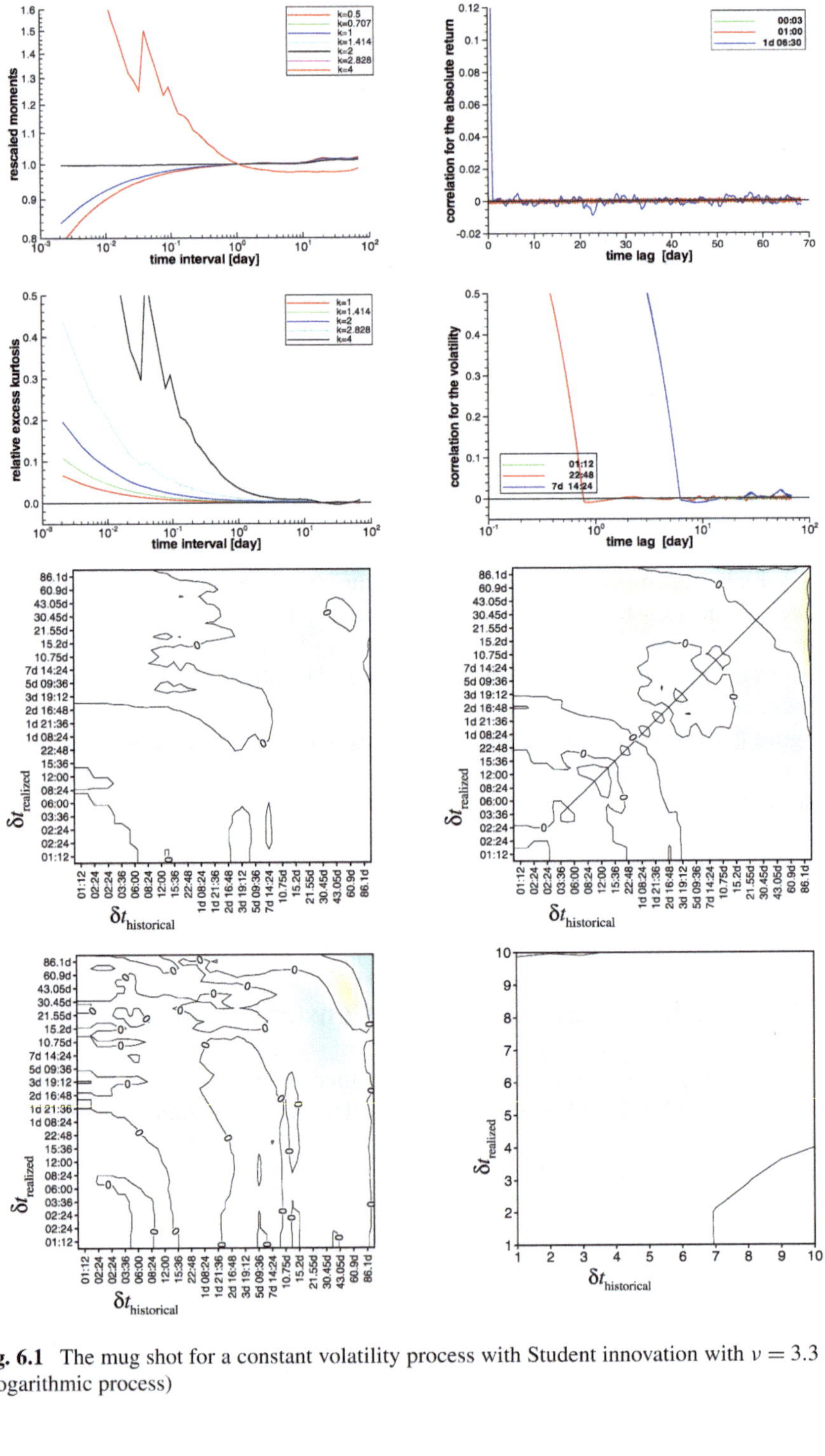

Fig. 6.1 The mug shot for a constant volatility process with Student innovation with $\nu = 3.3$ (for a logarithmic process)

series. Yet, because both process definitions are very similar, this seems a priori to be an elusive goal. The main difference is that the geometric process definition leads to well-defined values for the statistics on r_{diff}, including for fat-tailed innovations.

Using the geometric random walk model with a given present value $p(t)$, the (random) difference return $r_{\mathrm{diff}}[\Delta T]$ after n steps is

$$r_{\mathrm{diff}}[\Delta T] = p(t)\left\{\prod_{i=1}^{n}[1+\mu+\sigma\epsilon_i] - 1\right\} \tag{6.19}$$

with $\mu + \sigma\epsilon_i \geq -1$, $E[\epsilon] = 0$, and $E[\epsilon^2] = 1$, and ϵ_i is independent of ϵ_j (for $i \neq j$). Straightforward calculations lead to

$$\begin{aligned} -p(t) &\leq r_{\mathrm{diff}}[\Delta T], \\ E\big[p(t+\Delta T)\big] &= p(t)(1+\mu)^n \\ E\big[r_{\mathrm{diff}}[\Delta T]\big] &= p(t)\big\{(1+\mu)^n - 1\big\} \\ \mathrm{stdDev}\big(r_{\mathrm{diff}}[\Delta T]\big) &= E\big[\big(r_{\mathrm{diff}}[\Delta T] - E\big[r_{\mathrm{diff}}[\Delta T]\big]\big)^2\big]^{1/2} \\ &= \big(E\big[p(t+\Delta T)\big]\big)\big\{\big(1+\tilde{\sigma}^2\big)^n - 1\big\}^{1/2} \end{aligned} \tag{6.20}$$

with $\tilde{\sigma} = \sigma/(1+\mu)$. These relations are valid regardless of the distribution for ϵ (in particular for fat-tailed innovations) and regardless of the value for σ. When $n\sigma^2[\delta t] \ll 1$, assuming that the drift μ is close to zero, we have

$$\mathrm{stdDev}\big(r_{\mathrm{diff}}[\Delta T]\big) \simeq E\big[p(t+\Delta T)\big]\sqrt{n}\tilde{\sigma} \tag{6.21}$$

$$\mathrm{Skew}\big[r_{\mathrm{diff}}[\Delta T]\big] \simeq \frac{1}{\sqrt{n}}E\big[\epsilon^3\big] + 3\frac{n-1}{\sqrt{n}}\tilde{\sigma}. \tag{6.22}$$

Note that the variance has the same behavior as for a logarithmic random walk with Gaussian innovation given in Eq. (6.14); specifically, there is diffusive regime for short times and an exponential regime when $n\log(1+\tilde{\sigma}^2) \geq 1$ (see Sect. 6.5). Besides, the skewness is given by the skew of the innovation distribution for small time ΔT. With growing time ΔT, the second term increases and leads to positive skew. This term is created by the product of the random variables and is similar to the skewness for the logarithmic random walk (for normal innovations). Given the scaling of both terms with n, the skew always becomes positive for moderate value of n.

6.5 Long Time Properties of the (Constant Volatility) Random Walk Process

Both the logarithmic return and the relative return processes become unrealistic for long time ΔT. This follows from the variances given in (6.14) for logarithmic returns

$$E\big[\big(r_{\text{diff}}[\Delta T] - E\big[r_{\text{diff}}[\Delta T]\big]\big)^2\big] = \big(E\big[p(t+\Delta T)\big]\big)^2\big\{e^{n\sigma^2} - 1\big\}$$

and in (6.20) for relative returns

$$E\big[\big(r_{\text{diff}}[\Delta T] - E\big[r_{\text{diff}}[\Delta T]\big]\big)^2\big] = \big(E\big[p(t+\Delta T)\big]\big)^2\left\{\left(1+\left(\frac{\sigma}{1+\mu}\right)^2\right)^n - 1\right\}.$$

In both cases, the variance grow exponentially fast for sufficiently large n and/or σ. For the logarithmic random walk, the exponential growth is due to the exponential mapping, associated with a sufficient probability to have large logarithmic returns. This exponential regime can be driven by a combination of long horizons and/or large volatility estimates (in particular in presence of jumps, extremal events, or noise). Such growth can quickly lead to meaningless numbers, where a penny stock with a large volatility would have a variance exceeding the world market capitalization after a few years. Let us emphasize that most of the statistics on the difference returns become meaningless in the exponential regime. Moreover, in a portfolio, a single asset in the exponential regime will dominate all other positions. Clearly, some ingredients are missing in these simple diffusion models, and these are the economic constraints that will limit the growth of a given stock. One possibility to build processes that are also realistic for long time horizons would be to add a small mean reversion by using an Ornstein–Uhlenbeck term, but with a mean that would grow at a rate given by a risky rate (say in order to describe stocks).

On the other hand, with usual volatility figures, the situation is pretty well-behaved unless we let the horizon grow large. For most "normal" time series, the typical volatility is in the range of 10 to 20 % per year. For a 20 % annualized volatility, the volatility at the daily scale is $\sigma[1 \text{ day}] \simeq 1.25$ %, well within the region $\sigma^2[\Delta T] \ll 1$. With this volatility, the cross-over to the exponential behavior occurs for a horizon $\Delta T \sim 25$ years. This shows that, as long as the volatility is below 100 %, and as the risk horizons are up to a few years, the results are in the usual square root of time diffusive regime.

The "naive" variance scaling is given by the usual linear formula $\sigma^2[\Delta T] = \Delta T/\delta t \sigma^2[\delta t]$, while the exact variance scaling has been derived in Eq. (6.14) for logarithmic returns and normal innovations, and in Eq. (6.20) for relative returns. Let us emphasize that the difference between the square root scaling approximation and the exponential scaling is quickly becoming large when $\sigma^2[\Delta T] \gg 1$. In the exponential regime, the variance is becoming meaningless as ΔT grows, as well as most other statistics. Clearly, this is a limitation of the current models.

6.6 Geometric Process: Many Assets, Constant Volatility

The generalization for multivariate processes, with constant volatility, is fairly straightforward. We first explain the situation in one time-step and then for many time-steps. Unlike for the logarithmic random walk, the covariance is a nontrivial function of the time horizon for a the geometric random walk.

6.6.1 One Time-Step

Assume that we have N assets. A matrix R of size $N \times k_{\max}$ describes the dependencies (at the horizon of one time-step) between the assets and the $k_{\max}$ "sources of randomness". The multivariate geometric random walk process is defined as follows:

$$p_\alpha(t) = p_\alpha(t - \delta t)\bigl(1 + r_{\text{rel},\alpha}(t)\bigr) \tag{6.23}$$
$$r_{\text{rel},\alpha}(t) = \mu_\alpha + \sum_{k=1}^{k_{\max}} R_{\alpha,k}\epsilon_k(t) \qquad \text{with} \quad r_{\text{rel},\alpha}(t) > -1$$

where $\alpha \in [1, \dots, N]$ indexes the assets. The innovations $\epsilon_k(t)$ are normalized with

$$\begin{aligned} E[\epsilon_k(t)] &= 0 \\ E\bigl[\epsilon_k^2(t)\bigr] &= 1 \end{aligned} \tag{6.24}$$

and $\epsilon_k(t)$ is independent from $\epsilon_{k'}(t')$ for $k \neq k'$ or $t \neq t'$. This last technical condition is needed so that $E[f(\epsilon_k(t)) \cdot g(\epsilon_{k'}(t'))] = E[f(\epsilon_k(t))] \cdot E[g(\epsilon_{k'}(t'))]$ (when the indexes are different).

The covariance matrix of r_{rel} for one time-step is given by

$$E\bigl[(r_{\text{rel}} - \mu) \cdot (r_{\text{rel}} - \mu)^T\bigr] = \Sigma[\delta t] = R \cdot R' \tag{6.25}$$

so that R is essentially a square root of Σ (replacing the univariate volatility σ).

6.6.2 Many Time Steps

The multivariate process definition (6.23) can be iterated for n steps to obtain the aggregation formula

$$p_\alpha(t + \Delta T) = p_\alpha(t) \prod_{i=1}^{n} \Bigl\{1 + \mu_\alpha + \sum_k R_{\alpha,k}\epsilon_k(t_i)\Bigr\} \tag{6.26}$$

with $\Delta T = n\delta t$ and $t_i = t + i\delta t$. The relative return for the α asset is

$$r_{\text{rel},\alpha}[\Delta T] = \prod_{i=1}^{n} \Bigl\{1 + \mu_\alpha + \sum_k R_{\alpha,k}\epsilon_k(t_i)\Bigr\} - 1 \tag{6.27}$$

and the mean relative return is given by $E[r_{\text{rel},\alpha}[\Delta T]] = \{1 + \mu_\alpha\}^n - 1$. A convenient expression for computing the moments is

$$r_{\text{rel},\alpha} - E[r_{\text{rel},\alpha}] = (1 + \mu_\alpha)^n \left(\prod_{i=1}^{n} \Bigl\{1 + \sum_k \tilde{R}_{\alpha,k}\epsilon_k(t_i)\Bigr\} - 1 \right) \tag{6.28}$$

and with $\tilde{R}_{\alpha,k} = R_{\alpha,k}/(1+\mu_\alpha)$. The term $(1+\mu_\alpha)^n$ is due to the growth induced by the drift. The second term is independent of the drift, after its effect is absorbed in a rescaling of the dependency matrix R.

Denote by $\Sigma[\Delta T]$ the covariance matrix after n time steps. This covariance is a function of the 1-step covariance $\Sigma[\delta t] = \Sigma$. The covariance between asset α and β after n steps can be computed using the independence properties (6.24) of the innovations (and dropping the ΔT dependency):

$$\begin{aligned}
\Sigma_{\alpha,\beta}[\Delta T] &= E\big[\big(r_{\text{rel},\alpha} - E[r_{\text{rel},\alpha}]\big)\cdot\big(r_{\text{rel},\beta} - E[r_{\text{rel},\beta}]\big)\big] \\
&= (1+\mu_\alpha)^n(1+\mu_\beta)^n\left(E\left[\prod_{i=1}^{n}\Big\{1+\sum_k \tilde{R}_{\alpha,k}\epsilon_k(t_i)\Big\}\right.\right. \\
&\quad \left.\left.\times\prod_{j=1}^{n}\Big\{1+\sum_{k'}\tilde{R}_{\beta,k'}\epsilon_{k'}(t_j)\Big\}\right]-1\right) \\
&= (1+\mu_\alpha)^n(1+\mu_\beta)^n\left(E\left[\prod_{i=1}^{n}\Big\{1+\sum_k \tilde{R}_{\alpha,k}\epsilon_k(t_i)\Big\}\right.\right. \\
&\quad \left.\left.\times\Big\{1+\sum_{k'}\tilde{R}_{\beta,k'}\epsilon_{k'}(t_i)\Big\}\right]-1\right) \\
&= (1+\mu_\alpha)^n(1+\mu_\beta)^n\left(\prod_{i=1}^{n}\Big\{1+\sum_k \tilde{R}_{\alpha,k}\tilde{R}_{\beta,k}\Big\}-1\right) \\
&= (1+\mu_\alpha)^n(1+\mu_\beta)^n\big(\big(1+(\tilde{R}\cdot\tilde{R}')_{\alpha,\beta}\big)^n-1\big).
\end{aligned}$$

Therefore, the covariance after n time steps is

$$\Sigma_{\alpha,\beta}[\Delta T] = (1+\mu_\alpha)^n(1+\mu_\beta)^n\big(\exp\big(n\log\big(1+(\tilde{R}\cdot\tilde{R}')_{\alpha,\beta}\big)\big)-1\big). \tag{6.29}$$

Although it is fairly easy to calculate each entry of $\Sigma[\Delta T]$ from the above formula, the matrix Σ is not a simple function of the matrix R (as it is the case for a logarithmic random walk). When R is small, this expression can be approximated by $\Sigma_{\alpha,\beta}[\Delta T] = n(1+\mu_\alpha)^{n-1}R_{\alpha,k}\cdot(1+\mu_\beta)^{n-1}R_{\beta,k}$, which is a square, and with the familiar leading square root of time scaling. For Monte Carlo simulations, this reduction allows us to draw directly aggregated random returns at the scale ΔT using the matrix $\sqrt{n}(1+\mu_\alpha)^{n-1}R_{\alpha,k}$, without having to estimate the square root of $\Sigma[\Delta T]$. Still in the small R approximation, the correlation between $r_{\text{rel},\alpha}$ and $r_{\text{rel},\beta}$ can be computed and is independent of ΔT. However, in general, the square root of $\Sigma[\Delta T]$ should be computed numerically if needed.

6.7 Enforcing the Condition $r_{\rm rel} > -1$ or $r_{\rm rel} \geq -1$

In a univariate Monte Carlo simulation for a geometric process, the random return $r_{\rm rel}$ is given by

$$r_{\rm rel} = \mu + \sigma\epsilon \qquad \text{with} \quad r_{\rm rel} > -1 \quad \text{or} \quad r_{\rm rel} \geq -1. \tag{6.30}$$

A simple way to enforce the condition $r_{\rm rel} > -1$ is to use a rejection scheme, namely to reject the draws when $\epsilon \leq (\mu - 1)/\sigma$. Denoting by $P(\epsilon)$ the cumulative distribution corresponding to the density $p(\epsilon)$, the acceptance rate of this algorithm is $1 - P((\mu - 1)/\sigma)$. If σ is small enough such that $P((\mu - 1)/\sigma)$ is small, the algorithm is efficient. In general, for any symmetric distribution for ϵ, the acceptance rate is bounded below by $1/2$. In order to be consistent with the assumed properties for the distribution for ϵ, the truncated distribution $p(\epsilon)$ should be modified accordingly so that $E[\epsilon] = 0$ and $E[\epsilon^2] = 1$.

The process can be slightly modified to allow for bankruptcy. The condition becomes $r_{\rm rel} \geq -1$, and the zero price is an absorbing state: $p(t') = 0$ for $t' \geq t$ where t is the first time such that $r_{\rm rel}(t) = -1$. For this modified process, the condition can be enforced by replacing all draws with $\epsilon < (\mu - 1)/\sigma$ by $\epsilon = (\mu - 1)/\sigma$. This is equivalent to add a discrete mass $P((\mu - 1)/\sigma)$ at $r_{\rm rel} = -1$, similar to a finite probability for bankruptcy. As with the rejection scheme, the distribution for the innovations should be modified to be consistent with its assumed properties. Notice also that it is not possible to include naturally bankruptcy in a logarithmic random walk, as such an event occurs at minus infinity (i.e., at $\log(0)$).

In the multivariate case, the situation is more complicated. A draw for all assets can be accepted only when

$$r_{{\rm rel},\alpha} = \mu_\alpha + \sum_k R_{\alpha,k}\epsilon_k(t) > -1 \qquad \forall\alpha$$

leading to the acceptance rate that decreases exponentially with the number of time series. In a large scale computation, this could be a too large decrease of the algorithm efficiency. A practical algorithm is therefore to replace all values such that $\mu_\alpha + \sum_k R_{\alpha,k}\epsilon_k(t) < -1$ by -1 and interpreting such event as the bankruptcy of the asset α when this occurs. The subsequent correlations for the affected assets are then zero.

6.8 Skewness

Skewness is a quantity depending on the return definition, and for empirical statistics, the choice of the definition is a matter of taste. Due to the convexity of the mapping between relative and logarithmic returns, the skew measures are influenced by the selected return definition (i.e., skew(relative return) > skew(log return)). A related issue is the choice for the skew estimator, as the empirical time series for the returns exhibit fat-tailed distributions with exponents in the range of 4 to 7 (depending on the tail measure, time series, and available sample). Clearly, measuring the skew with the usual third reduced moment is dangerous because it is close to the limit for having an infinite value.

The first part of this section compares empirical skewness for a large set of stocks, evaluated with the usual estimator and with a robust L-estimator, using logarithmic and relative return definitions. All the analyses are done at time scales ranging from 1 day to 1 year, so as to have an overview of the multiscale properties of the skewness. A similar study along these lines but for a broader set of statistics can be found in [167].

The empirical data set is made of 777 stock time series part of the S&P 500, EuroStoxx and SMI indexes, and with a length between 2000 and 7000 days (the same set is used in Chap. 14). The relative or logarithmic returns are computed at the time horizons of 1, 2, 5, 10, 21, 65 130, and 260 days. The skews are computed with the textbook reduced moment estimator and with a ratio τ_L of robust L-estimator [82] given by

$$\tau_L = l_3 / l_2 \tag{6.31a}$$

$$l_2 = \frac{1}{2}\binom{N}{2}^{-1} \sum_j \left\{ \binom{j-1}{1} - \binom{N-j}{1} \right\} r_{[j]} \tag{6.31b}$$

$$l_3 = \frac{1}{3}\binom{N}{3}^{-1} \sum_j \left\{ \binom{j-1}{2} - 2\binom{j-1}{1}\binom{N-j}{1} + \binom{N-j}{2} \right\} r_{[j]} \tag{6.31c}$$

where $r_{[j]}$ is the j ordered statistics of the total sample of size N. This definition can be computed quite efficiently as it requires to order the data (an $N \log(N)$ operation).

For a given return time interval, the distribution of the skews over the set of time series is represented as a box plot. Figure 6.2 displays the box plots for the skew as a function of the return time intervals. The median values of the skews computed for the components of the respective indexes are overlayed in color. The four graphs are for each combination of relative/logarithmic returns and reduced moment/robust L estimators. The top line shows the skews evaluated with the reduced third moment estimator. The large number of points outside of the box shows that it is fairly dangerous to draw conclusions based on this estimator applied to financial time series. The bottom line uses the robust skew estimator, with a better coherence across the various time series. The overall difference between the skews computed with both return definitions is due to the convexity of the mapping between the returns. This shows that the common "negative skew for stocks" stylized fact is indeed dependent of the return definition. When using a relative return definition, it should be changed for "mostly positive skews" occurs for stocks.

Monte Carlo simulations can be used for various processes (defined in the forthcoming sections), either using relative returns, or the more traditional definitions using logarithmic returns. For ARCH processes, the return definition used in the

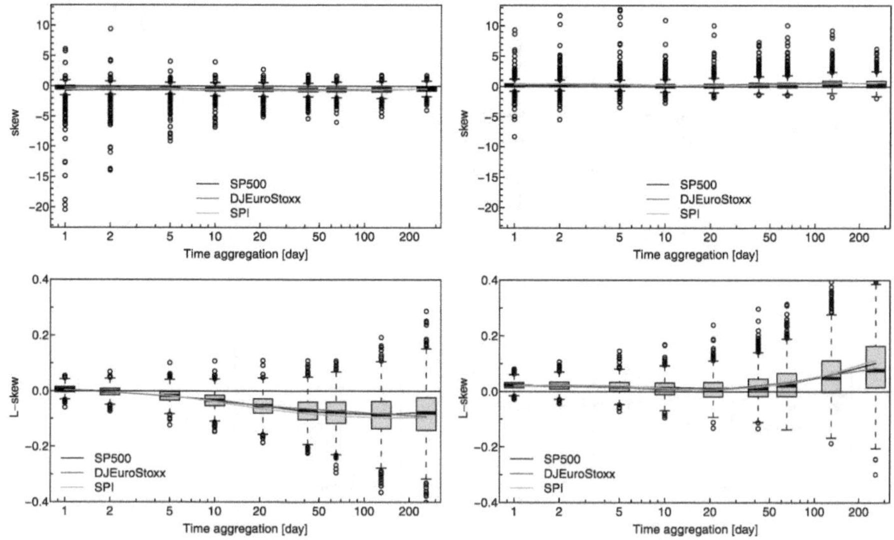

Fig. 6.2 The term structure of the skew, using different definitions and statistical estimators. The *horizontal axis* gives the time interval at which the returns are computed. The *top line* displays the skew computed with the third reduced moment, the *bottom line* with the robust L-skew estimator. The *left* (*right*) row corresponds to logarithmic (relative) returns

computation of the volatility is taken to be consistent with the price update. The parameters of the processes are adjusted so as to reproduce the broad features of a set of statistics across the data sample, like mean volatility, heteroscedasticity, leverage and robust kurtosis, and for time intervals ranging from 1 day to 1 year. The simulations are done so as to mimic the empirical data set, namely the same number of time series and for the same lengths are generated, with one fixed set of process parameters. Then, logarithmic and relative returns are used to compute the robust L-skews for each generated time series. The same box plot representation is used on the time series generated using a long-memory ARCH process with leverage, as shown on Fig. 6.3. The curves in color correspond to the median skews for simpler processes. Essentially, both lines of plots from the Monte Carlo simulations should be compared to the bottom line of the empirical plots in Fig. 6.2.

Clearly, the time aggregation only is creating the correct term structures for the skew, with a slowly increasing skewness for longer return time intervals. The details of the process and its parameters influence only weakly the skew. The skew values are nearly independent from the innovations distribution and from the heteroscedasticity structure (none, one time scale, or long memory). Only the leverage effect (negative correlation between past returns and future volatilities, see Chap. 14) reduces the skews as it creates larger moves on the downside (similar to crashes). For processes defined with logarithmic returns (top line), the skew is too high, and only the leverage effect is able to introduce the typical negative skew measured

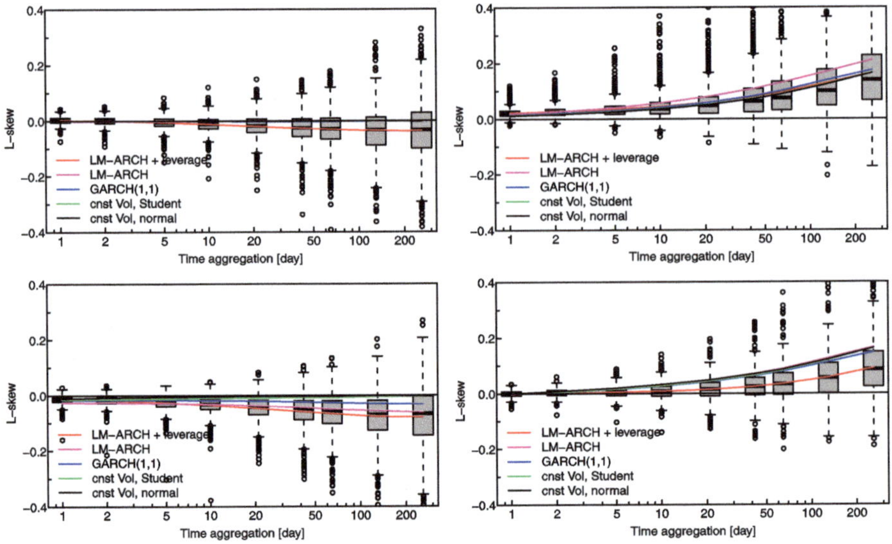

Fig. 6.3 The time structure of the skewness for several processes. The processes are defined with logarithmic returns (*top line*) and relative returns (*bottom line*). The statistical analysis for the generated time series is done for logarithmic return (*left*) and relative return (*right*), using the robust L-estimator for the skew. The *box plot* corresponds to the LM-ARCH process with leverage and for a fixed set of parameters, and with generated time series equivalent in number and lengths with the empirical data set used in Fig. 6.2

with logarithmic returns. Using relative returns in the process equations (bottom line) makes the skew time structure close to the empirical one. Indeed, the agreement with the best available process (long memory ARCH + leverage + fat-tailed innovation) is quite impressive (see [167] for more statistics). Let us emphasize that using a skewed innovation distribution can alter only the behavior for short time interval, while the long-term behavior is induced by the dynamics of the process.

Increasing the discretization step δt (or the volatility) makes the situation different. The relevant variable is $\sigma^2 \delta t$, namely the volatility at the scale of the simulation step. When $\sigma^2 \delta t \geq 1$, the situation changes as the skewness of the innovations becomes important, and the discretization matters. This is the case in long-term strategic simulations for insurance firms or pension funds where the Monte Carlo step is of the order of one year. With the usual geometric random walk setting and a symmetric distribution for the innovations, the logarithmic returns have a zero skew, while the relative returns have a positive skew (due to the exponential mapping). This feature of the discretization is at odd with the empirical skewness, which is negative at one year for the logarithmic returns. This difference is leading to overly positive previsions at a scale of a few years. For such a large discretization step, the skewness of the innovations should be properly calibrated.

6.9 The Broader Perspective So Far ...

In today algorithmic applications, an important concern is the robustness of a computation with respect to the input data. A realistic portfolio can contain several thousands of positions. With these sizes, it is likely that some time series will contain large moves (genuine or noise), leading to large volatility estimates. For a logarithmic random walk, the simulated prices in a Monte Carlo simulation are given essentially by $\exp(\sigma\epsilon)$, and the exponential amplifies large draws and large volatilities. The exponential mapping exacerbates the sensitivity to large events, whether genuine outliers or noise. Let us emphasize that even when using normal innovations in a process, the exponential mapping can lead to prices that are unrealistic. Clearly, removing the exponential mapping will make large-scale computations more robust.

Another important problem is option pricing, where slow progress is being made on pricing in a realistic framework. The basic option pricing model was set in 1973 by Black, Scholes, and Merton, in an idealized world involving continuous time and Wiener processes. Yet, a realistic set up needs to include heteroscedasticity and fat-tailed innovations. This is a much tougher framework, with slow progresses mainly due to [55, 80], and recently due to [40]). Compared to the original continuous-time option framework, some properties are not preserved in such more realistic world, for example, a unique replication. The work of Christoffersen et al. is an important contribution, separating cleanly the issues related to the equivalent martingale measure from the utility function. The proposed framework is a logarithmic random walk with a finite time step, while the possible specifications for the mean and for the volatility are very general. This allows for general Heteroscedastic dynamics like GARCH, and the distribution for the innovations can in principle be fairly arbitrary. Yet, the existence of an equivalent martingale measure depends on the characteristic function for the innovation distribution. For a logarithmic process, this is equivalent to the existence of $E[\exp(\sigma\epsilon)]$. In practice, this condition is quite restrictive and essentially allows only for normal innovations and clearly excludes fat-tailed distributions. In [116], the framework developed in [40] is adapted to geometric processes. In this setting, the distribution for the innovations can have fat-tails, while still allowing for a broad class of volatility dynamics. This option pricing framework is presented in Chap. 16.

For analytical computations, it is desirable to have the aggregated distribution after n steps, at least for Gaussian innovations. In the same direction but for numerical computations, if possible, it is efficient to replace an n-step simulation by a 1-step simulation with an aggregated variance. For logarithmic Gaussian random walks, this aggregation can be done exactly by using the stability of the normal distribution under the sum of random variables. For a geometric random walk, the natural aggregated variable is $r_{\text{diff}}[\Delta T]$ where a product of random variables appears. For small σ (and for small enough n such that $n\sigma \ll 1$), the expression can be expanded at first order in σ, leading to a simple sum which can be used to approximate the variance. A better method is to use the exact expression for the variance of $r_{\text{diff}}[\Delta T]$, leading to the aggregated variance (6.20) at the horizon ΔT. Yet, the probability distribution of $r_{\text{diff}}[\Delta T]$ cannot be derived, even in the Gaussian case, because expression (6.19) involves a product of random variables.

The inability to derive the law for aggregated random variables for process defined as a geometric random walk is a clear drawback. Yet, such constant volatility processes are no less than an important idealization. With the familiar logarithmic random walk, the modification of most of the ingredients used to build the process breaks this analytical tractability, as, for example, changing the distribution of the increments or adding an ARCH heteroscedasticity. Therefore, in practice, this analytical tractability is nice but does not hold for more realistic processes. With geometric random walk, such properties are lost even for the simplest process. As anyway numerical simulations should be used for more realistic processes, this is already irrelevant.

The availability of cheap computers and massive amount of financial data is clearly changing the requirements on models, either to describe empirical time series or to price complex instruments. The empirical properties of the financial data can be explored in great details, and the simplest random walk with constant volatility and Gaussian innovations is nowadays not accurate enough to be satisfactory. Yet, with the increasing accuracy and complexity of the processes, the analytical tractability has to be abandoned. Analytical results give a much stronger standpoint, but when not available, Monte Carlo simulations can be used to extract information about a process or to price an instrument. This broad evolution took place over the last 20 years, with the decreasing importance of solvable models in favor of more realistic ones.

The change from a logarithmic to a geometric discretization is a further step in the above landscape. For the simplest constant volatility and normal innovations model, the change from logarithmic to geometric processes makes the model not (exactly) solvable. As our goal is anyway to incorporate fat-tailed innovations and heteroscedasticity, the analytical tractability is of a lesser concern. Yet, the paradigm shift taking place when modifying the discrete process equations from a logarithmic to a geometric process allows us to solve elegantly a number of issues related to fat tails, robustness, skewness, and option pricing. The issue of the logarithmic versus geometric processes is explored further in Sect. 7.8 for the case of fairly general ARCH processes.

Chapter 7
ARCH Processes

The core idea of the ARCH (Auto Regressive Conditional Heteroscedastic) process is to have an effective volatility depending on the recent returns. This structure creates a feed-back loop where one large return increases the volatility that will in turn lead to subsequent larger returns. This is captured be the equation

$$\sigma_{\text{eff}}(t) = \sigma_{\text{eff}}\bigl[r(t), r(t-\delta t), r(t-2\delta t), \ldots)\bigr] \tag{7.1}$$

where the effective volatility is depending on the magnitude of the past returns $r(t - j\delta t)$. This simple mechanism was introduced by [25, 59, 61] and leads to cluster of volatility. The underlying intuition is that large moves in the market trigger other market participants to trade, hence creating subsequent volatility. The minimal model that captures this structure is the celebrated GARCH(1, 1) process which is based on one exponential moving average (EMA). The basic equations for this process are easy to modify in order to reproduce more, or more accurately, the observed stylized facts, and there is a bewildering array of ARCH processes. The subsequent problem is not to create new processes, but to select the equations that best capture the empirical regularities.

The next subsection discusses the basic GARCH(1, 1) process and its properties. This process is the archetype of a good model: it captures the basic mechanism leading to heteroscedasticity while being also very parsimonious. Yet, its structure and properties are fairly degenerate. Lifting this degeneracies is important to better understand the origin of the properties of the model and to generalize them in the right direction. This route leads us to the multicomponent ARCH processes, with the particular cases of the long-memory LM-ARCH and market-component Mkt-ARCH processes.

7.1 GARCH(1, 1)

The historical [25, 61] and common way to write the GARCH(1, 1) volatility equation is

$$\sigma^2_{\text{eff}}(t) = \alpha_0 + \alpha_1 r^2(t) + \beta_1 \sigma^2_{\text{eff}}(t-\delta t) \tag{7.2}$$

G. Zumbach, *Discrete Time Series, Processes, and Applications in Finance*,
Springer Finance, DOI 10.1007/978-3-642-31742-2_7,

with three parameters α_0, α_1, and β_1. This equation can be rewritten as follows:

$$\sigma_1^2(t) = \mu\sigma_1^2(t-\delta t) + (1-\mu)r^2(t) \tag{7.3a}$$

$$\begin{aligned}\sigma_{\text{eff}}^2(t) &= \sigma_\infty^2 + (1-w_\infty)\big(\sigma_1^2(t) - \sigma_\infty^2\big) \\ &= (1-w_\infty)\sigma_1^2(t) + w_\infty\sigma_\infty^2\end{aligned} \tag{7.3b}$$

with three parameters σ_∞, w_∞, τ. The relation between the characteristic time τ of the exponential and the decay factor μ of the exponential moving average (EMA) is given by $\mu = \exp(-\delta t/\tau)$. The equivalence between both sets of equations is proved by expressing (7.3b) for σ_1 and then inserting into (7.3a). By rearranging the terms, the volatility equation for a GARCH(1, 1) process is obtained, with the mapping between the coefficients

$$\begin{aligned}\alpha_0 &= w_\infty\sigma_\infty^2(1-\mu) \\ \alpha_1 &= (1-w_\infty)(1-\mu) \\ \beta_1 &= \mu.\end{aligned} \tag{7.4}$$

The inverse mapping is

$$\begin{aligned}\sigma_\infty^2 &= \frac{\alpha_0}{1-\alpha_1-\beta_1} \\ w_\infty &= 1 - \frac{\alpha_1}{1-\beta_1} \\ \mu &= e^{-\delta t/\tau} = \beta_1.\end{aligned} \tag{7.5}$$

In the form (7.3a), (7.3b), the GARCH(1, 1) process appears with one "internal variable" that can be interpreted as a historical volatility σ_1 measured by an EMA at the time horizon $\tau = -\delta t/\ln(\mu)$. The effective volatility σ_{eff} is given by the mean term, plus the difference between the historical volatility and the mean volatility, weighted by $1 - w_\infty$. In the second form, the effective volatility is a convex combination of the mean volatility σ_∞ and the historical volatility σ_1. With these equations, the three parameters have a direct interpretation for the time series, in contrast to the standard α, β parameterization in (7.2).

Although equivalent, the form (7.2) for the equation has two drawbacks. First, it is less amenable to an intuitive understanding of the role of the different terms and their economic or behavioral counterparts. Second, the natural generalization of (7.2) leads to the GARCH(p, q) form. Most studies find that GARCH(p, q) does not improve much over GARCH(1, 1) for reasonably small values of p and q. On the other hand, in the form (7.3b), the GARCH equations lead to the introduction of more EMA volatility components $\sigma_2, \sigma_3, \ldots$, with increasing time horizons $\tau_2, \tau_3, \ldots$. With a few components, a large span of time intervals can be covered while introducing only a minimal number of parameters.

The GARCH(1, 1) process has been estimated on many time series, mostly using a log-likelihood estimator on daily data. In the second parameterization, the parameters take fairly typical values. The value for σ_∞ is depending on the mean volatility on the data sample used to estimate the process, and this is clearly a parameter depending on the time series. A simple moment estimator is $\sigma_\infty^2 = \sum_t r(t)^2$ (see [150] for a full discussion of the estimation in different coordinates, and the restriction on the two-dimensional surface corresponding to the in-sample volatility). The value for w_∞ is typically around 0.1, with a fairly narrow range between 0.05 to 0.15. This means that the dynamics of the process is dominated by the historical EMA part as measured by σ_1, and the mean volatility gives a fairly weak return toward a constant value. The characteristic time τ of the EMA ranges from one to a few weeks. Such values allow us to obtain a good estimator of the short-term level of volatility, but with enough returns included in the EMA. Using the above relation $\beta_1 = \mu \simeq 1 - \delta t/\tau$, the parameter β_1 is in the range 0.8 to 0.95. With these values, the correlation time τ_{corr} as given by (7.8) is of the order of 2 to 6 months. Using the relation $\mu_{\text{corr}} = \alpha_1 + \beta_1 \simeq 1 - \delta t/\tau_{\text{corr}}$, we obtain the common observation that $\alpha_1 + \beta_1 \simeq 1$. Notice that this value originates in $\tau_{\text{corr}} \gg \delta t$, but the decay of the correlation is always exponential. This discussion explains intuitively why very similar parameter values are obtained for all financial time series. Essentially, this similarity results from simple statistical considerations (for τ) and from the quantitative importance of the dynamics of the volatility for all time series (for w_∞). Regardless of the time series, decent overall values for the GARCH(1, 1) parameters are $\tau = 10$ (business) days, $w_\infty = 0.1$, and σ_∞ given by a sample moment. And there is no need for 6 decimal figures in the parameter values to get a fairly good model.

In Sect. 7.1.1, the k-step volatility forecast formula for the GARCH(1, 1) process is computed:

$$\mathcal{F}\left[\Delta t = k\delta t; \sigma_{\text{eff}}^2\right](t) = \sigma_\infty^2 + \mu_{\text{corr}}^{k-1}\left(\sigma_{\text{eff}}^2(t) - \sigma_\infty^2\right). \tag{7.6}$$

This formula shows explicitly the exponential mean reversion toward σ_∞ for large time intervals Δt. The characteristic time for the mean reversion is $\tau_{\text{corr}} = -\delta t / \ln(\mu_{\text{corr}})$ with

$$\mu_{\text{corr}} = \alpha_1 + \beta_1 = \mu + (1 - w_\infty)(1 - \mu) = 1 - w_\infty(1 - \mu). \tag{7.7}$$

For $w_\infty(1 - \mu)$ small, an order of magnitude for the correlation time is

$$\tau_{\text{corr}} \simeq \frac{\delta t}{w_\infty(1 - \mu)} \simeq \frac{\tau}{w_\infty} \tag{7.8}$$

showing that a large correlation time is obtained when the coupling constant w_∞ is small. These formula are derived in the next section.

The lagged correlation for r^2 can be computed analytically [77]; it decays exponentially fast with the same characteristic time. Financial data display clearly a slow decay for the lagged correlation, and this discrepancy is the main shortcoming of the GARCH(1, 1) process. The long-memory processes introduced in Sect. 7.9 are constructed as a minimal extension of GARCH(1, 1) that remedies this deficiency.

For $w_\infty = 0$, the affine GARCH(1, 1) equations reduce themselves to the linear I-GARCH(1) process

$$\sigma^2_{\text{eff}}(t) = \mu\sigma^2_{\text{eff}}(t - \delta t) + (1 - \mu)r^2(t) \tag{7.9}$$

with one parameter μ. As shown by [77], the lagged correlation for the square return of an I-GARCH(1) process also decays exponentially. For the I-GARCH(1) process, the parameter value $\tau = 16$ (business) days or $\mu = 0.94$ gives good results for most time series, as was suggested for risk evaluations. The larger value for τ compared to the GARCH value compensates for the absence of mean volatility (that acts like a volatility component at an infinite τ).

This is our first encounter with pairs of processes that are closely related, and their similarities will become a recurring theme in this chapter. In the present case, the first member is the I-GARCH(1) process depending on one parameter. The equation for the variance is bilinear in the returns, or equivalently the variance is linear in the squared returns. The second member is the GARCH(1, 1) process, depending on three parameters. The equation for the variance is bilinear in the returns and with an additive term, or equivalently the variance is affine in the squared returns. The difference in the number of parameters comes from the mean variance σ_∞ and its coupling constant w_∞ in the affine processes. We will return a few times on the topic of linear versus affine ARCH processes, in particular in Sect. 7.4.

Similarly to GARCH(1, 1) and I-GARCH(1), the general multicomponent processes presented below come in pairs: an affine and a linear (for the variance) processes. They have the same parameters, except that the affine processes have two more parameters corresponding to the mean volatility σ_∞ and the corresponding "coupling constant" w_∞. This generalizes the GARCH(1, 1) and I-GARCH(1) construction, but some properties like the "mean reversion" should be carefully reexamined.

The widely used RiskMetrics formula [109, 154] for risk estimation corresponds to the I-GARCH(1) process, with a given value for the parameter μ. The most common choices, using daily data, are $\mu = 0.94$ and $\mu = 0.97$, corresponding respectively to exponential characteristic time $\tau \sim 16$ and $\tau \sim 32$ business days. In order to compute a simple and robust risk estimation for wide range of assets across different classes, it is important to have a simple linear process. This makes possible to have a one parameter process that can accommodate all securities. Because the mean volatility is clearly asset dependent, such a simple approach is not possible if affine equations are used. There has been numerous suggestion to improve risk estimates by using a GARCH(1, 1) process, but this is unmanageable as this would introduce at least n parameters for a portfolio with n risk factors (indeed n^2 or n^4 if a full multivariate extension is used!). A better path is to generalize the I-GARCH(1) equations so as to capture the long memory present in the empirical data, while still having a linear process [154]. As the lagged correlations for the volatility (or return square) show a logarithmic decay with a similar characteristic time for most assets, a process with the correct long memory can be built while still having no dependency on each particular time series. The relationship between processes and risk estimations is discussed in details in Chap. 15.

The mug shot for the GARCH(1, 1) process (with logarithmic return) is given in Fig. 7.1. The simulation time increment is 3 minutes. The parameters are $\tau = 1$ day and $w_\infty = 0.10$, leading to a correlation time of 10.0 days. The single characteristic time and the exponential decay of the lagged correlations can be clearly observed. The mug shot Fig. 7.2 shows the same process and parameter values, but with relative returns. The overall behavior is similar, with one characteristic time. Yet, some differences appear for long time intervals, in the lagged correlation and in the historical/realized volatility correlation.

So far, the presentation of the GARCH(1, 1) process concerns the implication of the volatility equation. The other important ingredient in the model is the distribution for the innovations $p(\epsilon)$, which must have a zero mean and unit variance but is otherwise unspecified. Let us give a first discussion about $p(\epsilon)$, while the detailed investigations are postponed to Chap. 13 after a good model for the volatility dynamics has been found. A convenient hypothesis is to assume a normal distribution, as this allows for simple log-likelihood estimation of the process parameters. When applied to financial time series, the empirical residuals are found to be distributed with a wider distribution, with large innovations occurring more frequently than what a normal distribution would imply. A simple solution is to use a distribution with heavy tails, like a Student t-distribution with ν degrees of freedom. This distribution improves the description of empirical time series by the process, with the new parameter ν to be estimated. This simple model includes the most important features observed in the empirical data, namely the diffusion of the prices (captured in the random walk for p), the clusters of volatility (captured in the GARCH dynamics), and the frequent occurrences of large events (captured by $p(\epsilon) \sim$ Student). Indeed, the GARCH(1, 1) process with Student innovations provides for an excellent model beyond a simple Gaussian random walk and a good stepping stone toward more accurate processes.

As an alternative to the normal distribution, more complex distributions for the innovations have been proposed by [36, 84, 105] in the families of α-stable and tempered stable distributions. For innovations with α-stable distributions, the tails are very heavy with $\nu < 2$, too large compared to the empirical time series. The tempered stable distributions offer a better adequation to the data, albeit at the price of increasing complexity. Following the scheme proposed by [55] for option pricing, these models are stable under the transformation from the physical measure to the risk-neutral measure. Yet, the option pricing scheme presented in Chap. 16 is simpler, allows for general ARCH processes, and is easier for the numerical implementation.

Finally, the time increment δt for the ARCH processes must be finite. The construction of the equivalent continuum time processes meets some serious difficulties, both related to the volatility equation and to the distribution of the innovations. The obstructions in the continuum construction are discussed at the end of Sect. 7.16 in the context of the multicomponent ARCH processes. In turn, the absence of the corresponding differential stochastic equations for the process leads to difficulties in the pricing of derivatives when the textbook continuous-time approach is followed.

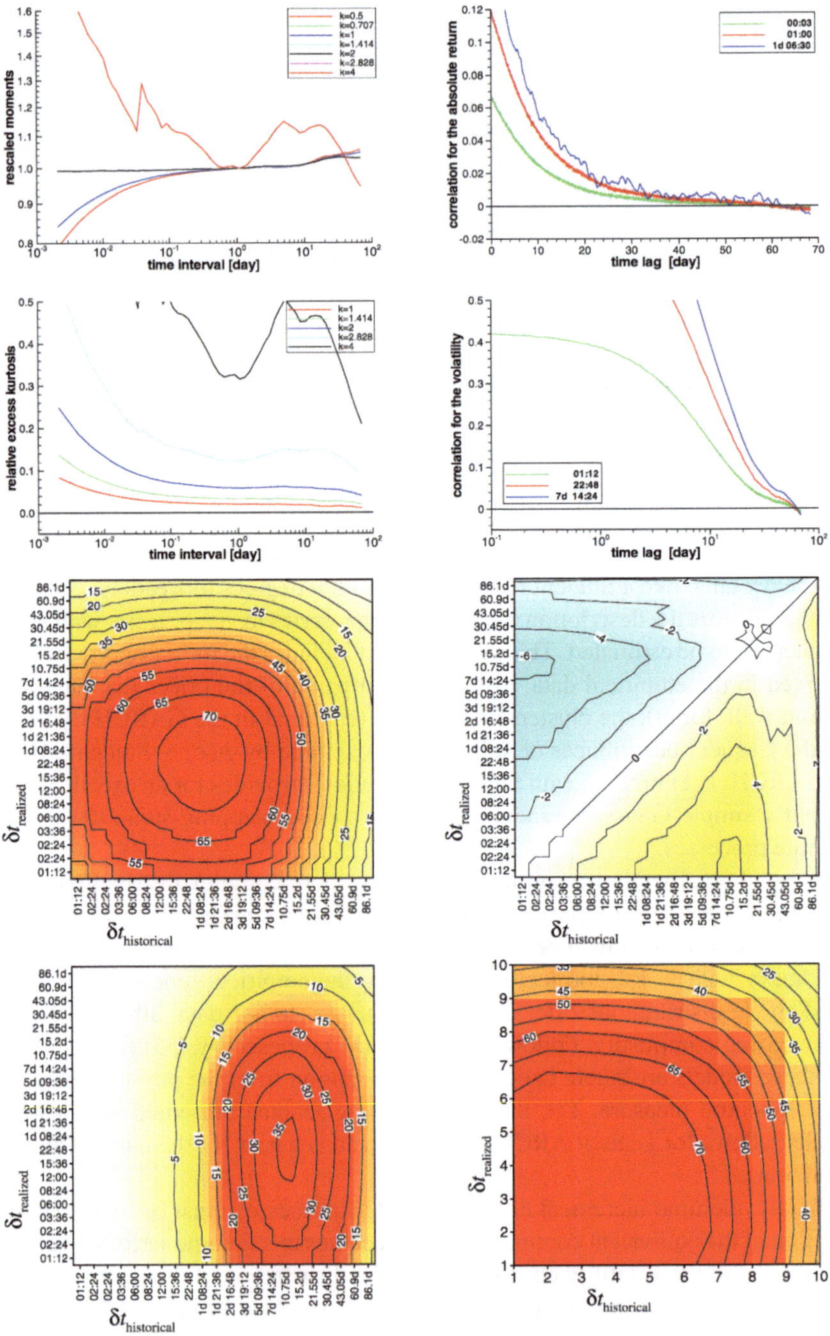

Fig. 7.1 The mug shot for the GARCH(1, 1) process (with logarithmic return)

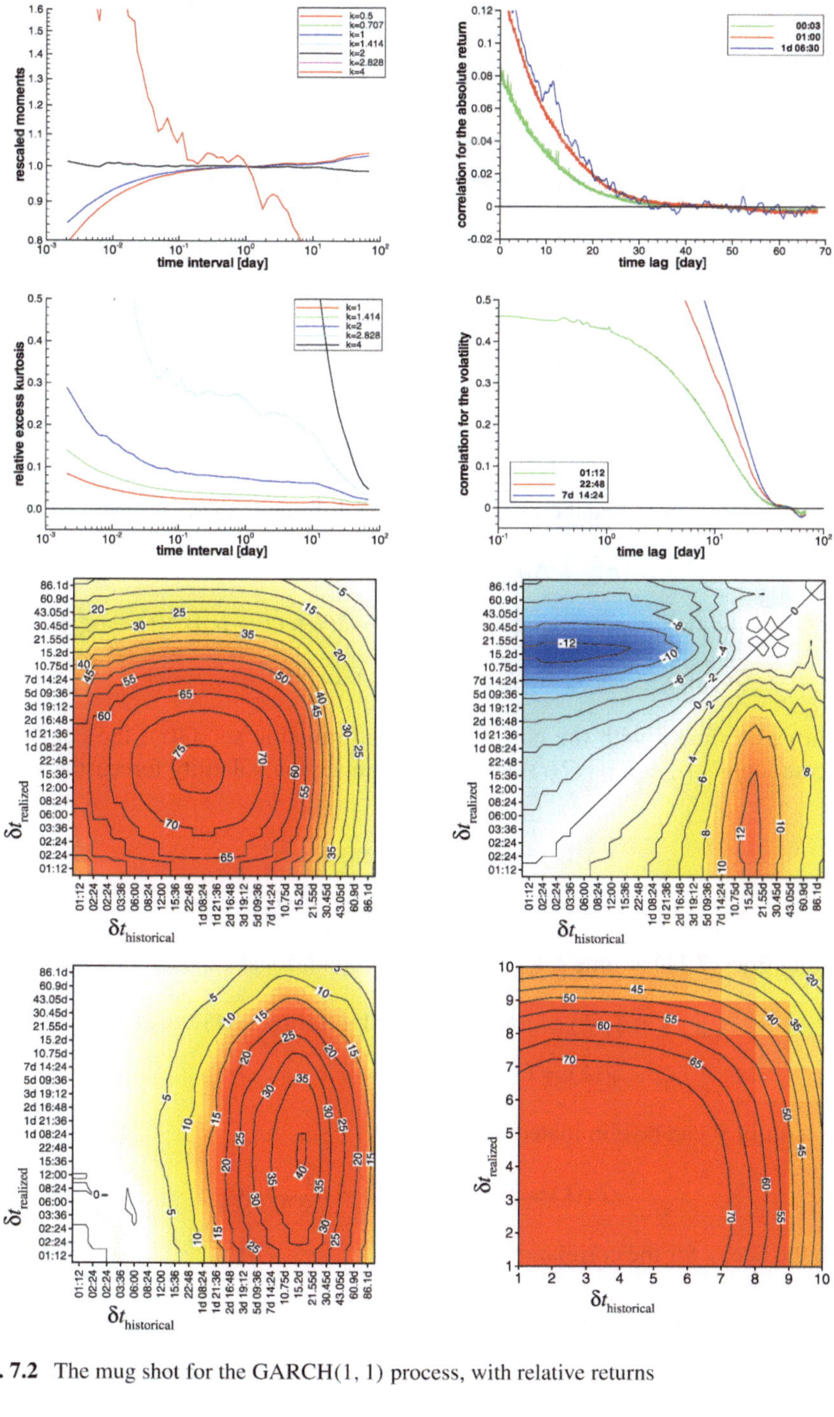

Fig. 7.2 The mug shot for the GARCH(1, 1) process, with relative returns

A proper extension to finite time increment of the option pricing framework is presented in Chap. 16, allowing us to use general ARCH processes with a broad class of distributions for the innovations.

7.1.1 *Volatility Forecast for the* GARCH(1, 1) *Process**

The equations for the GARCH(1, 1) process are simple enough to allow for an explicit evaluation of the volatility forecast. This subsection presents in details the computation; the evaluation is extended in Sect. 7.6 to the multicomponent ARCH processes following similar lines.

The unconditional expectations of (7.3a) and (7.3b) lead to

$$E\left[r^2\right] = E\left[\sigma_1^2\right] = E\left[\sigma_{\text{eff}}^2\right] = \sigma_\infty^2. \tag{7.10}$$

The process equations can be expressed for the difference to the means:

$$\begin{aligned} r^2(t+\delta t) &= \left(\sigma_\infty^2 + \Delta\sigma_{\text{eff}}^2(t)\right)\epsilon^2(t+\delta t) \\ \Delta\sigma_1^2(t) &= \sigma_1^2(t) - \sigma_\infty^2 = \mu\Delta\sigma_1^2(t-\delta t) + (1-\mu)\left(r^2(t) - \sigma_\infty^2\right) \\ \Delta\sigma_{\text{eff}}^2(t) &= \sigma_{\text{eff}}^2(t) - \sigma_\infty^2 = (1-w_\infty)\Delta\sigma_1^2(t). \end{aligned} \tag{7.11}$$

For a given time t with the corresponding information set $\Omega(t)$, the conditional expectations $E_t[\cdot] = E[\cdot \mid \Omega(t)]$ needed to evaluate the volatility forecast at $t + k\delta t$ are

$$\begin{aligned} \delta_1(k) &= E_t\left[\sigma_1^2(t+k\delta t) - \sigma_\infty^2\right] \\ \gamma(k) &= E_t\left[\sigma_{\text{eff}}^2(t+k\delta t) - \sigma_\infty^2\right]. \end{aligned} \tag{7.12}$$

Writing Eqs. (7.11) at time $t + k\delta t$ and taking conditional expectations lead to

$$\begin{aligned} \delta_1(k) &= \mu\delta_1(k-1) + (1-\mu)\gamma(k-1) \\ \gamma(k) &= (1-w_\infty)\delta_1(k). \end{aligned}$$

The term in γ can be eliminated, leading to

$$\delta_1(k) = \mu_{\text{corr}}\delta_1(k-1) = \mu_{\text{corr}}^k\delta_1(0) \tag{7.13}$$

with the decay for the correlation given by

$$\mu_{\text{corr}} = \mu + (1-w_\infty)(1-\mu) = 1 - w_\infty(1-\mu).$$

Equivalently, both decay parameters are related by the symmetrical expression $1 - \mu_{\text{corr}} = w_\infty(1-\mu)$. As $\mu < \mu_{\text{corr}} \leq 1$, the correlation time $\tau_{\text{corr}} = -\delta t / \ln(\mu_{\text{corr}})$

is larger than τ. When the characteristic time τ of the exponential is large compared to δt, the correlation time is

$$\tau_{\mathrm{corr}} \simeq \tau / w_{\infty}. \tag{7.14}$$

The volatility at time $t + k\delta t$, conditional on the information set at t, is given by

$$E_t\big[r^2(t + k\delta t)\big] = E_t\big[\sigma_{\mathrm{eff}}^2\big(t + (k-1)\delta t\big)\big] = \sigma_\infty^2 + \mu_{\mathrm{corr}}^{k-1}\big(\sigma_{\mathrm{eff}}^2(t) - \sigma_\infty^2\big). \tag{7.15}$$

7.1.2 Computation for the Lagged Correlation*

The computation of the lagged correlation is very similar to the derivation of the volatility forecast, except for the magnitude, as shown by using the conditional expectation

$$\begin{aligned}
&E\big[\big(r^2(t) - \sigma_\infty^2\big)\big(r^2(t + k\delta t) - \sigma_\infty^2\big)\big] \\
&\quad = E\big[\big(r^2(t) - \sigma_\infty^2\big)E_t\big[\big(r^2(t + k\delta t) - \sigma_\infty^2\big)\big]\big] \\
&\quad = \mu_{\mathrm{corr}}^{k-1} E\big[\big(r^2(t) - \sigma_\infty^2\big)\Delta\sigma_{\mathrm{eff}}^2(t)\big] \\
&\quad = \mu_{\mathrm{corr}}^{k-1} E\big[\big(r^2(t) - \sigma_\infty^2\big)\big\{\mu\Delta\sigma_{\mathrm{eff}}^2(t - \delta t) + (1 - w_\infty)(1 - \mu)\big(r^2(t) - \sigma_\infty^2\big)\big\}\big] \\
&\quad = \mu_{\mathrm{corr}}^{k-1}\big\{\mu E\big[\Delta\sigma_{\mathrm{eff}}^4\big] + (1 - w_\infty)(1 - \mu)E\big[\big(r^2 - \sigma_\infty^2\big)^2\big]\big\}.
\end{aligned} \tag{7.16}$$

Therefore, the lagged correlation is

$$\begin{aligned}
\rho_k &= \rho\big(r^2(t), r^2(t + k\delta t)\big) \\
&= \mu_{\mathrm{corr}}^{k-1}\left\{(1 - w_\infty)(1 - \mu) + \mu\frac{E[\Delta\sigma_{\mathrm{eff}}^4]}{E[(r^2 - \sigma_\infty^2)^2]}\right\} \\
&= \mu_{\mathrm{corr}}^{k-1}\rho_1.
\end{aligned} \tag{7.17}$$

As for the volatility forecast, the lagged correlation decays exponentially fast with the characteristic time τ_{corr}. The remaining task is to evaluate the correlation magnitude given by ρ_1, which is depending on the two quartic expectations that are related to the fluctuations of the volatility.

Equations (7.11) can be combined to obtain an equation involving $\Delta\sigma_1^2$ only:

$$\Delta\sigma_1^2(t) = \Delta\sigma_1^2(t - \delta t)\big\{\mu_{\mathrm{corr}} + (\mu_{\mathrm{corr}} - \mu)\big(\epsilon^2 - 1\big)\big\} + \sigma_\infty^2(1 - \mu)\big(\epsilon^2 - 1\big). \tag{7.18}$$

The excess kurtosis for the innovation is denoted by χ:

$$\chi = E\big[\big(\epsilon^2 - 1\big)^2\big]. \tag{7.19}$$

Using the process constraint $E[\epsilon^2] = 1$ on the residuals allows us to obtain

$$E\big[\Delta\sigma_1^4\big] = E\big[\Delta\sigma_1^4\big]\big\{\mu_{\text{corr}}^2 + (\mu_{\text{corr}} - \mu)^2\chi\big\} + \sigma_\infty^4(1-\mu)^2\chi \tag{7.20}$$

and therefore

$$E\big[\Delta\sigma_1^4\big] = \sigma_\infty^4 \frac{(1-\mu)^2\chi}{1-\mu_{\text{corr}}^2 - (\mu_{\text{corr}} - \mu)^2\chi}. \tag{7.21}$$

The numerator in ρ_1 can be computed using the last equation in (7.11):

$$E\big[\Delta\sigma_{\text{eff}}^4(t)\big] = (1-w_\infty)^2 E\big[\Delta\sigma_1^4\big]. \tag{7.22}$$

For the denominator, the expected fluctuations of r^2 is needed:

$$\begin{aligned} E\big[\big(r^2 - \sigma_\infty^2\big)^2\big] &= E\big[\sigma_{\text{eff}}^4\big](\chi+1) - \sigma_\infty^4 \\ &= \big(\sigma_\infty^4 + E\big[\Delta\sigma_{\text{eff}}^4\big]\big)(\chi+1) - \sigma_\infty^4 \\ &= \sigma_\infty^4\chi + E\big[\Delta\sigma_{\text{eff}}^4\big](\chi+1). \end{aligned} \tag{7.23}$$

The volatility fluctuations is composed of two terms, with $\sigma_\infty^4\chi$ that originates in the distribution of ϵ, whereas the term $E[\Delta\sigma_{\text{eff}}^4](\chi+1)$ is induced by the GARCH dynamic. Equations (7.20), (7.22), and (7.23) can be combined to obtain an expression in terms of the process parameters:

$$E\big[\big(r^2 - \sigma_\infty^2\big)^2\big] = \sigma_\infty^4\chi \frac{1 - 2\mu\mu_{\text{corr}} + \mu^2}{1-\mu_{\text{corr}}^2 - (\mu_{\text{corr}} - \mu)^2\chi}. \tag{7.24}$$

Simple algebra leads to the ratio

$$\frac{E[\Delta\sigma_{\text{eff}}^4(t)]}{E[(r^2 - \sigma_\infty^2)^2]} = \frac{(\mu_{\text{corr}} + \mu)^2}{1 - 2\mu\mu_{\text{corr}} + \mu^2} \tag{7.25}$$

which is not depending on χ, but only on the GARCH dynamic. Finally, the lag one correlation is

$$\rho_1 = (\mu_{\text{corr}} - \mu)\frac{1 - \mu\mu_{\text{corr}}}{1 - 2\mu\mu_{\text{corr}} + \mu^2} \tag{7.26}$$

and is therefore also independent of χ. The value for the lag one correlation is plotted on Fig. 7.3 for various values of the parameters. The curves make clear that w_∞ is the main parameter controlling the magnitude of the correlation. Given the empirical lagged correlation for the square returns, this figure makes easy to parameterize a GARCH(1, 1) process with the desired characteristics.

This computation depends on (7.18) where only $\Delta\sigma_1$ appears. This allows us to compute $E[\Delta\sigma_1^4]$ in closed form and to have an explicit expression for ρ_1. With a multicomponent ARCH process, this computation seems not anymore possible, and only the shape of the lagged correlation can be obtained.

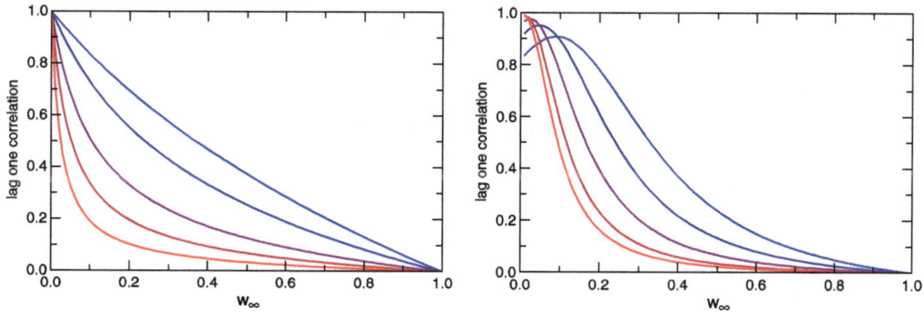

Fig. 7.3 The lag one correlation for the square return for the GARCH(1, 1) process. In the *left panel*, τ is constant for *each curves*, with $\tau = 1, 2, 5, 10, 21$ days and *colored* respectively from *blue* to *red*. In the *right panel*, τ_{corr} is constant for *each curves*, with $\tau_{\text{corr}} = 5, 10, 21, 42, 63$ days and *colored* respectively from *blue* to *red*

7.2 I-GARCH(2)

When written in the form (7.3a), (7.3b), the GARCH(1, 1) equation calls for the natural extension

$$\sigma_1^2(t) = \mu_1 \sigma_1^2(t - \delta t) + (1 - \mu_1) r^2(t) \tag{7.27a}$$

$$\sigma_2^2(t) = \mu_2 \sigma_2^2(t - \delta t) + (1 - \mu_2) r^2(t) \tag{7.27b}$$

$$\sigma_{\text{eff}}^2(t + \delta t) = (1 - w)\sigma_1(t)^2 + w\sigma_2(t)^2(t). \tag{7.27c}$$

This process is linear (in the variances and square returns), with two components. As the equations are very similar to I-GARCH(1), we name this process I-GARCH(2). It has three parameters τ_1, τ_2, w (or μ_1, μ_2, w). Roughly, σ_2 plays a similar role as the mean volatility, with the difference that σ_2 is computed from a long EMA on the past return and therefore adjusts to the past of the time series. This contrast from the parameter σ_∞, which is static (as obtained from a historical sample moment estimator, for example). Clearly, more volatility components can be added in the process, and these extensions are presented later in this chapter. Notice that [52] introduced a similar two-component process but with an affine term. An empirical study on volatility forecasts presented in [152] finds that the I-GARCH(2) performs better than the GARCH(1, 1) process, precisely because the mean volatility is difficult to estimate.

7.3 EGARCH(1, 1)

Most of the processes studied in the literature have a quadratic structure $\sigma^2 \sim r^2$. The reason for this choice is to have some analytical tractability, as, for example, in the computation of unconditional expectations, volatility forecast, or lagged correlation. More general analytical structures can be easily written, for example, with $\sigma^p \sim |r|^p$, but in general any analytical computations become impossible[1] for $p \neq 2$.

An interesting variation in the algebraic structure of the process is the Exponential GARCH process, or EGARCH, that has been introduced by [114]:

$$r(t+\delta t) = \sigma_{\text{eff}}(t)\epsilon(t+\delta t) \tag{7.28a}$$

$$\sigma_{\text{eff}}(t) = \sqrt{\frac{\delta t}{1\text{y}} e^{h(t)}} \tag{7.28b}$$

$$h(t) = (1-\beta_1)\ln(\sigma_\infty) + \alpha_1\left(\frac{|r(t)|}{\sigma_{\text{eff}}(t-\delta t)} - E[|\epsilon|]\right) + \beta_1 h(t-\delta t). \tag{7.28c}$$

The three process parameters are $\ln(\sigma_\infty)$, α_1, and β_1 with the constraint $\beta_1 < 1$. The parameter $\ln(\sigma_\infty)$ fixes the mean annualized volatility (but σ_∞ is not equal to the mean volatility!). The constant $E[|\epsilon|]$ depends on the probability distribution for the residuals ϵ. Essentially, this model is "like" GARCH(1, 1), but for the logarithm of σ. The prefactor in (7.28b) scales the volatility from 1 year to the time increment of the process δt. This allows us to have parameter values at the more intuitive 1 year scale.

The "mug shot" for this process is given in Fig. 7.4. Overall, it is quite similar to GARCH(1, 1), as both processes are depending on a single time scale. This similarity shows that the time scale structure is more important than the algebraic structure of the process equations. For this reason, we will focus on the (quadratic) multicomponent ARCH processes as they allow us to better capture the empirical properties of financial data, while preserving some analytical tractability.

7.4 Linear Versus Affine Processes

Many process properties are depending on whether the process equations for the variance are *linear* or *affine*. With linear process equations, like I-GARCH, the squared volatility is a linear combination of the historical squared volatility and the squared returns. Directly rooted in the linear structure of the equations are that the initial conditions set the long term mean volatility and that the variance of the volatility diverges. With affine process equations, a constant is added to the combination of historical squared returns. The simplest example of affine process equation is the GARCH(1, 1) process which includes in the equation for $\sigma^2(t)$ an additive constant term, in this book denoted by $w_\infty \sigma_\infty^2$ but mostly denoted α_0 or ω in the literature. This constant fixes the mean volatility (with some constraints on the parameters to ensure the existence of the mean). For the GARCH(1, 1) affine process

[1] In principle, the choice of the exponent, or more generally the algebraic structure of the processes, should not be a matter of analytical convenience but dictated by the empirical properties of the financial time series, or be rooted in some microstructure properties of the markets or the trading rules.

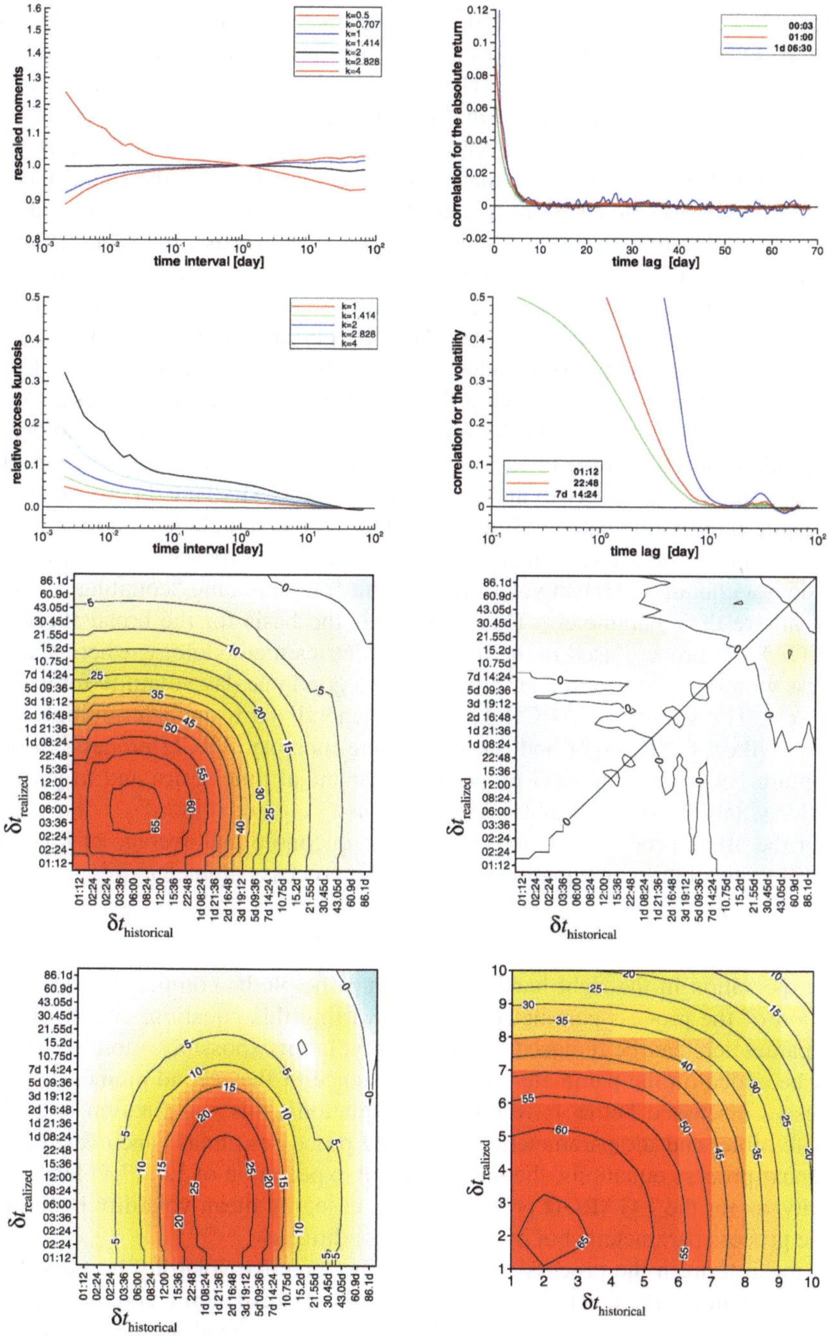

Fig. 7.4 The mug shot for the EGARCH(1, 1) process

equations, regardless of the initial conditions, the long-term mean volatility converges to a constant related to the process parameters. For the I-GARCH(1) and GARCH(1, 1) processes, the above key statistical properties originate in the linear or affine structure of the equations. These two models have been studied rigorously by [115], who shows in particular that the probability density of the volatility for the (linear) I-GARCH process converges to a singular distribution with all the mass at zero. This difference between linear and affine equations for the I-GARCH and GARCH processes extends to most ARCH processes which come in pairs with linear and affine versions (a subtle case is the FIGARCH process, see Sect. 7.12).

Because the differences between linear and affine processes are deeply rooted in their mathematical structure, it is plausible that some statistical properties of more complex processes are similar to the I-GARCH(1) and GARCH(1, 1) processes, respectively (with possibly some added constraints on the parameter values for the existence of the mean). Yet, proving those properties for generic processes is not elementary. In [152], very long Monte Carlo simulations are used to explore the asymptotic properties of the linear and affine long-memory processes. As expected, they are similar respectively to the I-GARCH(1) and GARCH(1, 1) processes.

Despite their mathematical difficulties and possible singular asymptotic properties, linear processes are very interesting for forecasting purposes, precisely because they do not contain the mean volatility (and the corresponding "coupling constant" w_∞) among their parameters. This property is the basis for the broad success of the I-GARCH process used in the basic RiskMetrics methodology, where a simple process with one fixed parameter provides for a good one day volatility forecast for all assets. The simple I-GARCH(1) can be generalized to include multiple components, like I-GARCH(2) and the linear long-memory ARCH process, in order to capture better the empirical properties of financial time series and to get more accurate volatility forecasts and risk estimations.

For the affine process equations, taking the unconditional average of Eqs. (5.2) and (7.1), one obtains an equation involving the process parameters ϑ:

$$\sigma_\infty^2 = E\big[\tilde{\sigma}^2[\vartheta, \delta t]\big]. \tag{7.29}$$

The expectation in the right-hand side can in principle be computed, leading to a function of the process parameters ϑ. By inverting this equation, one of the process parameters can be replaced by the mean volatility (possibly, more constraints must be added on the parameters for the existence of the second moment). As the processes discussed below have quadratic equations, this computation is analytically feasible, and the parameterization of the processes already includes σ_∞. For the linear process equations, the unconditional expectation of Eq. (7.1) leads to an identity, as for the I-GARCH process. In this case, the mean volatility is not fixed by the process parameters, but by the initial conditions.

A mean term in the process equation for the effective volatility introduces two constants, namely the mean volatility σ_∞ and the corresponding coupling constant w_∞. For long-term Monte Carlo simulations, a well-defined asymptotic distribution is necessary, and therefore affine processes must be used. But for most other situations, a linear process is perfectly suitable.

A *different* property is the return to the mean after a volatility spike and the inclusion of the "return to the mean" in a forecast. The subtlety of the "return to the mean" is that this behavior can be induced by a mean volatility parameter or by volatility components at long time horizons. The difference can be understood by comparing the I-GARCH(1), GARCH(1, 1), and I-GARCH(2) processes. The I-GARCH(1) process has no "mean reversion", the GARCH(1, 1) process has "mean reversion" induced by the mean volatility σ_∞^2, and the I-GARCH(2) process has "mean reversion" at characteristic times between τ_1 and τ_2 induced by the long-horizon volatility σ_2. In general, for a process with multiple time horizons, the long-term components act as a mean for the short-time horizons (independently of a possible mean volatility parameter). Notice that the GARCH(1, 1) process is singular as the "mean reversion" is completely provided by the mean volatility. In [157], the comparison of the 1 day forecasts between the I-GARCH(1), GARCH(1, 1), and I-GARCH(2) shows that this "reversion toward the longer term" is quantitatively important, as the last two processes are much better that I-GARCH(1). Yet, I-GARCH(2) is better than GARCH(1, 1), indicating that the constant mean volatility term is not the relevant factor. Similarly, results for the long-memory processes show no differences between linear and affine versions, as all include a similar "mean reversion" at the 1 day forecast horizon. In short, the presence of long-horizon volatility components is an efficient mechanism to induce the "return to the mean" at short to intermediate time horizons. This property is different from having well-behaved asymptotic properties for long times, which are controlled by the presence of a mean volatility σ_∞ in the equations.

7.5 Multicomponent ARCH Processes

The GARCH(1, 1) process as written in the form (7.3a) and (7.3b) leads naturally to a generalization with more volatilities σ_k measured at time horizons τ_k, and with σ_{eff} as a convex combination of these (historical) volatilities and σ_∞. As the volatilities are measured over increasing characteristic times, the corresponding returns can be either taken at the smallest time horizon δt or over increasing time span $n\delta t$. A rich crop of processes can be built along these lines, using either a microscopic or aggregated definition for the historical volatilities, a linear or affine structure for σ_{eff}, and different choices for the characteristic times τ_k and the corresponding weights w_k. The basic ingredients are presented here, while the detailed recipes for various processes are given in the following subsections.

Given a set of time horizons τ_k, the corresponding decay factors are

$$\mu_k = e^{-\delta t/\tau_k} \qquad k = 1, \dots, k_{\max}. \tag{7.30}$$

The *microscopic* historical volatility is computed with

$$\sigma_{\text{mic},k}^2(t) = \mu_k \sigma_k^2(t - \delta t) + (1 - \mu_k) r^2(t) \tag{7.31}$$

and r is measured at the finest scale δt regardless of τ_k. The *aggregated* historical volatility is

$$\begin{aligned}
\sigma^2_{\mathrm{agg},k}(t) &= \mu_k \sigma^2_k(t-\delta t) + (1-\mu_k) r[l_k \delta t]^2(t) \\
r[l_k \delta t](t) &= \frac{x(t) - x(t - l_k \delta t)}{\sqrt{l_k}}
\end{aligned} \tag{7.32}$$

where $l_k \geq 1$ are integers. The aggregated returns $r[l_k \delta t]$ are measured over the time horizon $l_k \delta t$ and scaled back to the time horizon δt. With this choice, all the volatilities are scaled to the process increment δt. This convenient choice allows for a direct comparison of the process parameters. The rationale for both formula is to use all available information for the microscopic definition, in the spirit of the quadratic variations, whereas the aggregated definition captures the behavior of traders that use data at lower frequency when their characteristic time is longer.

The effective volatility σ_{eff} is a convex combination of the historical volatilities

$$\begin{aligned}
\sigma^2_{\mathrm{eff}}(t) &= \sigma^2_\infty + (1 - w_\infty) \sum_{k=1}^{k_{\max}} \chi_k \big(\sigma^2_k(t) - \sigma^2_\infty\big) \\
&= \sum_{k=1}^{k_{\max}} w_k \sigma^2_k(t) + w_\infty \sigma^2_\infty \\
w_k &= (1 - w_\infty)\chi_k.
\end{aligned} \tag{7.33}$$

The parameters are such that $\sum \chi_k = 1$, $0 < \chi_k < 1$, and $\sum w_k + w_\infty = 1$, $0 \leq w_\infty < 1$. The linear processes are obtained for $w_\infty = 0$.

Many processes can be built using these bricks, with different choice for the characteristic times τ_k, weights χ_k, coupling w_∞ for the mean volatility σ_∞, as well as the choice of a microscopic or aggregated definition for the volatility σ_k. The correct balance has to be found between the complexity, number of parameters, and accuracy of a process compared to the empirical facts, with the optimal choice depending on the desired application. Mainly two interesting routes can be followed. The first one is to take multicomponent but simples functions for τ_k and χ_k. The functions are chosen so as to reproduce the overall multiscale stylized facts and in particular the long memory observed in the lagged correlations. The processes along this line are the long-memory processes, with LM-Mic-Lin-ARCH (long-memory-microscopic-linear-ARCH), LM-Agg-Lin-ARCH (long-memory-aggregated-linear-ARCH), LM-Mic-Aff-ARCH (long-memory-microscopic-affine-ARCH), and LM-Agg-Aff-ARCH (long-memory-aggregated-affine-ARCH). The second route is to take a few characteristic times τ_k and to adjust the corresponding weights χ_k so that the detailed structure found in the historical/realized volatility correlation can be reproduced. The family of market processes Mkt-ARCH are constructed along this line, with the same microscopic/aggregated and linear/affine variants.

7.6 General Considerations on Volatility Forecast

For the multicomponent (quadratic) ARCH volatility processes, a forecast for the volatility can be computed analytically. More precisely, at time t, given the information set $\Omega(t)$, the forecast $\mathcal{F}[\sigma^2_{\text{eff}}]$ for the time $t+\Delta t$ is given by the conditional average

$$\mathcal{F}\big[\Delta t; \sigma^2_{\text{eff}}\big](t) = E\big[\sigma^2_{\text{eff}}(t+\Delta t) \,\big|\, \Omega(t)\big] \tag{7.34}$$

with $\Delta t = k\delta t$. Then, using iteratively the process equations, the right-hand side can be expressed as a function of the variables contained in the information set $\Omega(t)$. Note that this is possible by simple analytical means because the above equations are quadratic in the returns. General nonlinear processes can be easily written, for example, by replacing the square by a power p, but this analytical tractability is then lost. The forecast for the integrated volatility between t and $t+\Delta t$ is given by

$$\overline{\mathcal{F}}\big[\Delta t; \sigma^2_{\text{eff}}\big](t) = \sum_{j=0}^{m-1} \mathcal{F}\big[j\delta t; \sigma^2_{\text{eff}}\big](t) \tag{7.35}$$

with $\Delta t = m\delta t$. This is indeed the interesting quantity to forecast as it is relevant for portfolio management, option pricing, and risk assessment. The forecast of the integrated volatility is also very interesting for process estimations: on historical data, the realized volatility can be computed, and the parameters of the process can be optimized to give the best forecast for the realized volatility. A simple L^2 distance $d[\Delta t, \theta]$ gives a convenient functional to measure the forecast error

$$d^2[\Delta t, \theta]\Big(\sqrt{\overline{\mathcal{F}}[\sigma^2_{\text{eff}}]}, \sigma_{\text{rel}}\Big) = \sum_t \Big(\sqrt{\overline{\mathcal{F}}[\Delta t, \sigma^2_{\text{eff}}](t)} - \sigma_{\text{rel}}[\Delta t](t)\Big)^2. \tag{7.36}$$

This quantity is usually called the RMSE (root mean square error). The forecast depends on the process parameters θ, and therefore the distance functional $d[\theta]$ can be minimized to produce the best forecast. This procedure is an alternative to the usual estimate of the process parameters by the maximization of a log-likelihood functional.

If the emphasis is on volatility forecast, then estimating the parameters by minimizing the forecast error is appropriate, whereas if the emphasis is on the data generating process, a log-likelihood estimate is better suited. The salient features of the log-likelihood estimate and forecast error estimate are summarized in Table 7.1.

In particular, when optimizing on the forecast error, the explicit form of the forecast error needs to be chosen (i.e., the distance functional between the realized and forecast volatilities), but no distributional assumption on the residues is needed. The most common choice is the above L^2 (or RMSE) distance for the quadratic realized volatility.

Table 7.1 Log-likelihood versus forecast error for the estimation of the parameter values in a process

	Log-likelihood	Forecast error
Hypothesis	Distributional assumption on the residues	Choice of the realized volatility and distance functional
Sensitivity to the different time scales	Mainly at the smallest time scale δt	Mainly at all time scales up to Δt
Can be evaluated for	Any processes	quadratic process equations and forecast models

7.7 Volatility Forecast for the Microscopic ARCH Processes

For the microscopic multicomponent ARCH processes, the EMA can be "unwinded", leading to simple expressions for the effective volatility and for the volatility forecast. After elimination of the (unwinded) σ_k volatilities, the effective volatility takes the form

$$\sigma_{\text{eff}}^2(t) = w_\infty \sigma_\infty^2 + \sum_{i\geq 0} \lambda(i) r^2(t - i\delta t) \qquad \lambda(i) > 0 \tag{7.37}$$

with $\sum_{i\geq 0} \lambda(i) = 1 - w_\infty$. The historical weights $\lambda(i)$ are related to the component weights w_k by

$$\lambda(i) = \sum_{k=1}^{k_{\max}} w_k (1 - \mu_k) \mu_k^i.$$

The same computation can be done in a volatility forecast, using the weights derived explicitly in the next section, leading to

$$\overline{\mathcal{F}}[\Delta T; \sigma_{\text{eff}}^2](t) = \widetilde{\sigma^2}[\Delta T](t) = \frac{\Delta T}{\delta t} \left\{ \lambda_\infty(n) \sigma_\infty^2 + \sum_{i\geq 0} \lambda(n, i) r^2(t - i\delta t) \right\} \tag{7.38}$$

with $\Delta T = n\delta t$. In this form, the weight $\lambda(n, i)$ can be computed a priori, leading to simple and efficient volatility forecasts. This explicit forecast with $w_\infty = 0$ is used for risk estimation in Chap. 15, and Fig. 15.1 compares the weights for the I-GARCH and LM-ARCH processes. In particular, the forecasts induced by a long-memory process correspond to our intuition that short-term forecasts are dominated by the recent past while long-term forecasts need to put more weights on the distant past.

*7.7.1 Explicit Equations for the Volatility Forecasts**

For the multicomponent microscopic affine LM-Mic-Aff-ARCH(n) and linear LM-Mic-Lin-ARCH(n) processes, the iterative equations for the conditional average take a simple form that is derived in this section. The line of this computation follows the derivation for the GARCH(1, 1) process given in Sect. 7.1.1. Using the process equations for the LM-Mic-Aff-ARCH process, we obtain:

$$\begin{aligned} E\big[\sigma_k^2(t+j\delta t)\,\big|\,\Omega(t)\big] &= \mu_k E\big[\sigma_k^2\big(t+(j-1)\delta t\big)\,|\,\Omega(t)\big] \\ &\quad + (1-\mu_k)E\big[\sigma_{\mathrm{eff}}^2(t+j\delta t)\,\big|\,\Omega(t)\big] \qquad (7.39) \\ E\big[\sigma_{\mathrm{eff}}^2\big(t+(j+1)\delta t\big)\,\big|\,\Omega(t)\big] &= \sigma_\infty^2 + \sum_k w_k\big\{E\big[\sigma_k^2(t+j\delta t)\,|\,\Omega(t)\big] - \sigma_\infty^2\big\}. \end{aligned}$$

Introducing the new variables

$$\begin{aligned} \delta_k(j) &= E\big[\sigma_k^2(t+j\delta t)\,\big|\,\Omega(t)\big] - \sigma_\infty^2 \\ \gamma(j) &= E\big[\sigma_{\mathrm{eff}}^2(t+j\delta t)\,\big|\,\Omega(t)\big] - \sigma_\infty^2 \end{aligned} \qquad (7.40)$$

the conditional average equations are reduced to

$$\begin{aligned} \delta_k(j) &= \mu_k\delta_k(j-1) + (1-\mu_k)\gamma(j) \\ \gamma(j+1) &= \vec{w}'\cdot\vec{\delta}(j) \end{aligned} \qquad (7.41)$$

where $\vec{w}$ is the vector of weights w_k, and similarly for $\vec{\delta}$ and $\vec{\mu}$. For the LM-Mic-Lin-ARCH(n), the computations are identical except for the term σ_∞^2, which is absent. We can introduce the diagonal matrix $M_{k,p} = \delta_{k,p}\,\mu_k$ with $\delta_{k,p}$ the Kronecker symbol ($\delta_{k,p} = 1$ if $k = p$, zero otherwise). Equations (7.41) can be combined

$$\vec{\delta}(j) = \big\{M + (\vec{1}-\vec{\mu})\vec{w}'\big\}\vec{\delta}(j-1) \qquad (7.42)$$

where $\vec{1}$ denotes the constant vector $\vec{1}_k = 1$. This equation can be iterated j times:

$$\vec{\delta}(j) = \big\{M + (\vec{1}-\vec{\mu})\vec{w}'\big\}^j\vec{\delta}(0) \qquad (7.43)$$

and $\vec{\delta}(0)$ is in the information set. This expression relates $\mathcal{F}[\sigma_{\mathrm{eff}}^2]$ linearly to the $\sigma_k^2(t)$. For γ, (7.41) can be expressed as

$$\gamma(j+1) = \vec{w}'(j)\cdot\vec{\delta}(0) \qquad (7.44)$$

with the coefficients $w_k(j)$ given by the recursive equation

$$\begin{aligned} w'(j+1) &= w'(j)\big\{M + (\vec{1}-\vec{\mu})\vec{w}'(0)\big\} \\ \vec{w}(0) &= \vec{w}. \end{aligned} \qquad (7.45)$$

The equation can be expressed for the components

$$w_k(j+1) = \mu_k w_k(j) + w_k(0)\sum_p w_p(j)(1-\mu_p). \qquad (7.46)$$

The sum of the coefficients can be computed

$$\sum_k w_k(j) = (1 - w_\infty) \sum_k w_k(j-1) = (1 - w_\infty)^{j+1} \tag{7.47}$$

and the "coupling constant" for the mean volatility in the forecast is $w_\infty(j)$ with

$$w_\infty(j) = 1 - (1 - w_\infty)^{j+1}. \tag{7.48}$$

Therefore, the coefficients $w_k(j)$ can be evaluated a priori, and the forecast for the effective volatility computed by a simple scalar product. We can now express the above forecast equations in the form (7.38). The iterative equation for σ_k^2 can be unwinded in order to express the volatility with the lagged returns and exponential weights:

$$\sigma_k^2(t) = (1 - \mu_k) \sum_{i=0}^{\infty} \mu_k^i r^2(t - i\delta t). \tag{7.49}$$

In practice, the sum over the lags needs be cut off at some $i_{\max}$, and the formula becomes

$$\sigma_k^2(t) = \frac{1 - \mu_k}{1 - \mu_k^{i_{\max}}} \sum_{i=0}^{i_{\max}-1} \mu_k^i r^2(t - i\delta t). \tag{7.50}$$

This form is introduced in definition (7.33) for σ_{eff}^2 to obtain Eq. (7.37) with the weights

$$\lambda(i) = \sum_{k=1}^{k_{\max}} w_k \frac{1 - \mu_k}{1 - \mu_k^{i_{\max}}} \mu_k^i \tag{7.51}$$

The same substitution in the forecast equations leads to (7.38) with

$$\lambda(n, i) = \sum_{k=1}^{k_{\max}} \frac{1}{n} \sum_{j=0}^{n-1} w_k(j) \frac{1 - \mu_k}{1 - \mu_k^{i_{\max}}} \mu_k^i \tag{7.52a}$$

$$\lambda_\infty(n) = 1 - \sum_i \lambda(n, i) \tag{7.52b}$$

and with $w_k(j)$ given by the iterative equations (7.45). The integrated process is defined with the constraint $\sum_{k=1}^{k_{\max}} w_k = 1$, and the coefficients $\lambda(i)$ obey

$$\sum_{i=0}^{i_{\max}-1} \lambda(i) = \sum_{i=0}^{i_{\max}-1} \lambda(n, i) = 1. \tag{7.53}$$

This computation is interesting for three purposes. First, it allows us to evaluate very efficiently the forecasts for the processes LM-Mic-Aff-ARCH(n) and

LM-Mic-Lin-ARCH(n) with a closed formula. Secondly, the effective weights $\vec{w}(j)$ can be used in nonquadratic forecasting models, as a parsimonious parameterization of the weightings that are used for each component. Thirdly, the same weights can be used as an approximation for aggregated processes (see below).

This computation gives also the shape of the lagged correlation, similarly to Sect. 7.1.1. The magnitude of the lagged correlation is given by the expectation of fourth moments. The corresponding analytical computation is presented in Sect. 7.1.2 for GARCH(1, 1), but seems not possible for a multicomponent process.

For the processes with an aggregated volatility, like the LM-Agg-Aff-ARCH(n) and LM-Agg-Lin-ARCH(n) processes, the forecast cannot be expressed in a simple scalar product. This happens because the returns at different time horizons enter into the equations, and therefore the forecast depends on the recent price history. Yet, the evaluation of the conditional average for the volatilities is fairly straightforward when computed iteratively, and the numerical implementation of the equations, even though cumbersome, presents no particular difficulty. A simpler solution for the aggregated models is to use the forecasting weights w_k for the volatility components σ_k as derived for the microscopic model. Although not rigorous, these weights capture the main properties of the forecast. Such a solution is used for options in Chap. 16 in order to compute a variance forecast up to the option expiry.

7.8 Geometric ARCH Processes

Section 7.5 introducing multicomponent ARCH processes is based on the usual logarithmic random walk discretization. As discussed in Chap. 6, a geometric discretization presents important advantages, in particular with respect to the inclusion of fat-tails in the distribution of the innovations. The expression of the multicomponent ARCH processes in this setting is simply:

$$\sigma_k^2(t) = \mu_k \sigma_k^2(t-\delta t) + (1-\mu_k) r^2(t), \qquad k = 1, \ldots, k_{\max} \tag{7.54a}$$

$$\sigma_{\text{eff}}^2(t) = \sum_{k=1}^{k_{\max}} w_k \sigma_k^2(t) + w_\infty \sigma_\infty^2 \qquad \text{with} \quad \sum_{k=1}^{k_{\max}} w_k + w_\infty = 1 \tag{7.54b}$$

$$r_{\text{rel}}[\delta t](t+\delta t) = \sigma_{\text{eff}}(t)\, \epsilon(t+\delta t) \tag{7.54c}$$

$$p(t+\delta t) = p(t)\big(1 + r_{\text{rel}}(t+\delta t)\big). \tag{7.54d}$$

The meaning of the various terms are identical from the previous sections, except for using the relative return r_{rel}. Equation (7.54d) involving the price is modified by the formal replacement $r = r_{\log} \to r_{\text{rel}}$. Yet, the modification in (7.54d) breaks analytical tractability. For example, the parameter σ_∞ fixes the mean volatility but is not equal to the mean volatility. The unspecified return r in (7.54a) can be either the microscopic or aggregated return, and either the logarithmic or relative return. This leaves clearly many possible processes to be explored. Monte Carlo simulations for this process with the return in (7.54a) taken as an aggregated logarithmic return points to long-term stability issues. This is not the case when this return is taken as an aggregated relative return, consistently with the other equations.

After n steps, the relative return is

$$r_{\mathrm{rel}}[\Delta T] = \prod_{i=1}^{n} \left[1 + \sigma_{\mathrm{eff}}(t_{i-1})\epsilon(t_i)\right] - 1. \tag{7.55}$$

For volatility forecasts, the expected variance $E[r^2_{\mathrm{rel}}[\Delta T] \mid \Omega(t)]$ needs to be computed, but the product breaks the exact analytical computations.

7.9 Long-Memory ARCH Processes

The goal of the long-memory ARCH processes is to capture in a minimal way the long memory observed in the volatility lagged correlation. This long memory is observed on all the mug shots and is characterized more precisely in Chap. 12. The idea of this process is to use a large number of components so as to be close to a "continuum limit" but with a simple parametric form for the weights. The basic time structure l_k of the process is given by the geometric series $l_k = \rho^{k-1}$. In order to approximate a continuum limit, ρ is chosen close to 1. In practice, $\rho = 2$ or $\rho = \sqrt{2}$ is close enough so as to be an irrelevant parameter. The characteristic times for the EMA are

$$\tau_k = l_k \tau_1 \qquad k = 1, \ldots, n \tag{7.56}$$

and the historical EMA volatility can be computed using either the microscopic or the aggregated definition.

The weights χ_k that define the effective volatility are taken as a simple function of the characteristic times. In practice, a function $\chi_k(\tau)$ that decays as the lagged correlation essentially reproduces the desired lagged correlation, with either a power law or a logarithmic shape. The logarithmic shape for χ_k reproduces better the empirical lagged correlation

$$\chi_k = c\left(1 - \frac{\ln \tau_k}{\ln \tau_0}\right) \tag{7.57}$$

with the condition $\tau_n < \tau_0$. The constant c is such that $\sum_k \chi_k = 1$. A power law decay is a popular choice

$$\chi_k = c\rho^{-(k-1)\lambda} = c\left(\frac{1}{l_k}\right)^{\lambda} = c'\left(\frac{\tau_1}{\tau_k}\right)^{\lambda}. \tag{7.58}$$

Such heteroscedasticity structures have three parameters: the lower cut-off τ_1, the upper cut-off τ_n, and the decay of the weights given either by the characteristic time τ_0 (logarithmic decay) or the exponent λ (power law). As usual, the affine process has two more parameters given by w_∞ and σ_∞. The constant w_∞ fixes the vol-of-vol and the overall level of correlation (the correlation increases with decreasing w_∞, as for the GARCH(1, 1) process). By using the microscopic or aggregated definition of the volatility, different processes can be constructed. This family of processes has been explored in [152] with respect to volatility forecast and in [156] with respect to their time reversal properties. The characteristics of the volatility lagged correlations are investigated in Chap. 12, with the conclusion that the logarithmic shape gives the best overall description of the financial time series.

As for the GARCH(1, 1) process, the optimal parameters take fairly similar values regardless of the time series. The characteristic decay time τ_0 of the long memory is of the order of a few years, with a typical value around 2 to 4 years. The upper cut-off τ_n is of the order of 1 year, and the process properties are fairly insensitive to this value. For the affine process, w_∞ is of the order of $w_\infty \simeq 0.05$. Compared to the GARCH(1, 1) process, the smaller value can be understood by the long-term volatility components taking some part of the mean volatility. As the limiting case $w_\infty = 0$ is singular in the long run, an interesting question is why the empirical time series are systematically fairly close to this singular limit? The answer probably lies in the market structures and the human attitude toward profit and risk, but is not known at this time.

The mug shots for the microscopic and aggregated long-memory ARCH are given respectively in Figs. 7.5 and 7.6. Both processes reproduce well probability distributions and lagged correlations. In particular, the volatility lagged correlations have a decay that is very similar to the empirical data for both processes. Yet the difference between them is clearly observed in the historical/realized volatility correlation: the asymmetry is very weak for the microscopic process, but the aggregated process has the correct size compared to the empirical data. The volatility graining shows also a clear difference between both processes, with the same respective symmetry and asymmetry. This shows that the granularity used in the definition of the volatility σ_k is an important ingredient and that an aggregated definition of the volatility should be used to reproduce the stylized facts.

7.10 The Long-Memory Affine Pure-Aggregated ARCH Process

This process has been introduced by [26], and the name in this book is chosen to be consistent with the other ARCH processes. The idea of the "pure-aggregated" ARCH process LM-Aff-PureAgg-ARCH is to measure the volatility at the scale τ_k by $\sigma_k^2 = r^2[\tau_k]$. The motivation is that traders take their decisions based on price moves, and therefore the volatility σ_k influencing the next price at a give time scale τ_k should be related to the magnitude of the return at the corresponding horizon. The effective volatility for the LM-Aff-PureAgg-ARCH process is given by

$$\sigma_{\text{eff}}^2(t) = w_\infty \sigma_\infty^2 + (1 - w_\infty) \sum_{l=1}^{l_{\max}} \chi_l r^2[l\delta t](t). \tag{7.59}$$

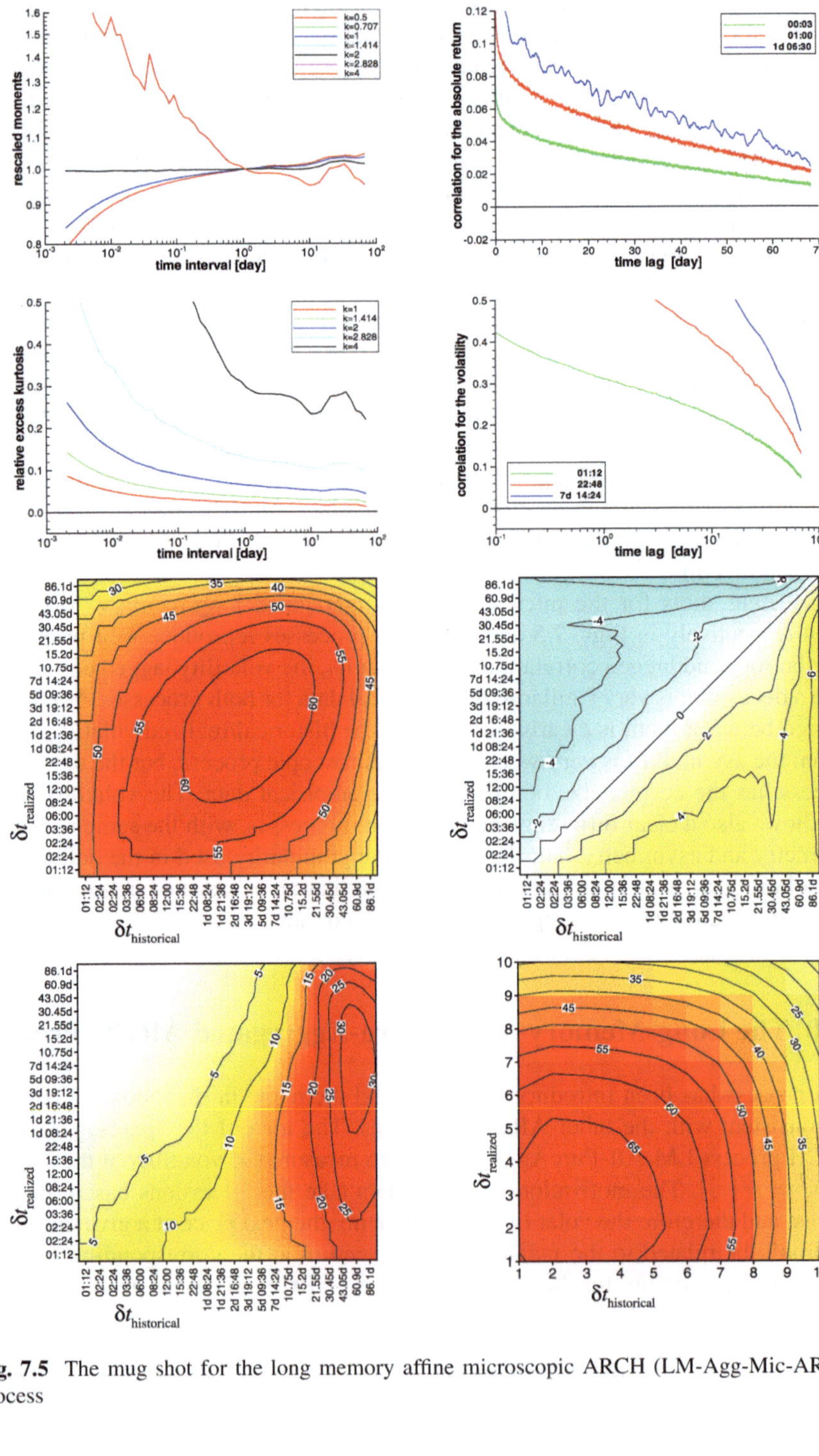

Fig. 7.5 The mug shot for the long memory affine microscopic ARCH (LM-Agg-Mic-ARCH) process

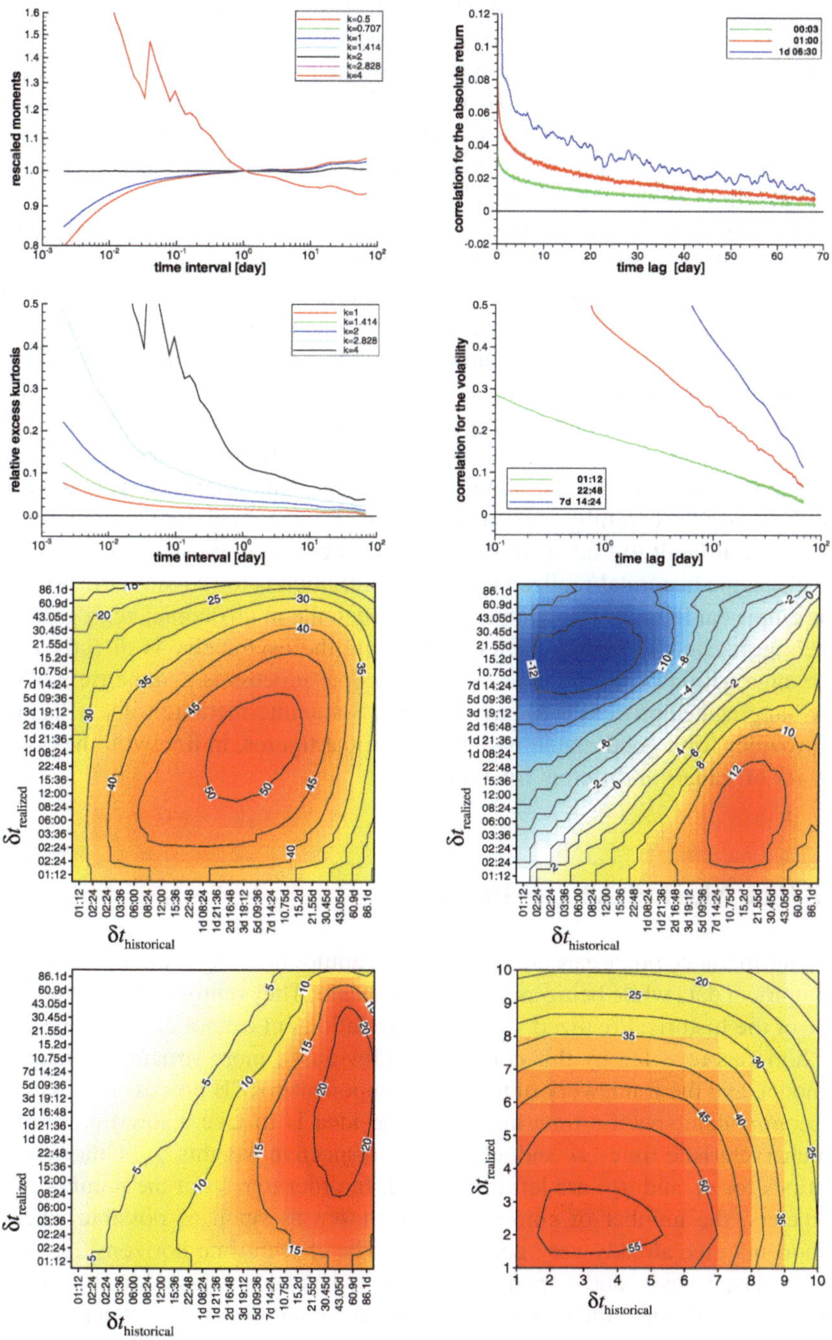

Fig. 7.6 The mug shot for the long memory affine aggregated ARCH (LM-Agg-Agg-ARCH) process

The sum over the lags runs over all integer values up to the upper cut-off l_{max}. This is different from the above long-memory processes where the lags are taken as a geometric series. Otherwise, this model is essentially equivalent to the limit $\tau_0 \to 0$ of the LM-Aff-Agg-ARCH process. It has the same parameters as the microscopic or the aggregated processes, only the aggregation is taken as large as possible. The weights are taken as a simple decaying functions (power law, logarithmic decay), but clearly other choices are possible.

As the model is formulated, the computational time is proportional to the cut-off (whereas for the LM-Aff-Agg-ARCH, it is proportional to the log of the cut-off). As we investigate large cut-off compared to the process increment δt, this difference makes the simulation time much (much) longer. On the other hand, thanks to its simpler structure, the model has some analytical tractability, whereas models with EMAs seem more complicated (see [26]).

The mug shot for the pure-aggregated long-memory ARCH process is given in Fig. 7.7. This process reproduces some empirical features like the slow decay of the lagged correlation and the anomalous scaling of the moment. The agreement could be better for carefully chosen parameters, but the much larger computational time makes difficult to find accurate values for them. Yet, regardless of the parameter values, the historical/realized volatility correlation is too asymmetric compared to the empirical figures. The comparison between the microscopic, aggregated, and "pure"-aggregated processes shows clearly that the asymmetry is directly related to the aggregation of the returns: the microscopic process is symmetric while the "pure"-aggregated process is too asymmetric. Some intermediate value of aggregation provides for a good match with the empirical figures, in line with the possible behavior of traders at different time horizons.

7.11 The Market-ARCH Processes

The long-memory processes are designed to capture the long memory observed in the empirical data while being as simple as possible. They reproduce well the overall shape of the historical/realized volatility correlation but cannot capture the specific time horizons revealed by this correlation. Obviously, more structure is needed in the model, and the framework of the multicomponent ARCH process is rich enough to accommodate specific time horizons. The idea is to take a few time horizons with characteristic times τ_k and adjust their respective weights χ_k. Otherwise, the equations for σ_k and σ_{eff} are left unchanged. In order to restrict the number of free parameters, the number of components is chosen as small as possible, and four components give already very good results. The chosen time horizons $2\tau_k$ are 3.8 hours, 1 day, 1 week, and 1 month, and only τ_1 has been adjusted to the empirical data. This set of time scale grows roughly like a geometric series with a progression of 5. The remaining parameters are $w_\infty = 0.082$, $\chi_1 = 0.53$, $\chi_2 = 0.16$, $\chi_3 = 0.14$, and $\chi_4 = 0.17$. The resulting mug shot is shown in Fig. 7.8. The overall structure is correct, the inhomogeneity is clearly present, but the inhomogeneity structure is too

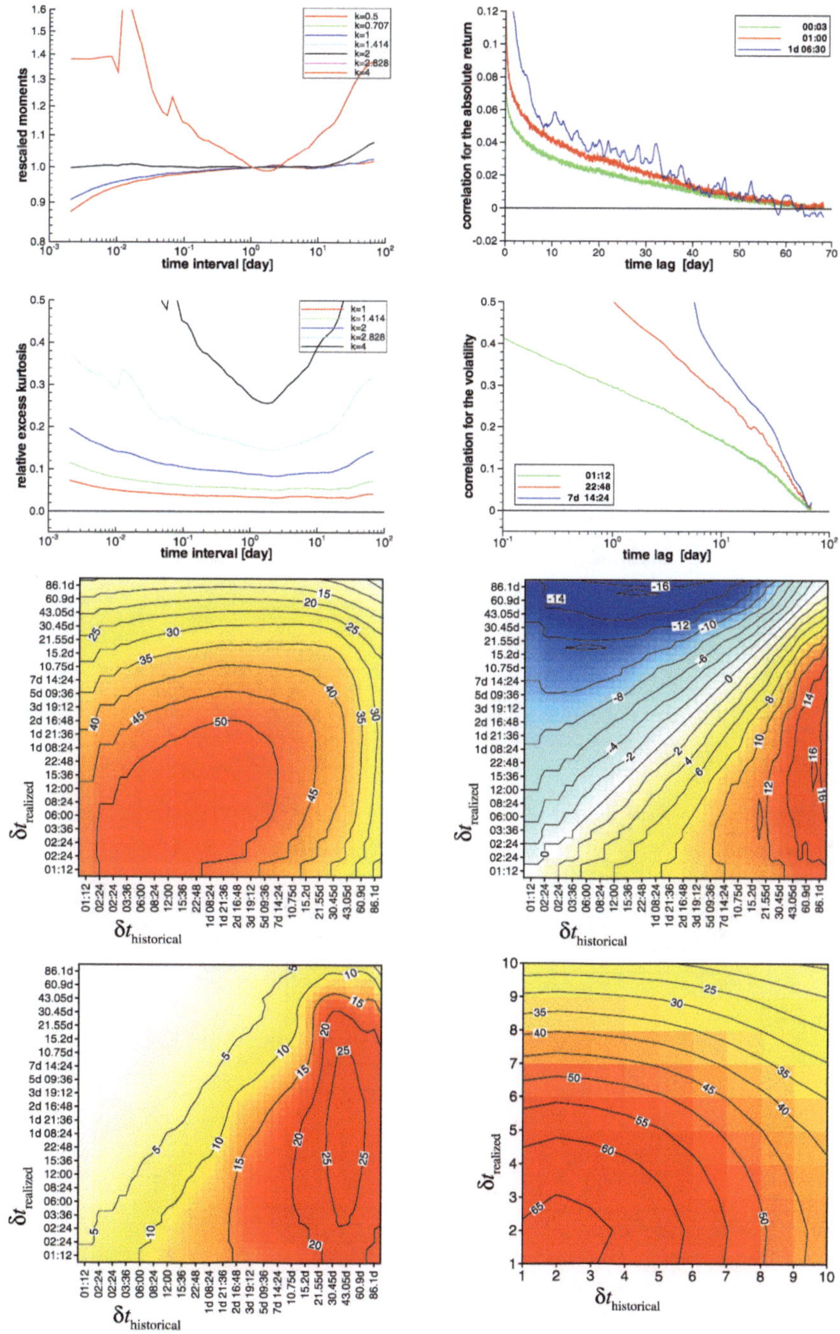

Fig. 7.7 The mug shot for the long memory affine "pure"-aggregated ARCH (LM-Agg-PureAgg -ARCH) process (parameter set 2)

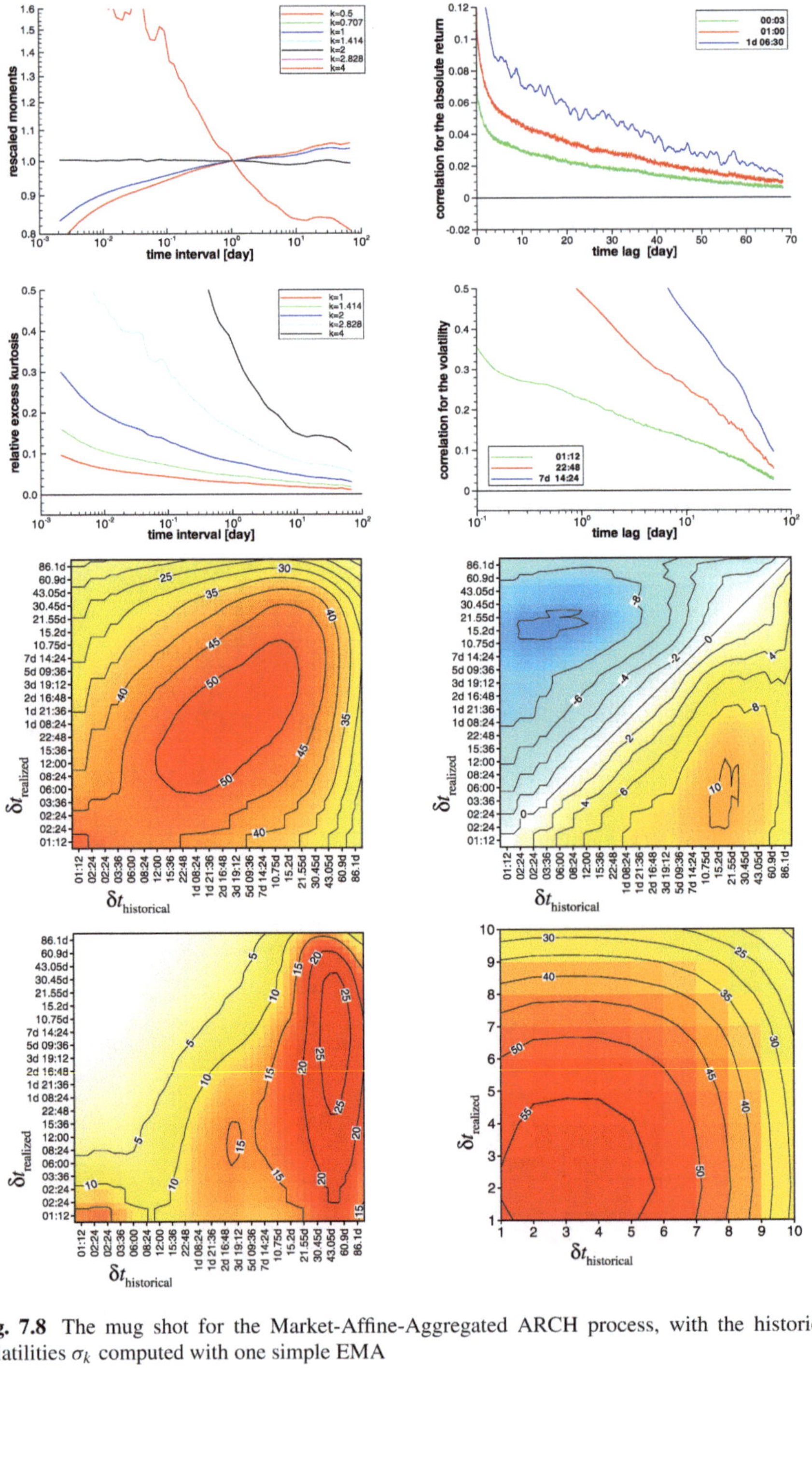

Fig. 7.8 The mug shot for the Market-Affine-Aggregated ARCH process, with the historical volatilities σ_k computed with one simple EMA

soft. For example, by changing the parameters in this process, it is not possible to reproduce the strong dip around 16 hours.

In order to get sharper market components, a slight improvement of the process is needed. Instead of measuring the historical volatilities with a simple EMA of r^2, a Moving Average (MA) operator is used because it has a more rectangular kernel (see Sect. 2.5). The value $m = 32$ has been used for the computations.

As for the long-memory processes, the historical volatilities can be measured with the microscopic or the aggregated return squared. This leads to the microscopic and aggregated definition of the Market ARCH processes, and their respective mug shots are displayed in Figs. 7.9 and 7.10. The parameters have been adjusted on the USD/CHF empirical figure. The agreement between the empirical data and these processes is superb. The differences between Figs. 7.8 and 7.10 show that a fairly rectangular kernel must be used to describe the memory of each market component. From this difference between the processes we could infer that the market participants forget the past quickly beyond their time frames.

Similarly to the long-memory processes, the microscopic measure of historical volatilities leads to a symmetric historical/realized volatility correlation, while the aggregated measure of volatility displays the correct amount of asymmetry. The interesting point is that some structural properties of the empirical figures can be reproduced only by some structures in the process equations, regardless of the parameter values. This allows us to select in a structural way among the wealth of processes that can be constructed, as opposed to a log-likelihood comparison, which is sample dependent and relies on optimal values for the parameters. A structural selection is clearly a much stronger standpoint.

The parameters for the market ARCH processes have been chosen so as to reproduce the historical/realized volatility correlation. As a by product, the other figures on the mug shot are also very similar to the empirical figures. For example, the volatility lagged correlation has the correct slow decay. This shows that some empirical properties can be reproduced by many processes and are in fact not very discriminating for process selection. On the other hand, the historical/realized volatility correlation proves to give a powerful diagnostic about many properties of a time series.

7.12 FIGARCH(p, d, m)

The Fractionally Integrated GARCH process has been introduced by Baillie, Bollerslev, and Anderson [14] in order to include a long memory in the volatility. The idea of the derivation is to note that, often, in empirical fits of GARCH(p, q) processes, the estimated coefficients are such that $\sum_{i=1} \alpha_i + \beta_i \simeq 1$. This may indicate that the GARCH process is misspecified, and an integrated process should be used instead. This diagnostic however can be misleading. Consider the simplest case of GARCH(1, 1), for which $\alpha_1 + \beta_1 = \mu_{\text{corr}} = \exp(-\delta t/\tau_{\text{corr}})$. This process has one exponential characteristic time, and regardless of the value of the parameters, is never close to a power law. If the estimated parameters are such that

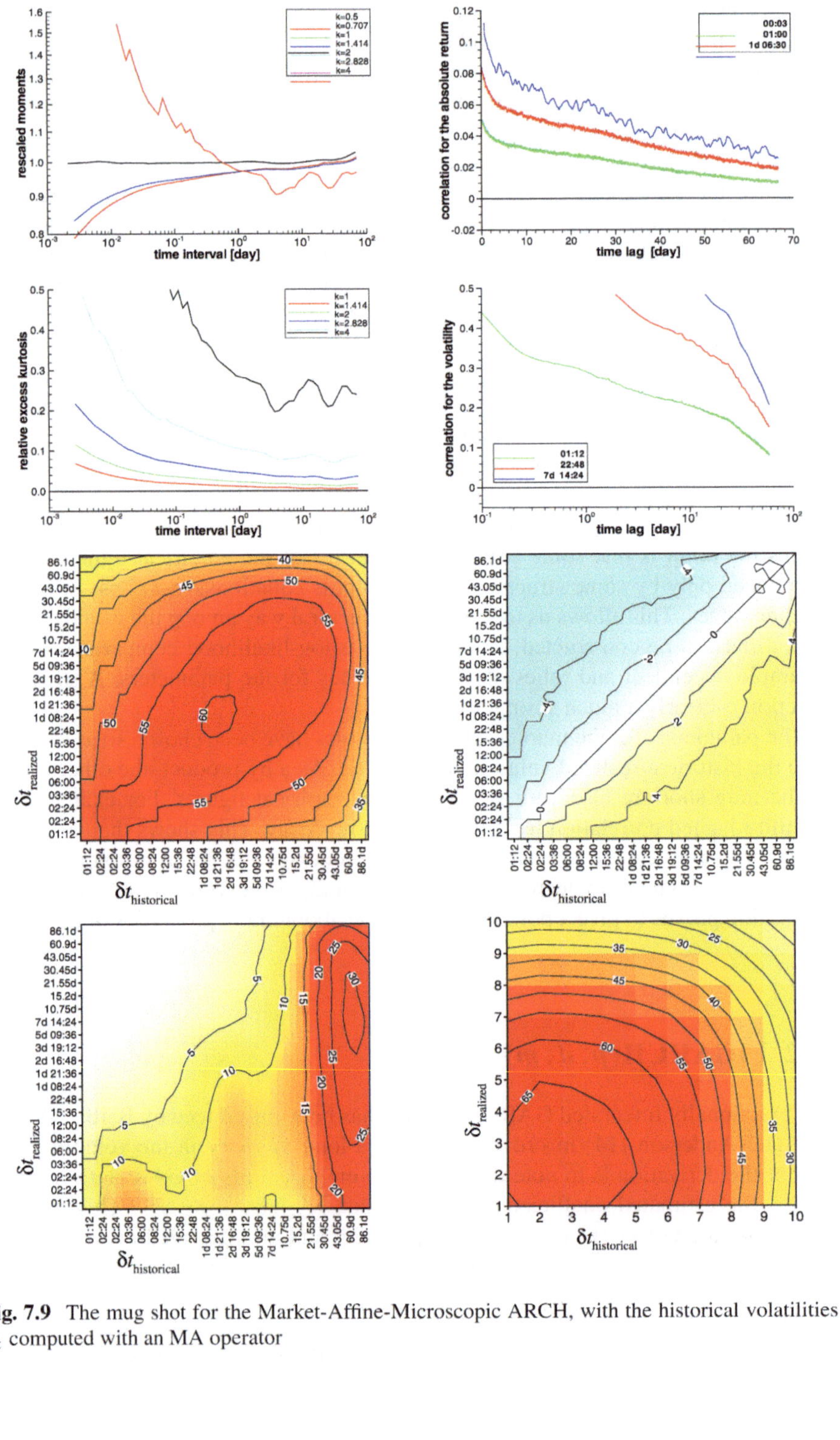

Fig. 7.9 The mug shot for the Market-Affine-Microscopic ARCH, with the historical volatilities σ_k computed with an MA operator

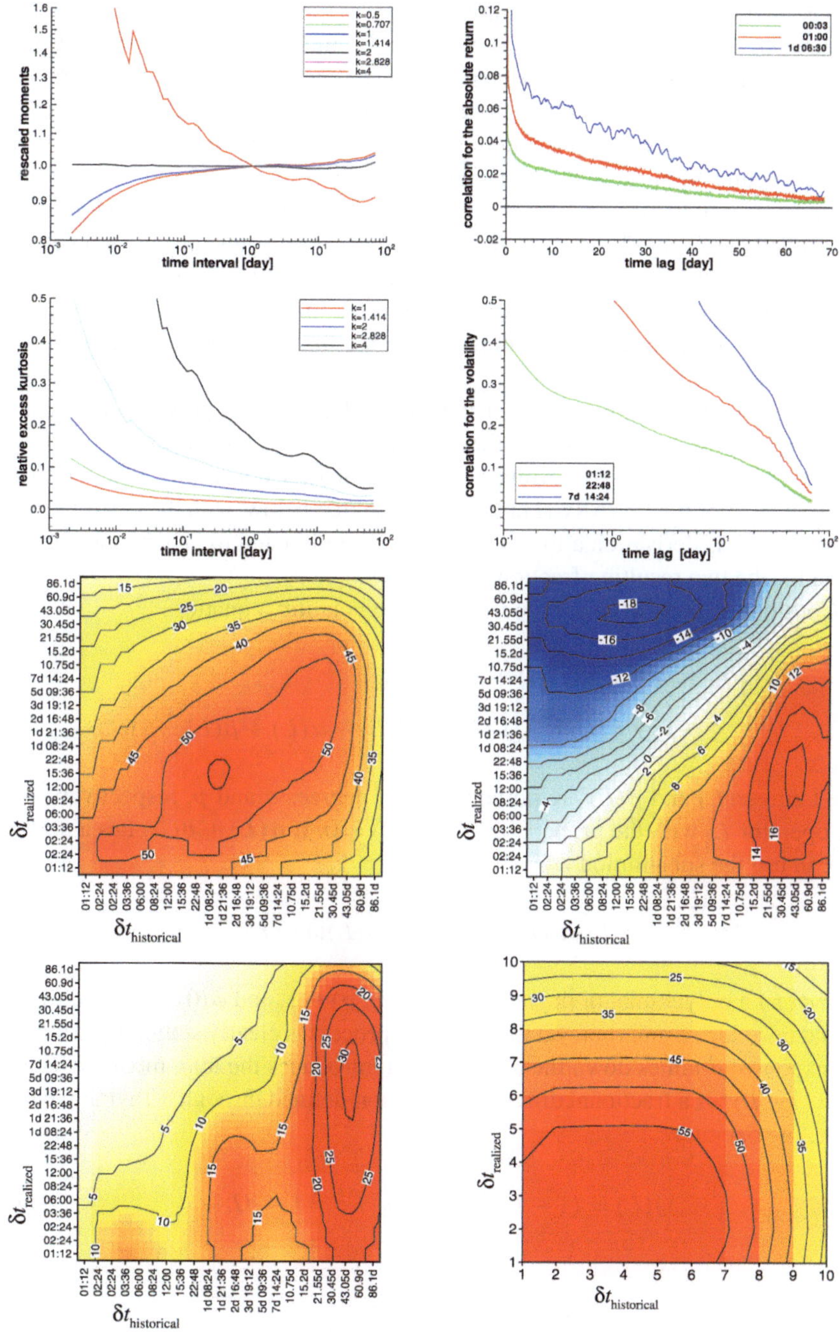

Fig. 7.10 The mug shot for the Market-Affine-Aggregated ARCH, with the historical volatilities σ_k computed with an MA operator

$\alpha_1+\beta_1=\mu_{\text{corr}}\simeq 1$, these values imply that $\delta t/\tau_{\text{corr}}$ is small, namely that $\tilde{\tau}_{\text{corr}}\gg\delta t$. When using high-frequency data for which δt is small, this is the generic case.

Beside, as shown in [150], a numerical estimate of the α_1 and β_1 parameters for a GARCH(1, 1) process can be misleading as there exist a nearly degenerate one-dimensional manifold of solutions. In particular, this manifold goes close to $\mu_{\text{corr}}\simeq 1$, and because it is nearly flat, it can fool an optimization algorithm to converge. Possibly, several empirical published results using GARCH(1, 1) have not properly converged. As show in [150], other coordinates for the parameters, as the one we are using, are better behaved, and the typical cost function (say a log-likelihood) shows a nice parabolic behavior. This improves much the convergence of numerical optimization algorithm and avoids that the algorithm returns with a spurious solution.

The practical implementation of the FIGARCH process presents an important difficulty related to the cut-off of the fractional difference operator, as pointed out by [141] and [41]. As both references are unfortunately not readily available, a simple derivation of the FIGARCH process is presented below, with the cut-off problem explained. This derivation allows us to fix the notation and to give an unambiguous name for the two resulting fractional processes.

Consider a GARCH(p,q) process with the volatility equation

$$\begin{aligned}\sigma^2_{\text{eff}}(t)&=\omega+\alpha(L)r^2(t)+\beta(L)\sigma^2_{\text{eff}}(t)\\&=\omega+\beta(L)\sigma^2_{\text{eff}}(t)+\bigl(1-\beta(L)+\alpha(L)+\beta(L)-1\bigr)r^2(t).\end{aligned}\tag{7.60}$$

The terms $\alpha(L)$ and $\beta(L)$ are polynomials of degrees q and p, respectively, of the lag operator L, with the property $\alpha(0)=\beta(0)=0$. If $\alpha(L)+\beta(L)$ has a unit root, we can use the substitution

$$1-\alpha(L)-\beta(L)\rightarrow\phi(L)(1-L)^d\tag{7.61}$$

with $\phi(L)$ a polynomial of degree $m=\max(p,q)-1$, and $\phi(0)=1$. A term $1-L$ would introduce a unit root, yet the volatility process is clearly stationary. Therefore, the exponent d tames down the unit root and introduces the long memory. The new term $(1-L)^d$ is a fractional difference operator, defined through a Taylor expansion in the lag operator

$$\delta_d(L)=(1-L)^d=\sum_{j=0}\delta_{d,j}L^j=1-dL+\cdots\tag{7.62}$$

$$\delta_{d,j}=\frac{(-1)^j\Gamma(d+1)}{\Gamma(j+1)\Gamma(d-j+1)}.$$

The fractional difference operator obeys the important identity $\delta_d(1)=0$ which will play a central role below. The coefficients of the Taylor expansion can be computed easily with the recursion

$$\begin{aligned} \delta_{d,j+1} &= \frac{j-d}{j+1}\delta_{d,j} \\ \delta_{d,0} &= 1. \end{aligned} \tag{7.63}$$

For large j, the coefficients $\delta_{d,j}$ decay as $j^{-(d+1)}$. This power law decay of the operator coefficients induces the long memory of the process. Using the above substitution, the volatility process becomes

$$\sigma^2_{\text{eff}}(t) = \omega + \beta(L)\sigma^2_{\text{eff}}(t) + \big(1 - \beta(L) - \phi(L)\delta_d(L)\big)r^2(t). \tag{7.64}$$

By taking the average of this equation and using $E[r^2] = E[\sigma^2_{\text{eff}}]$ and $\delta_d(1) = 0$, we obtain $\omega = 0$, and the mean volatility is not determined (we assume that the parameters are such that the processes are covariant stationary). Therefore, the FIGARCH(p, d, m) volatility process is given by the linear equation (for the variance)

$$\sigma^2_{\text{eff}}(t) = \beta(L)\sigma^2_{\text{eff}}(t) + \big(1 - \beta(L) - \phi(L)\delta_d(L)\big)r^2(t). \tag{7.65}$$

Moreover, the positivity of the left-hand side imposes that all the coefficients in the expansion of $1 - \beta(L) - \phi(L)\delta_d(L)$ are positive, giving further restriction on the coefficients of β and ϕ.

The practical implementation of this equation is quite subtle, because the expansion of the fractional difference operator has to be cut-off at some upper limit j_{max}. The cut-off of the infinite sum implies that the identity $\delta_d(1) = 0$ is not true anymore. Let us emphasize that the violation of this identity can be important for realistic values of the parameters. For example, taking $d = 0.25$ and $j_{\text{max}} = 1000$, we obtain $\delta_d(1) = 0.145$. For the differential operator *with a finite cut-off*, taking again the average of (7.64) and using $E[r^2] = E[\sigma^2_{\text{eff}}]$, we obtain that ω fixes the mean volatility

$$E\big[\sigma^2_{\text{eff}}\big] = \frac{\omega}{\phi(1)\delta_d(1)}. \tag{7.66}$$

Therefore, we define the affine FIGARCH process, or Aff-FIGARCH(p, d, m), by the equation

$$\sigma^2_{\text{eff}}(t) = \sigma^2_\infty\phi(1)\delta_d(1) + \beta(L)\sigma^2_{\text{eff}}(t) + \big(1 - \beta(L) - \phi(L)\delta_d(L)\big)r^2(t) \tag{7.67}$$

and the affine parameter σ_∞ fixes the mean volatility. On the other hand, in order to define a linear process Lin-FIGARCH from the FIGARCH equation (7.65), with a finite cut-off in the fractional difference operator, the identity $\delta_d(1) = 0$ has to be enforced. This is done by defining

$$\tilde{\delta}_d(L) = 1 + \gamma(d)\sum_{j=1}^{j_{\text{max}}}\delta_{d,j}L^j \tag{7.68}$$

and adjusting the constant $\gamma(d)$ so that $\tilde{\delta}_d(1) = 0$.

The simplest of the FIGARCH models are the Aff-FIGARCH$(1, d, 0)$ and Lin-FIGARCH$(1, d, 0)$ processes, derived from GARCH$(1, 1)$. These processes have respectively three parameters $(\sigma_\infty, \beta, d)$ and two parameters (β, d). Notice that the process Lin-FIGARCH has only one parameter less than Aff-FIGARCH, namely the mean volatility, and not two as with the other linear/affine pairs of processes. This difference between the linear and affine processes occurs because the weight of the mean volatility is controlled by $\delta_d(1)$ that is not taken as a free parameter, but given implicitly by $j_{\max}$ which is fixed a priori. Finally, let us add that the process discussed in the original literature [14] is the Aff-FIGARCH process.

The computational requirement of FIGARCH process is quite high. The number of operations required to evaluate one iteration of the FIGARCH process is dominated by the fractional difference, and this computational time grows linearly with the cut-off of this operator. This behavior contrasts sharply with the above long-memory processes whose computational time grows only as the logarithm of the cut-off. For the market-component ARCH, the computational time grows linearly with the number of components, and a few components are enough to model well the empirical data.

As a final note on the derivation of the FIGARCH model, one can notice that it is based on a mathematical observation, namely that $1 - \mu L \sim (1 - L)^\mu$. But it is not based on a model for the market participants or on an econometric approximation. As a result, the FIGARCH model is difficult to modify to incorporate more features, say, for example, the coarse grained versus fine grained volatility asymmetry, or the preferred time horizons in the volatility–volatility correlation. Despite the fact that it incorporates one specific form for the long memory, the FIGARCH model is less appealing, both because of the mathematical rigidity of the process equations, and of the lack of connection with our intuition about the market and its participants. Because of these two facts, there exists no interesting extension of FIGARCH. The mug shot for this process is given in Fig. 7.11.

7.13 The GARCH Processes with Trends: GARTCH*

In Sect. 3.9.2, the influence of the past price moves on the volatility is investigated. The simple explanation of the drift/trend effect is that market participants keep their positions in a drifting market, whereas clear price moves make them trades. Subsequently, the volatility decreases or increases, respectively. The (past) drift/trend market conditions are measured by the indicator $(rL[r])(t) = r[\delta t_r](t)L[\delta t_r; r[\delta t_r]](t) = r[\delta t_r](t)r[\delta t_r](t - \delta t_r)$, and the argument above leads to a positive correlation with the subsequent volatility. The empirical analysis shows that this is indeed the case. Moreover, the characteristic human time horizons can be observed on the correlation plot. The aim of this section is to add the required term(s) in an ARCH process in order to reproduce this stylized fact. Because ARCH processes are based on the returns, it is easy to add drift/trend term(s). In order to limit the complexity of the processes, the philosophy is to respect the

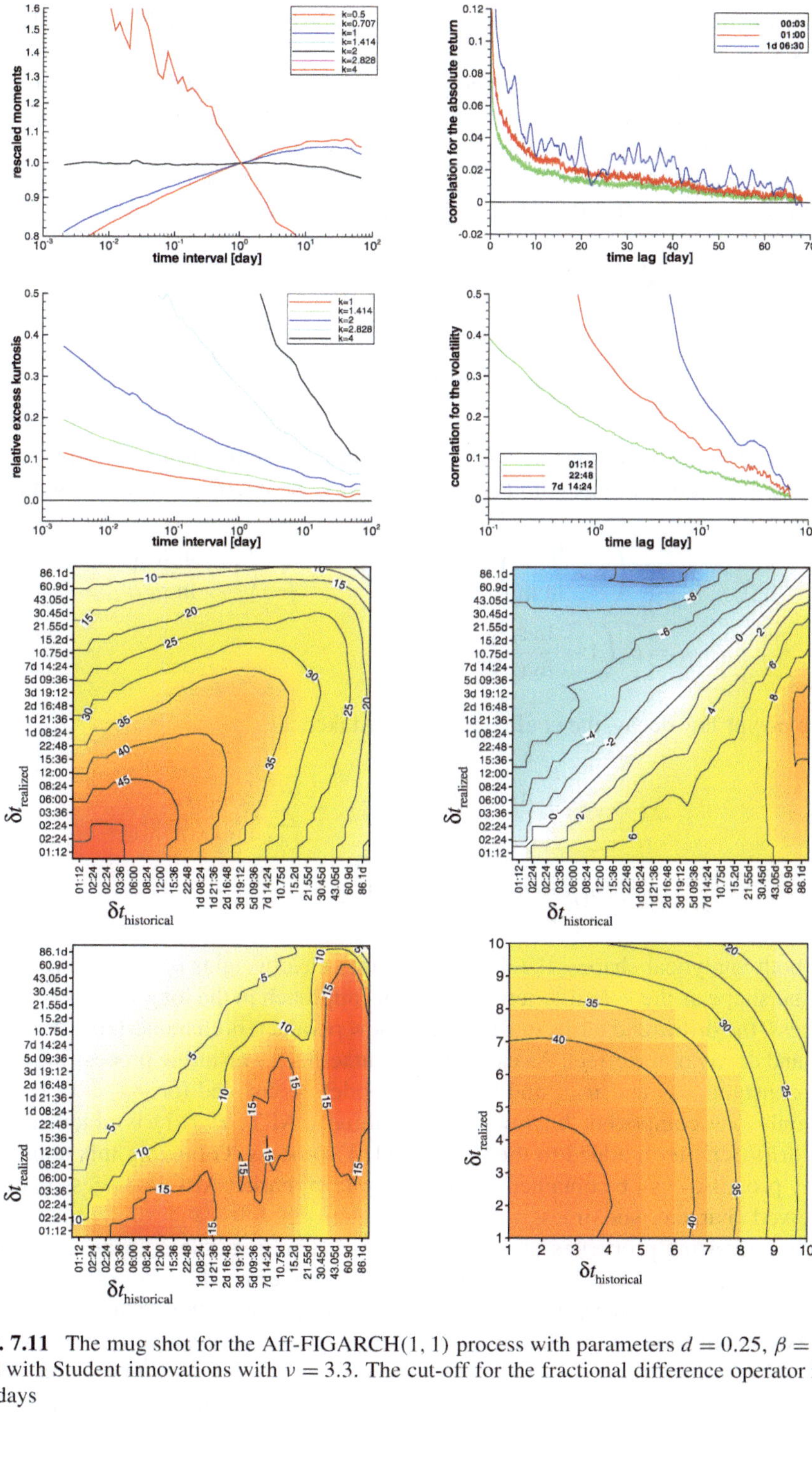

Fig. 7.11 The mug shot for the Aff-FIGARCH(1, 1) process with parameters $d = 0.25$, $\beta = 0.1$, and with Student innovations with $\nu = 3.3$. The cut-off for the fractional difference operator is at 68 days

basic structure of the process and to add a corresponding number of drift/trend terms.

For a multiple components process, the idea is to add a drift/trend term at all time horizons included in the process, with a similar structure between the volatility and the drift/trend coefficients. The effective volatility σ_{eff} for the long-memory (LM) affine (Aff) ARTCH (Auto-Regressive with Trend Conditional Heteroscedastic) processes LM-Aff-ARTCH(n), with n components, is

$$\sigma^2_{\text{eff}}(t) = \sum_{k=1}^{k_{\max}} w_k \sigma_k^2(t) + w_\infty \sigma_\infty^2 + \sum_{k=1}^{k_{\max}} \theta_k \cdot r[l_k](t) \cdot r[l_k](t - l_k \delta t) \tag{7.69}$$

$$w_k = (1 - w_\infty)\chi_k \qquad \text{with} \quad \sum_k \chi_k = 1$$

with the characteristic time $\tau_k = l_k \tau_1$ and the decay coefficient for the EMA $\mu_k = \exp(-\delta t/\tau_k)$. For a model with logarithmic decay, the coefficients are chosen as

$$\begin{aligned} \chi_k &= c\left(1 - \frac{\ln \tau_k}{\ln \tau_0}\right) \qquad \text{with} \quad 1/c = \sum_{k=1}^{k_{\max}} \left(1 - \frac{\ln \tau_k}{\ln \tau_0}\right) \\ \theta_k &= \theta_1 \left(1 - \frac{\ln \tau_k}{\ln \tau'_0}\right). \end{aligned} \tag{7.70}$$

With power law decay, the coefficients can be taken as

$$\begin{aligned} \chi_k &= c\rho^{-(k-1)\lambda} \qquad \text{with} \quad 1/c = \sum_{k=1}^{k_{\max}} \rho^{-(k-1)\lambda} \\ \theta_k &= \theta_1 \rho^{-(k-1)\lambda'} = \theta_1 \left(\frac{1}{l_k}\right)^{\lambda'}. \end{aligned} \tag{7.71}$$

For both analytical shapes, the drift/trend terms have weights θ_k that decay with a similar shape as the volatility terms. With this approach in the long-memory ARCH process, the drift/trend terms introduce two new parameters characterizing the intensity and decay of the effect. As usual, the mean terms for an affine process introduce two constants σ_∞ and w_∞, and the linear model is obtained for $w_\infty = 0$. By taking only one component $k_{\max} = 1$, the minimal GARTCH(1, 1) is obtained with one drift/trend term added to the GARCH(1, 1) process. Following this approach, many processes can be obtained as, for example, a market ARCH process with the observed characteristic times.

An interesting properties of these processes is that, regardless of the time structure and coefficients for the volatility and drift/trend terms, the process is quadratic in the returns. Therefore, the integrals required to compute the volatility forecasts can be evaluated, leading to iterative equations that can be implemented numerically. Beyond volatility forecasts, such a computation can be used to estimate process parameters by minimizing a cost function like the root mean square error between the forecasted volatility and the realized volatility. This approach is used in [157] to explore systematically the relationship between the process structure and the performance of the volatility forecast. The main outcome is that the forecast accuracies increase steadily by including successively the main empirical characteristics of the volatility. For example, the relative RMSE measure of forecast quality increases for the sequence of processes: "rectangular moving average" (7.2 %), I-GARCH(1) (10.7 %), I-GARCH(2) (13.5 %), LM-Lin-Mic-ARCH (14.2 %), and LM-Lin-Mic-ARTCH (15.7 %). This sequence shows well the successive improvement provided by the memory shapes converging to a slow decay, and finally by the trend term. Interestingly enough, the affine processes (like GARCH(1, 1)) have lower performances compared to a similar linear process (like I-GARCH(2), which has the same number of parameters).

The ARCH and trend structure can be generalized further, by including all the cross-terms between past returns:

$$\sigma_{\text{eff}}^2(t) = w_\infty \sigma_\infty^2 + (1 - w_\infty) \sum_{i,j=0}^{J-1} K(i, j) \cdot r(t - i\delta t) \cdot r(t - j\delta t) \tag{7.72}$$

where the kernel $K(i, j)$ measures the influence of the past returns over the volatility. This model has been proposed by [136] (with an added leverage term proportional to the return). All the microscopic and aggregated processes, with trend terms, can be reduced to this generic form. The difficulty with such an approach is the large number of parameters, $2 + J(J + 1)/2$, related to its lack of structure. The longest time horizon included in the model is given by J, and the number of parameters grows as J^2. For this reason, it is difficult to include long time horizons in this generic form without further simplifications. This model was studied recently by [38], with the interesting finding that $K(i, j)$ is dominated by the diagonal terms.

7.14 Sensitivity with Respect to the Estimated Parameters

The paper [157] includes a study of the out-of-sample sensitivity with respect to the process parameters as measured by the volatility forecasts. The idea is the following. Assume a process with a given set of parameters θ. The parameters can be estimated, say by minimizing a log-likelihood or a forecast error. A given subsample (t_1, t_2) leads to optimized parameters $\theta(t_1, t_2)$. Are the differences in the values of the parameters important when changing t_1 and t_2? The difficulty is that there is no natural metrics on the space of parameters, and a given process can be parameterized differently. For example, the GARCH(1, 1) process can be parameterized with $\alpha_0, \alpha_1, \beta_1$ in (7.2), or with $\sigma_\infty, w_\infty, \tau$ in (7.3a), (7.3b), or with $\sigma_\infty, w_\infty, \mu$ in (7.3a), (7.3b). As such, the problem is not well posed as the parameters θ cannot be compared.

The solution is given by the process it-self: the effective volatility σ_{eff} and the volatility forecasts provide for functions of the parameters θ which are invariant by reparametrizations. The paper [157] defines a measure of sensitivity with respect to changing data sets by

$$Q = \frac{1}{n} \sum_t \left\| \sigma_{\mathrm{eff}}[t - \Delta T, t](t) - \sigma_{\mathrm{eff}}[t_{\mathrm{start}}, t_{\mathrm{end}}](t) \right\| \tag{7.73}$$

where n is the number of terms in the sum. The effective variance $\sigma_{\mathrm{eff}}[t_1, t_2](t)$ is a one day forecast for the realized variance with the parameters estimated on the subsample from t_1 to t_2. Essentially, the sensitivity factor Q measures the difference in the forecasts when estimating the parameters in a trailing sample of length ΔT compared to the full sample, and with the forecasts computed, respectively, out-of-sample and in-sample. For large Q values, the process is more dependent on the choice of the subsamples and therefore less robust. The same definition can be used with the effective variance replaced by a volatility forecast.

An empirical study for the CHF/USD foreign exchange over a large set of ARCH processes shows that the mean volatility is a parameter that is difficult to estimate, in the sense that it strongly depends on the chosen estimation subsample. This dependency is measured of the sensitivity Q, which is systematically larger for all affine processes, roughly by a factor ~ 1.5 to ~ 2 when compared to the corresponding linear processes. For example, the I-GARCH(1), I-GARCH(2), and GARCH(1, 1) have respective sensitivities $Q = 0.11$ %, 0.15 %, and 0.28 %. These figures should be compared to the mean daily volatility on the sample of 0.6 % (equivalent to an annualized volatility of 10 %). Another case where the sensitivity of the mean volatility parameter can be observed is for the linear and affine long-memory processes: their forecast qualities are very similar, while the affine version has a larger sample sensitivity Q.

7.15 Long-Term Dynamics for the Mean Volatility*

The mean volatility σ_∞ is an important parameter for the affine ARCH processes, and the infinity subscript is a reference to the time horizon at which this parameter could be measured. Yet, the mechanism that sets the value of the mean volatility is still unexplained. One possibility is that the quote and trading process creates price fluctuations, which sets the mean volatility. In this explanation, the origin is microscopic as related to the elementary trades. Another possibility is that the fundamental economic shifts induce price moves, and the mean volatility is a measure of the underlying economic activities and/or imbalances. In this explanation, the mean volatility level is set by long-term economic factors and their fluctuations. A possible process which describes this behavior is to have a long-term stochastic volatility combined with a short-term ARCH process. The long-term stochastic volatility describes the fluctuation of the underlying economic factors, while the ARCH part captures the dynamic of the financial market and its participants. A simple process combining an exponential stochastic volatility with multiple component ARCH is the following:

$$
\begin{aligned}
\sigma_k^2(t) &= \mu_k \sigma_k^2(t-\delta t) + (1-\mu_k) r^2(t) \qquad k = 1, \ldots, k_{\max} \\
h(t) &= -\frac{\delta t}{\tau_{SV}} h(t-\delta t) + \gamma \sqrt{\frac{\delta t}{\tau_{SV}}} \epsilon_{SV}(t) \\
\sigma_{SV}(t) &= \sigma_\infty \exp\bigl(h(t)\bigr) \\
\sigma_{\text{eff}}^2(t) &= w_\infty \sigma_{SV}^2(t) + \sum_k w_k \sigma_k^2(t).
\end{aligned}
\tag{7.74}
$$

This process has one characteristic time τ_{SV} corresponding to the mean reversion for the stochastic volatility and $k_{\max}$ characteristic time for the EMA. The argument about the long-term economic fluctuations would lead to $\tau_{SV} \gg \tau_k$. With one component $k_{\max} = 1$, a combination of GARCH(1, 1) with a stochastic volatility is obtained. The mug shot for this process is given in Fig. 7.12, with the parameters $w_\infty = 0.1$, $\tau_{SV} = 50$ days, $\gamma = 0.1$ and EMAs with characteristic times ranging from 0.05 day to 25 days and with a power law decay for the weights with exponent $\nu = 0.15$. The correlation plot for the historical/realized volatility is particularly interesting, showing the asymmetry created by the multiple components ARCH for time interval up to ~3 weeks, and a more symmetrical part for longer time intervals. Essentially, this figure combines the plots for both processes in their respective range. The empirical mug shots do not show this tendency toward a symmetrical historical/realized volatility. Therefore, if such a mechanism for the long-term volatilities takes place, it is for longer time horizons, say above one year.

7.16 The Induced Dynamics for the Volatility

The multicomponent ARCH processes are stochastic processes for the return in which the volatilities are convenient intermediate quantities. As already emphasized when discussing ARCH processes, it is important to realize that the volatilities σ_k and σ_{eff} are useful and intuitive in formulating a model, but they can be completely eliminated from the ARCH equations in order to formulate the process only in terms of the lagged returns. Going in the opposite direction, we want in this section to eliminate the returns, namely to derive the equivalent process equations for the dynamics of the variance induced by a multicomponent ARCH process.

The microscopic multicomponent ARCH processes as defined in Sect. 7.5 are used as a starting point. The equation for σ_k can be rewritten as

$$
\begin{aligned}
d\sigma_k^2(t) &= \sigma_k^2(t) - \sigma_k^2(t-\delta t) \\
&= (1-\mu_k)\bigl\{-\sigma_k^2(t-\delta t) + \epsilon^2(t)\sigma_{\text{eff}}^2(t-\delta t)\bigr\} \\
&= (1-\mu_k)\bigl\{\sigma_{\text{eff}}^2(t-\delta t) - \sigma_k^2(t-\delta t) + \bigl(\epsilon^2(t)-1\bigr)\sigma_{\text{eff}}^2(t-\delta t)\bigr\}.
\end{aligned}
\tag{7.75}
$$

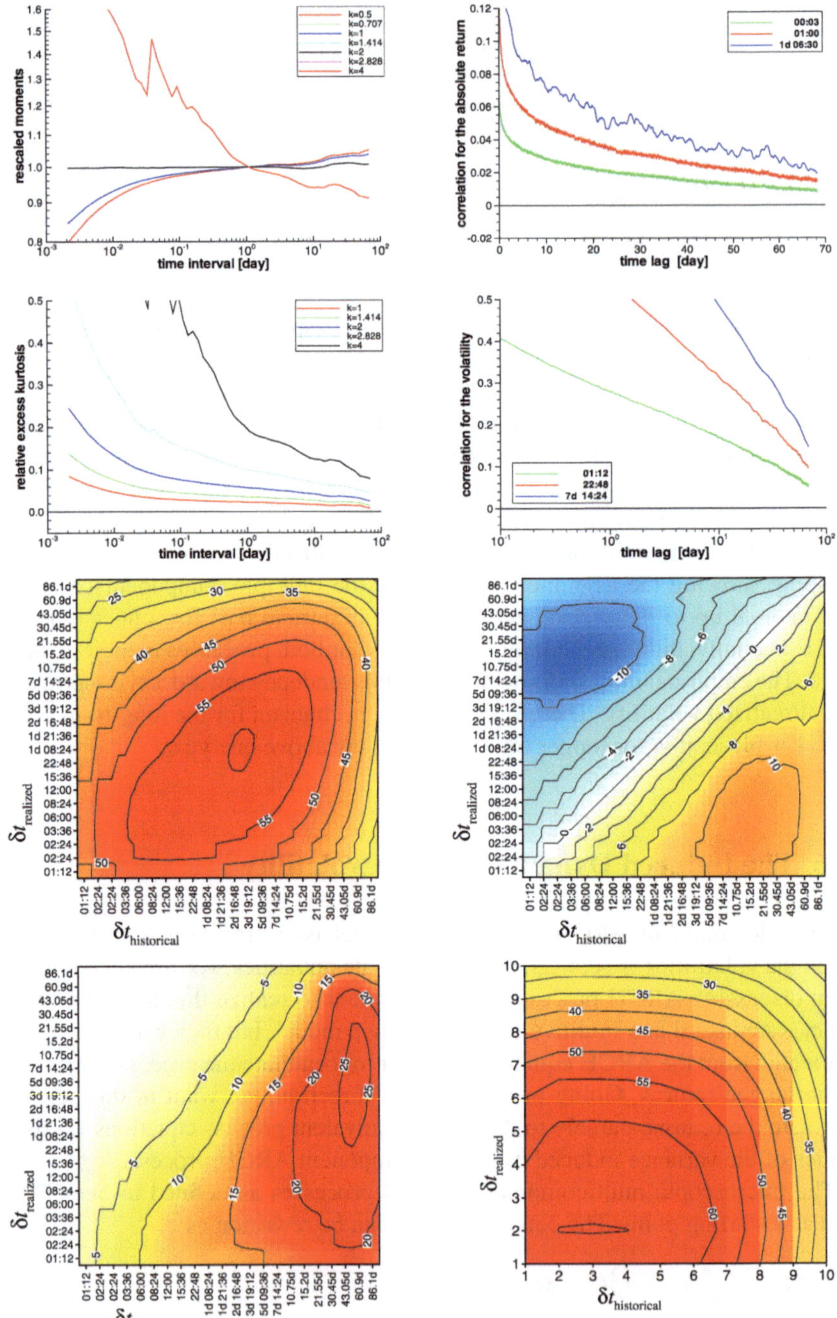

Fig. 7.12 The mug shot for the LM-Aff-Agg-ARCH process combined with a stochastic volatility

The equation can be simplified by introducing the annualized variances

$$\begin{aligned} v_k(t) &= 1y/\delta t\ \sigma_k^2(t) \\ v_{\text{eff}}(t) &= 1y/\delta t \sigma_{\text{eff}}^2(t) \end{aligned} \tag{7.76}$$

and a new random variable χ with

$$\chi(t) = \epsilon^2(t) - 1 \qquad \text{such that} \quad E\big[\chi(t)\big] = 0, \quad \chi(t) > -1. \tag{7.77}$$

Assuming that the time increment δt is small compared to the time scales τ_k in the model, the following approximation can be used:

$$1 - \mu_k = \frac{\delta t}{\tau_k} + O\big(\delta t^2\big). \tag{7.78}$$

In the present derivation, this expansion is used only to make contact with the more usual continuous-time form, but no terms of higher order are neglected. Exact expressions are obtained by replacing $\delta t/\tau_k$ by $1 - \mu_k$ in the equations below.

These notations and approximations allow us to write the equivalent volatility process in the form

$$v_{\text{eff}}(t) = \sum_k w_k v_k(t) + w_\infty v_\infty \tag{7.79a}$$

$$dv_k(t + \delta t) = \frac{\delta t}{\tau_k}\big\{v_{\text{eff}}(t) - v_k(t) + v_{\text{eff}}(t)\chi\big\} \tag{7.79b}$$

with the usual equation $v_k(t + \delta t) = v_k(t) + dv_k(t + \delta t)$ for the one step increment of the variances. These equations involve only the variances, and the innovations for the return are hidden in the new random variable χ. These are in some sense the stochastic volatility process equivalent to a multicomponent ARCH process.

The content of Eq. (7.79b) is the following. The term $\delta t\{v_{\text{eff}} - v_k\}/\tau_k$ gives a mean reversion toward the current effective volatility v_{eff} at a time scale τ_k. This structure is fairly standard, except for $v_{\text{eff}}(t)$ which is given by a convex combination of all the variances $v_k(t)$. Then, the random term is unusual. All the variances share the same random factor $\delta t\chi/\tau_k$, which has a standard deviation of order δt instead of the usual $\sqrt{\delta t}$ appearing in diffusion processes.

An interesting property of this equation is to enforce positivity for v_k through a somewhat peculiar mechanism. Equation (7.79b) can be rewritten as

$$dv_k = \frac{\delta t}{\tau_k}\big\{-v_k + (\chi + 1)v_{\text{eff}}\big\}. \tag{7.80}$$

Because $\chi \geq -1$, the term $(\chi + 1)v_{\text{eff}}$ is never negative, and as $\delta t v_k(t - \delta t)/\tau_k$ is smaller than $v_k(t - \delta t)$, this implies that $v_k(t)$ is always positive (even for a finite δt). Another difference with the usual random process is that the distribution for χ is not Gaussian. In particular, if ϵ has a fat-tailed distribution, as is required in order to have a data generating process that reproduces the properties of the empirical time series, the distribution for χ also has fat tails.

Let us emphasize that the derivation of the volatility process as induced by the ARCH structure involves only elementary algebra. Essentially, if the price follows an ARCH process (one or multiple time scales, with or without mean σ_∞), then the volatility follows a process according to (7.79a), (7.79b). The structure of this process involves a random term of order δt, and therefore it cannot be reduced to a usual random walk (and subsequently to take a continuum limit with $\delta t \to 0$). This is a key difference between the processes explored in this chapter and continuous-time finance.

The implications of Eqs. (7.79a), (7.79b) are important as they show a key difference between ARCH and stochastic volatility processes. This has clearly implications for option pricing, but also for risk evaluation. In a risk context, the implied volatility is a risk factor for any portfolio that contains options, and it is likely better to model the dynamics of the implied volatility by a process with a similar structure. The relationship of (7.79a), (7.79b) with processes for the forward variance and with at-the-money implied volatility in explored further in [158].

The derivation above is done for a finite time increment δt. The continuum limit of the GARCH(1, 1) process with normal innovations was investigated by [115] and [45]. Depending on the dependency on δt assumed for the coefficients, two limits can be constructed. The differences in the construction can be intuitively understood using the GARCH(1, 1) process formulation given in Eq. (7.3a). Assuming that $\mu_1 = 1 - \delta t/\tau_1$ and $w_\infty = \tau_1/\tau_C$ fixed (with τ_C the characteristic exponential time of the lagged correlation), a degenerate limit is obtained where the volatility process is deterministic (corresponding to $\alpha = 0$ in [45]). With the different assumptions $\mu_1 = 1 - \sqrt{\delta t/\tau_\alpha} + O(\delta t)$ and $w_\infty = \sqrt{\delta t \tau_\alpha}/\tau_C$, [115] obtained a nondegenerate limit where the volatility follows an Ornstein–Uhlenbeck-like diffusion with a second source of randomness (corresponding to $\alpha \neq 0$ in [45] and with $\tau_\alpha = 2/\alpha^2$). In the present formulation, the construction of this limit depends on the relation $v_{\text{eff}} - v_1 = O(\sqrt{\delta t})$, following the assumption $w_\infty = O(\sqrt{dt})$ and the presence of one component v_1 in the GARCH(1, 1) process. With multiple components, it is not possible to have $v_{\text{eff}} - v_k = O(\sqrt{\delta t})$ for all components, and it is therefore impossible to construct an interesting non degenerate limit with multiple components. Hence, the scaling used in (7.78) should be used, and (7.79b) reduces to the deterministic equation $dv_k(t) = \delta t/\tau_k\{v_{\text{eff}}(t) - v_k(t)\}$ in the continuum limit. Notice that the derivation of (7.79a), (7.79b) is done with a finite time increment δt, and the interesting empirical parameter values correspond to τ_1 be a small multiple of δt. For a fixed number of components, it is therefore not very natural to take a continuum limit of this volatility process.

7.17 Some Thoughts on the Small Values of w_∞

For all the affine ARCH processes studied in this book, the estimation for w_∞ gives always small values, ranging between 0.02 to 0.2. The upper values correspond to the GARCH(1, 1) process. For multicomponent ARCH processes, increasing the upper cut-off in the volatility components included in the model leads to a decrease of the value of w_∞. A realistic multiscale process corresponds to w_∞ between 2 to 10 %, regardless of the details of the model.

The process parameter w_∞ measures the weight of the mean volatility. For the processes discussed in this book, the mean volatility σ_∞ is a parameter given externally, essentially by the human activity and the economy. An estimate for the "naked" underlying variance is $w_\infty \sigma_\infty^2$, which is of the order of 2 to 10 % of the variance of a given time series. The complement $1 - w_\infty$ measures the importance of the endogenous feed-back loop in the volatility dynamics. The small values for w_∞ indicate that the largest part of the volatility dynamics originates in the trading mechanism, in the range of 90 to 98 %.

For $w_\infty = 0$, the equations reduce to the linear ARCH processes which are singular (see the discussion in Sect. 7.4, page 96). Roughly, the financial time series are described by processes which are as far as possible from the constant volatility process (i.e., $w_\infty = 1$), but just no singular (i.e., $w_\infty = 0$). A similar point was made by [29] following the analysis of the dynamics in the order queues, namely the markets operate close to criticality. This situation holds for all the time series analyzed by several authors and therefore should be generic. For example, all the stock time series share this property, regardless of the market where they are traded. This universality points to a deeper reason why the trading of an asset has to converge to such an equilibrium point. This reason likely originates in the free trading, the organization of the markets, the role and impact of market and limit orders, and the human behavior, but it remains to be found.

7.18 Which Simple ARCH Process and Parameter Values to Use Overall?

Many computations in finance require to have some volatility estimators, either for the current level of volatility or for a volatility forecast over some time span. Some examples are the estimation of innovations (see Chap. 13), the computation of market risks (see Chap. 15), or of implied volatilities (see Sect. 16.9). What is needed is a good and robust volatility estimator, but not the best and most accurate one. Other applications require Monte Carlo simulations and the generation of possible price paths. An example is the pricing of derivatives (see Chap. 16). For all these cases, our goal is to get the job done, with the minimal amount of chores corresponding to implementations, parameter estimations, testing, and selection.

The long-memory ARCH process is a perfect candidate for such applications. A logarithmic form (7.57) for the memory of the volatility is validated by the empirical behavior of most time series (see Chap. 12). This analytical form contains only one parameter τ_0 which fixes the decay, and a value of the order of one or a few years describes well most time series. Depending on the application, the linear or affine version of the process can be used.

For computing the current level of the volatility or a volatility forecast, the linear version is sufficient. Neglecting the affine term is validated by the small value for w_∞ that is found when estimating financial time series. For volatility forecasts, the forecast weights corresponding to the microscopic long-memory ARCH process are given by simple recursion equations in Sect. 7.7.1. These weights can be evaluated numerically in a few dozen lines in your favorite programming language. As the process is linear, there is no need to evaluate the mean volatility parameter σ_∞. Clearly, it is difficult to simplify further (i.e., one parameter and no estimation). Such an approach is used in Chap. 15 for the estimation of market risks.

For Monte Carlo simulations, the affine version of the process should be used in order to fix its asymptotic properties. The equations are barely more complicated, but this introduces two more parameters namely the mean volatility σ_∞ and the corresponding coupling constant w_∞. The mean volatility is depending on the time series and should be estimated using historical data. For this task, the variance of the available historical sample gives a very simple and explicit estimator for σ_∞^2. The coupling constant w_∞ is found to be small for all time series, and values between $w_\infty \simeq 0.05$ and 0.1 are decent default values. This parameter can be adjusted further in order to obtained the desired level of heteroscedasticity, for example, by calibrating the lagged correlation of the volatility. In the implementation of the process for Monte Carlo simulations, the aggregated version of the process is a bit more complex to program compared to the microscopic version, but it will be more realistic.

These two choices will deliver good volatility estimators, forecasts or simulation paths for time horizons ranging from one day to a few months. In this range, the long memory of the variance dominates the estimators, and it is captured efficiently by a logarithmic decay. Capturing correctly the heteroscedasticity is enough to obtain good performances with a minimal complexity. For longer horizons, other terms can contribute (for example, the leverage effect, see Chap. 14), and much more work is required in order to obtain accurate estimators or processes. An example describing stocks is given in [167] and for pricing options on stock indexes in [166].

The aficionados of the GARCH(1, 1) process might want to stick to their favorite tool. In this case, the parameterization given in (7.3a), (7.3b) gives a more intuitive set of parameters than the original one. Good default values are $\tau \simeq 5$ days and $w_\infty \simeq 0.2$, while σ_∞ can be estimated with a moment estimator on historical data as describe above. For daily data, these values correspond to $\alpha_1 = 0.145$ and $\beta_1 = 0.82$, comparable to the values given in the literature for many time series and obtained by a log-likelihood estimator. Compared to the long-memory ARCH process, the disadvantage of the GARCH(1, 1) process is that it delivers less accurate volatility estimators and gives forecasts in a more limited range.

At a pragmatic level, the propositions above will get most of the work done while spending a minimal amount of work. They could be used in a broad range of applications and for a broad range of financial time series. Finally, the outcome is weakly sensitive to the parameter values, making the scheme robust. Moving beyond these propositions allows one to obtain better results but requires substantially more work.

Chapter 8
Stochastic Volatility Processes

8.1 Introduction

The basic idea for the stochastic volatility processes is that the volatility is a genuine independent process with its own source(s) of randomness, whereas the return is a “slave” process with a magnitude given by the volatility. The structure of the volatility process should be such that the volatility stays positive and bounded. In a discrete-time formulation, the general structure of the equations is

$$\sigma_{\text{eff}}(t) = \sigma_{\text{SV}}\big[\Omega_\sigma(t), \epsilon_\sigma(t)\big] \tag{8.1a}$$

$$r[\delta t](t+\delta t) = \sqrt{\frac{\delta t}{1\text{y}}}\sigma_{\text{eff}}(t)\epsilon_r(t) \tag{8.1b}$$

On the right-hand side of (8.1a), the function σ_{SV} fixes the dynamics of the volatility. The information set Ω_σ includes the previous values for σ_{eff} and other possible internal states, but does not include the returns r. The volatility depends on one or several sources of randomness $\epsilon_{\sigma,k}$. The return is then a simple “slave” process with the amplitude given by the volatility, and with its source of randomness ϵ_r. The volatility is annualized, and the term $\delta t/1\text{y}$ scales the volatility from 1 year to the process increment δt.

The processes ϵ_r and ϵ_σ have zero means and unit variances. In order to obtain realistic pdf from the simulated processes, they should be chosen with fat-tailed distributions. They can be independent, but a (negative) correlation is often added to include the leverage effect (returns and volatility are negatively correlated for stocks),

$$\rho(\epsilon_r, \epsilon_{\sigma,k}) = \rho_k$$

with $\rho_k < 0$ for stock and stock indexes. In an option pricing context, a negative correlation is used to introduce a tilted smile in the option-implied volatility. Otherwise, the $\epsilon_{\sigma,k}$ are independent:

$$\rho\big(\epsilon^p_{\sigma,k}, \epsilon^q_{\sigma,k'}\big) = 0 \qquad \text{for} \quad k \neq k'.$$

G. Zumbach, *Discrete Time Series, Processes, and Applications in Finance*, Springer Finance, DOI 10.1007/978-3-642-31742-2_8,

With this setup for the processes, the return is essentially a slave process of the volatility, and there is no feed-back from the return to the volatility. With correlated increments, the return $r[\delta t]$ can have a contemporaneous dependency on the volatility σ_{eff} such that $\rho(r, \sigma_{\text{eff}}) \neq 0$. Otherwise, the ϵ_r do not influence the volatility σ_{eff}. In particular, a sequence of ϵ_r such that $r[\Delta T]$ is large (for example, ϵ_r are mostly negative) has no influence on the volatility. Therefore, due to the structure of the stochastic volatility process, it is not possible to capture the volatility granularity effect in this class of processes.

Depending on the function σ_{SV}, many stochastic volatility processes can be derived with one or multiple time scales, and with different algebraic structure. Similarly to the ARCH family of processes, the problem is not to invent new structures, but to select the relevant model(s) among the dozen of existing propositions. As the empirical studies have shown, an important feature of financial time series is the dynamics over multiple time scales, and our way on the SV models follows this empirical hint.

The salient features of these models follow from the splitting on the SV process into a cascade from volatility to return.

- There is no feedback from the return on the volatility.
  This structure is quite counter-intuitive with respect to our perception of the financial market where the prices are the source of public information. As the volatility is an independent variable, it acquires a "higher status" and corresponds, for example, to the current "mood" or consensus of the market with respect to the future uncertainty, or to a measure of the disagreement between market participants. Yet, the volatility is not published, neither as an information nor as a tradable quantity (this situation is somewhat changing for very liquid instruments with the availability of volatility indexes and variance swaps). Notice that this is dramatically different from an ARCH process, where the volatility is a convenient way to express the model but can be completely eliminated from the equations in order to express the next return as dependent only on the past returns.
- Because the volatility is not observable, the parameter estimation is fairly difficult.
  For example, we cannot compute simple statistics of the volatility that relates to the parameters of the volatility process, like generalized moment estimate. Instead, more complex methods should be used.
  For our purpose, a precise estimation of the parameter values is not needed. Instead, plausible values are used by expressing the process parameters in dimensionful units. These first-order values are good enough to obtain the generic properties of the processes.
- Because of the simpler structure for the equations, some properties can be derived analytically.
  In particular, the simple Heston [80] model can be solved, and option prices based on this process can be computed. It is important to be able to price efficiently options, with an underlying process more realistic than a simple Gaussian random

walk. Yet, the analytical tractability is lost for more realistic SV processes, in particular for the multiscale versions.

- The equations for the volatility dynamics are of the usual diffusion type, with the random innovation of size $\sqrt{\delta t}$. For the ARCH processes, the dynamics of the effective volatility is of a different type, as shown in Sect. 7.16.

As mentioned above, the volatility process must be such that the volatility stays positive. There are two simple ways to meet this constraint: (1) to use the logarithm of the volatility as the variable for the process or (2) to multiply the random term by a power of the volatility such that the stochastic part diminishes when the volatility decreases. These two solutions lead respectively to the exponential SV model and to the Heston model. They are explored in turn in the following sections, in their basic form with one time horizon, and with a multi-time scale structure. The other main ingredient entering into the volatility equation is a mean-reverting term such that the volatility stays bounded. This is done usually with a simple Ornstein–Uhlenbeck term, but other analytical forms are possible.

The literature on the stochastic volatility processes is very large, as well as their application to option pricing. A general reference on this class on processes is [137], see also [124].

8.1.1 Stochastic Time Processes

The core idea of a stochastic time process is to have a simple random walk for the price, but indexed by a stochastic time. The stochastic process for the time can be sophisticated enough to include heteroscedasticity. The random time $T(t)$ is indexed by the physical (or business) time t and is such that it is strictly growing: $dT > 0$. In a discrete formulation, the price increment dx follows

$$dx = \sigma_\infty \sqrt{\frac{dT}{1y}} \epsilon(T) \tag{8.2}$$

with σ_∞ the annualized constant volatility of the random walk and with the usual conditions on $\epsilon(T)$. This can be rewritten as

$$dx = \sigma_\infty \sqrt{\frac{dT}{dt}} \sqrt{\frac{dt}{1y}} \epsilon\big(T(t)\big) = \sigma(t) \sqrt{\frac{dt}{1y}} \epsilon(t) \tag{8.3}$$

with the stochastic volatility

$$\sigma(t) = \sigma_\infty \sqrt{\frac{dT}{dt}}. \tag{8.4}$$

In this form, the volatility measures essentially the "speed of time", namely $\sigma^2(t)$ gives the (stochastic) Jacobian of the transformation between the stochastic time T

and the physical time t. This corresponds to our intuition that large events lead to the "time flying faster", or equivalently to larger volatility.

Structurally, there is no difference between a stochastic time process and a stochastic volatility process. In particular, the fundamental limitation of the stochastic time depending on its own past but not on the returns restricts the statistical behaviors that can be produced. In a multivariate setting, the random-time formulation leads to difficulties related to the specific random times for each time series. Essentially, the processes have to be synchronized again, in effect formulating the processes as stochastic volatility processes in time t. Because of these difficulties, the presentation of the stochastic time processes is not pursued in this book. The interested readers can check the original references, in particular [13, 69, 96, 112].

8.2 Exponential Stochastic Volatility Process

8.2.1 One-Component Exponential Stochastic Volatility Process

The one-component discrete exponential stochastic volatility model is defined by

$$\begin{aligned}
\sigma_{\text{eff}}(t) &= \sigma_\infty e^{h(t)} \\
h(t) &= h(t-\delta t) + dh(t) \\
dh(t) &= -\frac{\delta t}{\tau} h(t-\delta t) + \gamma \sqrt{\frac{\delta t}{\tau}} \epsilon_\sigma(t).
\end{aligned} \tag{8.5}$$

The parameters of the process are σ_∞, τ, γ, the pdf for ϵ_r and ϵ_σ, and the correlation ρ between the two processes ϵ_r and ϵ_σ. The various terms are:

- σ_∞: This constant fixes the mean annualized volatility. Essentially, $\sigma_\infty \simeq E[\sigma_{\text{eff}}(t)]$ up to correction of order $\exp(\langle h^2 \rangle)$.
- τ: The characteristic time of return toward zero of $h(t)$.
- γ: The strength of the volatility noise, or the volatility of the log volatility.
- ϵ_σ: Random variable with zero mean and unit variance.
- ρ: The correlation between contemporaneous return and volatility, introducing a leverage effect.

Realistic first-order values for the parameters are $\tau \simeq$ a few days, $\gamma \simeq 1$, and $\rho = 0$ for foreign exchange time series. The corresponding mug shot is given in Fig. 8.1. From the structure of the equations, this process has one time horizon with an exponential decay of the correlation. The Monte Carlo simulations show clearly this feature. The next two models expand along the same algebraic structure but add multiple time scales in the process.

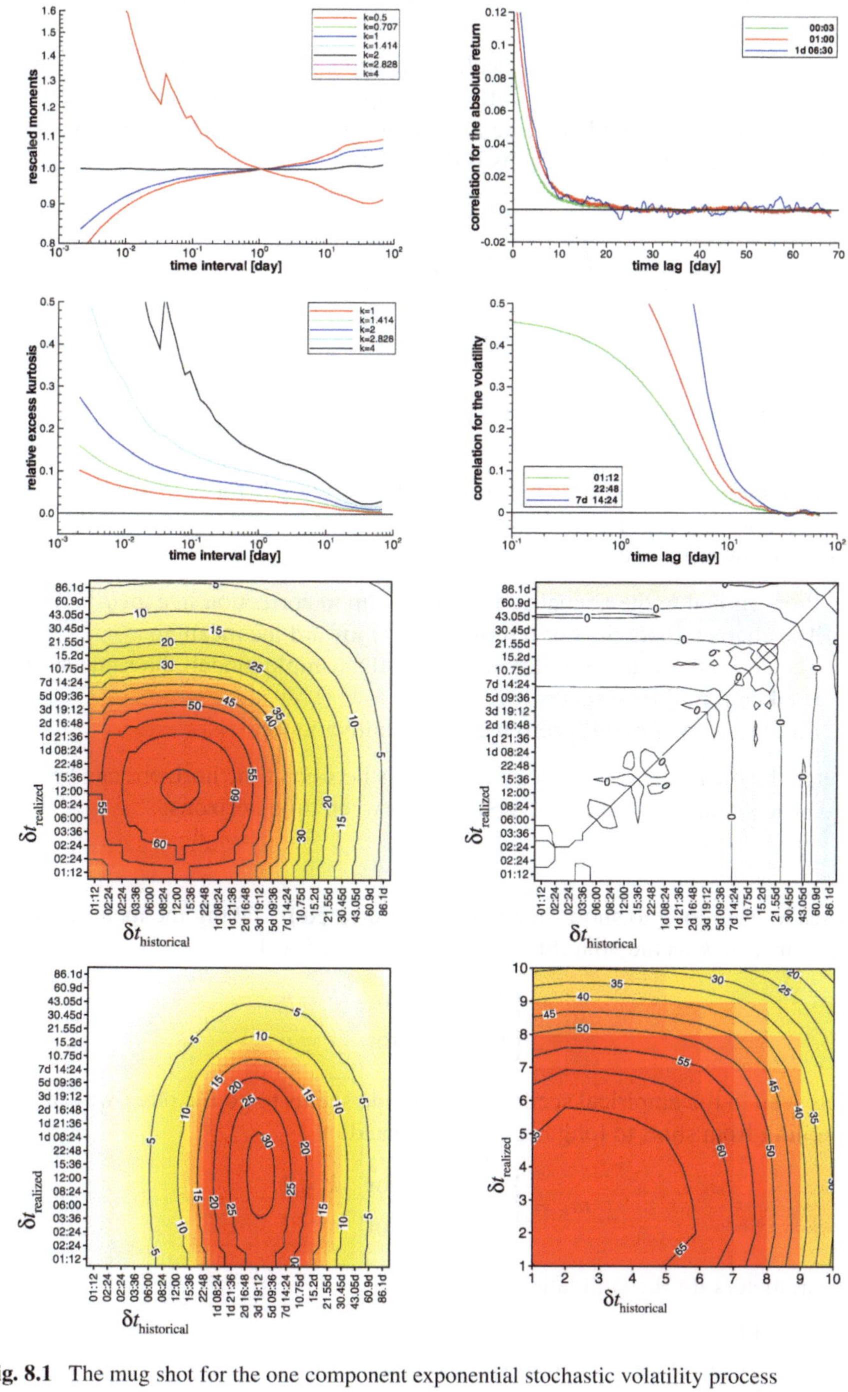

Fig. 8.1 The mug shot for the one component exponential stochastic volatility process

8.2.2 Long-Memory Exponential Stochastic Volatility Process

The generalization from one time scale to multiple time scales is straightforward, and fairly general models can be written. A simple structure with a minimal number of parameters is presented here. The long-memory exponential stochastic volatility model with n cascading components is defined by

$$\begin{aligned}
\tau_k &= \rho^{k-1}\tau_1 \qquad k = 1, \ldots, k_{\max} \\
\sigma_{\text{eff}}(t) &= \sigma_\infty \exp\left(\frac{1}{k_{\max}} \sum_k h_k(t)\right) \\
h_k(t) &= h_k(t-\delta t) + dh_k(t) \qquad k = 1, \ldots, k_{\max} \\
dh_k(t) &= -\frac{\delta t}{\tau_k}\big(h_k(t-\delta t) - m_k(t-\delta t)\big) + \frac{\gamma}{\sqrt{k_{\max}}}\sqrt{\frac{\delta t}{\tau_k}}\epsilon_{\sigma,k}(t).
\end{aligned} \tag{8.6}$$

The parameters of the process are σ_∞, τ_1, τ_n, γ, the pdf for ϵ_r and $\epsilon_{\sigma,k}$, and the mean reversion values m_k.

The various terms are:

- σ_∞: Proportional to the annualized volatility (up to correction in $\exp(\langle h^2\rangle)$).
- τ_k: The characteristic time of return for $h_k(t)$ toward the mean m_k.
- γ: The strength of the volatility noises. In this simple model, the same strength is taken for all time scales.
- $\epsilon_{\sigma,k}$: Random variables with zero mean and unit variance.

Through the mean terms m_k, various couplings between time horizons can be introduced. The simplest is to have no coupling between time horizons:

$$m_k = 0 \tag{8.7}$$

A cascade from long to short time horizons is created by taking the mean volatility at a given scale k as the volatility at the longer scale $k+1$:

$$m_k = \begin{cases} h_{k+1} & \text{for} \quad k < n \\ 0 & \text{for} \quad k = n \end{cases} \tag{8.8}$$

Yet, as seen in the empirical section, the cascade seems to go the other way around. A coupling from short to long horizons is created by

$$m_k = \begin{cases} 0 & \text{for} \quad k = 1 \\ h_{k-1} & \text{for} \quad k > 1 \end{cases} \tag{8.9}$$

The parameters for the simulations are:

- $\sigma_\infty = 0.11$
- $\tau_0 = 0.03125$ day, $\tau_n = 32$ day, $\rho = 2$ (i.e., $n = 12$)
- $\gamma = 0.07$

- $p(\epsilon_r)$ = Student with 3.3 degrees of freedom
- $p(\epsilon_\sigma)$ = Student with 3 degrees of freedom

Figure 8.2 corresponds to no coupling between time scales as given by (8.7), and Fig. 8.3 to a cascade from long to short time horizons as given by (8.8). Albeit the lagged volatility correlations are much closer to the empirical figures, the asymmetry in the historical/realized volatility correlation is always missing. This indicates that these processes are time-reversal invariant, regardless of the cascade structure created by the coupling between time horizons. Therefore, these processes cannot reproduce some empirical features, despite the multiple time horizons and their couplings.

8.3 Heston Stochastic Volatility Process

8.3.1 One-Component Heston Stochastic Volatility Process

The Heston model is a stochastic volatility model, where the noise term for the variance dynamic is modulated by the volatility. The volatility equation for the process is

$$\sigma_{\text{eff}}^2(t) = \sigma_{\text{eff}}^2(t-\delta t) - \frac{\delta t}{\tau}\left(\sigma_{\text{eff}}^2(t-\delta t) - \sigma_\infty^2\right) + \gamma\sqrt{\frac{\delta t}{\tau}}\sigma_\infty\,\sigma_{\text{eff}}(t-\delta t)\epsilon_\sigma(t) \quad (8.10)$$

where $\sigma_{\text{eff}}^2(t)$ is an annualized variance, σ_∞^2 fixes the mean annualized variance, τ sets the mean reversion time, and γ fixes the mean volatility of the variance (the vol-of-vol). The random term is essentially $\sigma_{\text{eff}}\epsilon_\sigma$. For small variance, its intensity decreases, and the mean reversion dominates, pulling back the volatility toward σ_∞. This is the mechanism that enforces the positivity of the volatility. This argument can be made rigorous in the continuous version of the model; in the discrete case, the positivity must be enforced. As our time increment is very small, the details of the positivity test do not introduce significant distortion in the distributions.

The main interest of the model is that an exact solution for the European option price can be computed when the residues ϵ_r and ϵ_σ are normal [80]. The solution is expressed as an integral of a complex function, which presents some numerical difficulties related to the complex cut of the logarithm function and to an oscillatory integrant. As this model incorporates a nontrivial dynamic for the volatility and fat-tailed price distributions, the option prices are more realistic than using a simple Gaussian random walk for the underlying prices.

The one-component Heston process has one characteristic time scale τ, and this leads to an exponential decay of the correlations. The mug shots are given in Figs. 8.4 and 8.5 for Gaussian and Student innovations, respectively. The parameters are $w_\infty = 0.11$ (annualized), $\tau = 4$ day, $\gamma = 1.65$, and $\nu = 4$ for Student innovations. With Gaussian innovations, the return and volatility distributions do not have enough fat tails, while Student innovation leads to more realistic distributions. The next process keeps the basic idea of the model to have a positive volatility, but adds a multi-time scale structure.

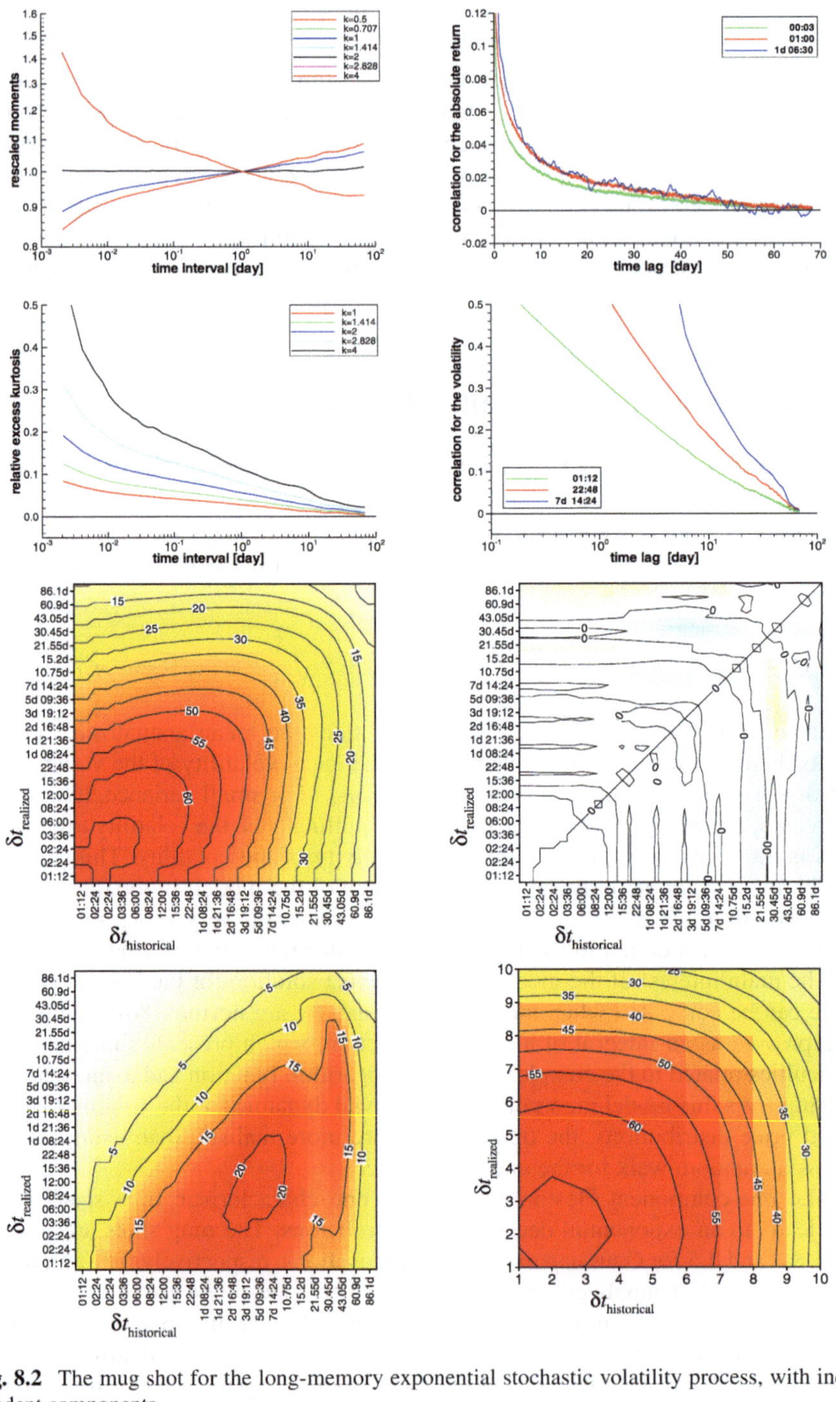

Fig. 8.2 The mug shot for the long-memory exponential stochastic volatility process, with independent components

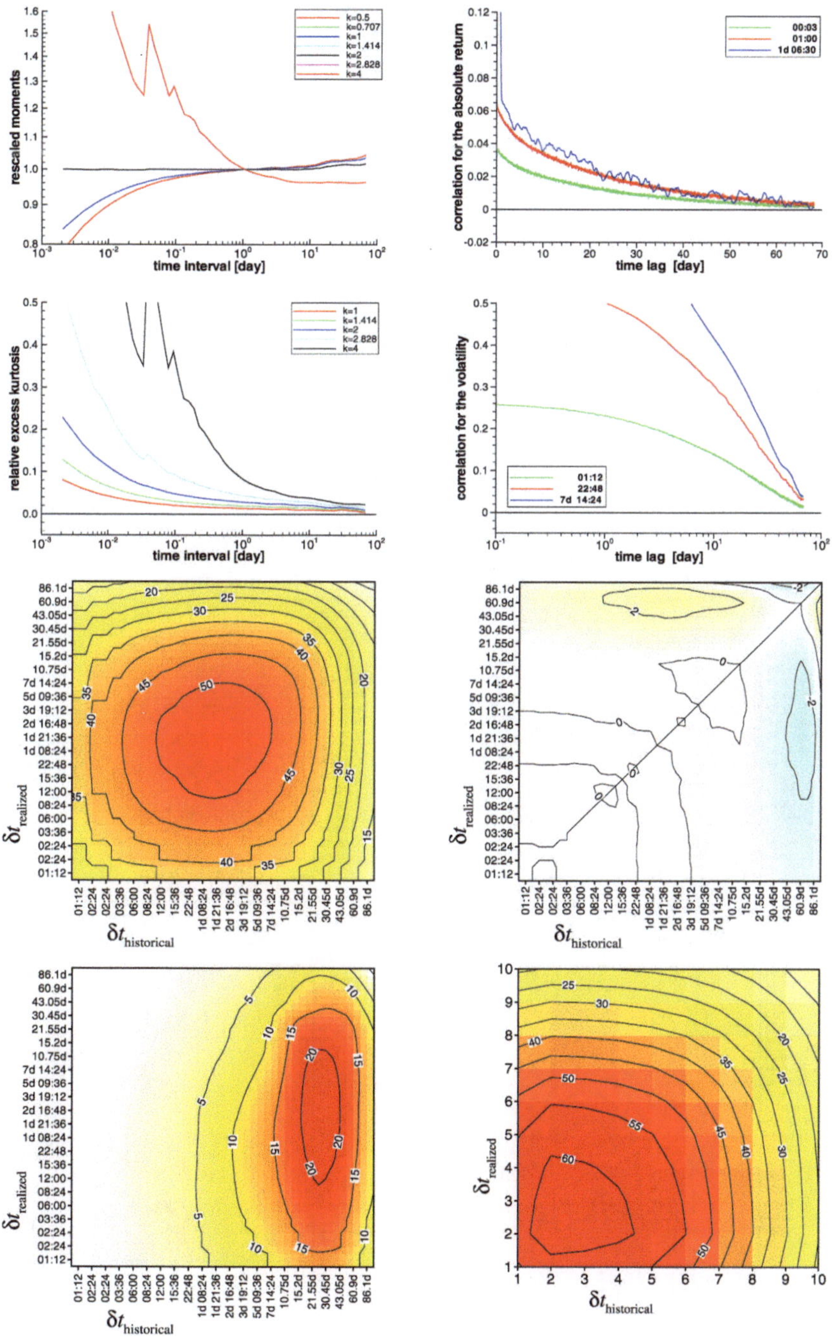

Fig. 8.3 The mug shot for the long-memory exponential stochastic volatility process, with a cascade from long to short time horizons

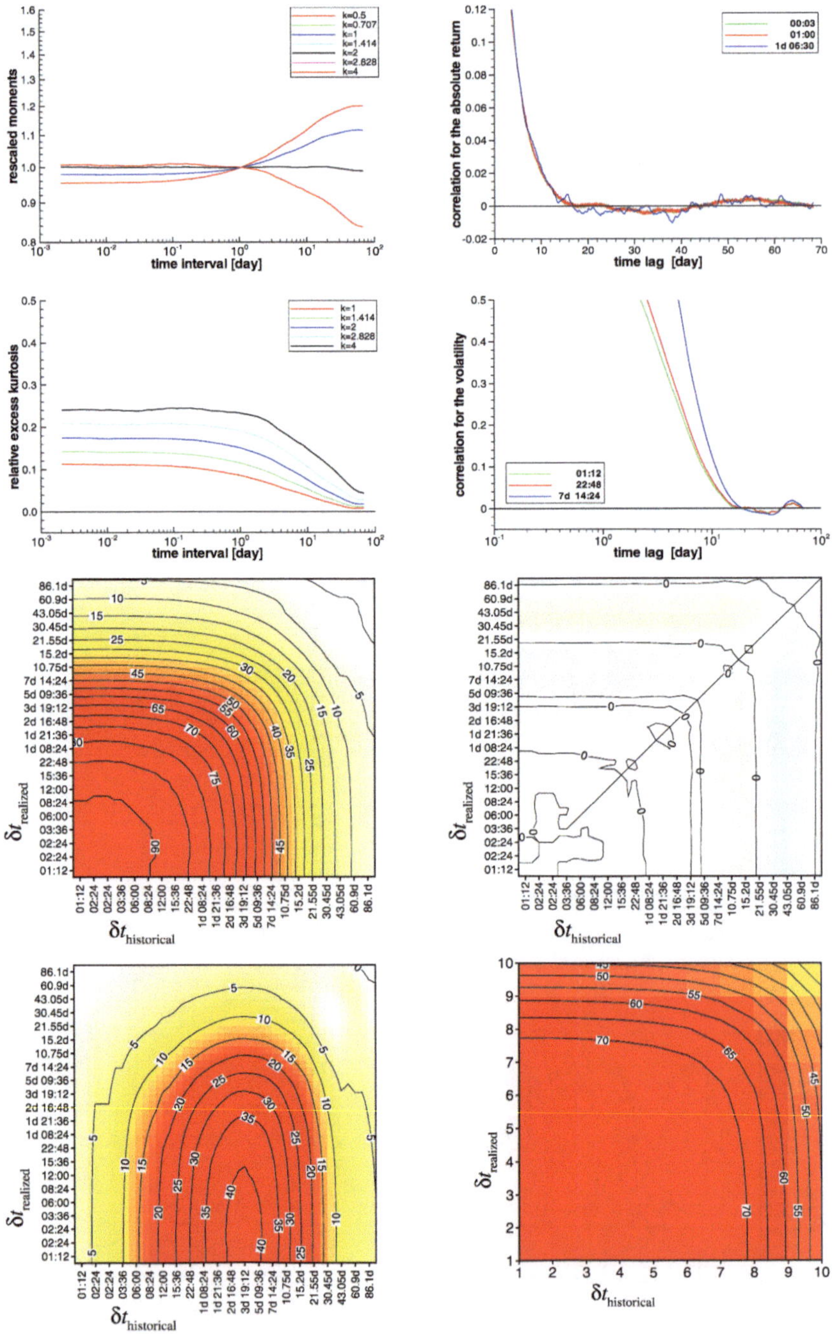

Fig. 8.4 The mug shot for the (one-component) Heston stochastic volatility process, with Gaussian innovations

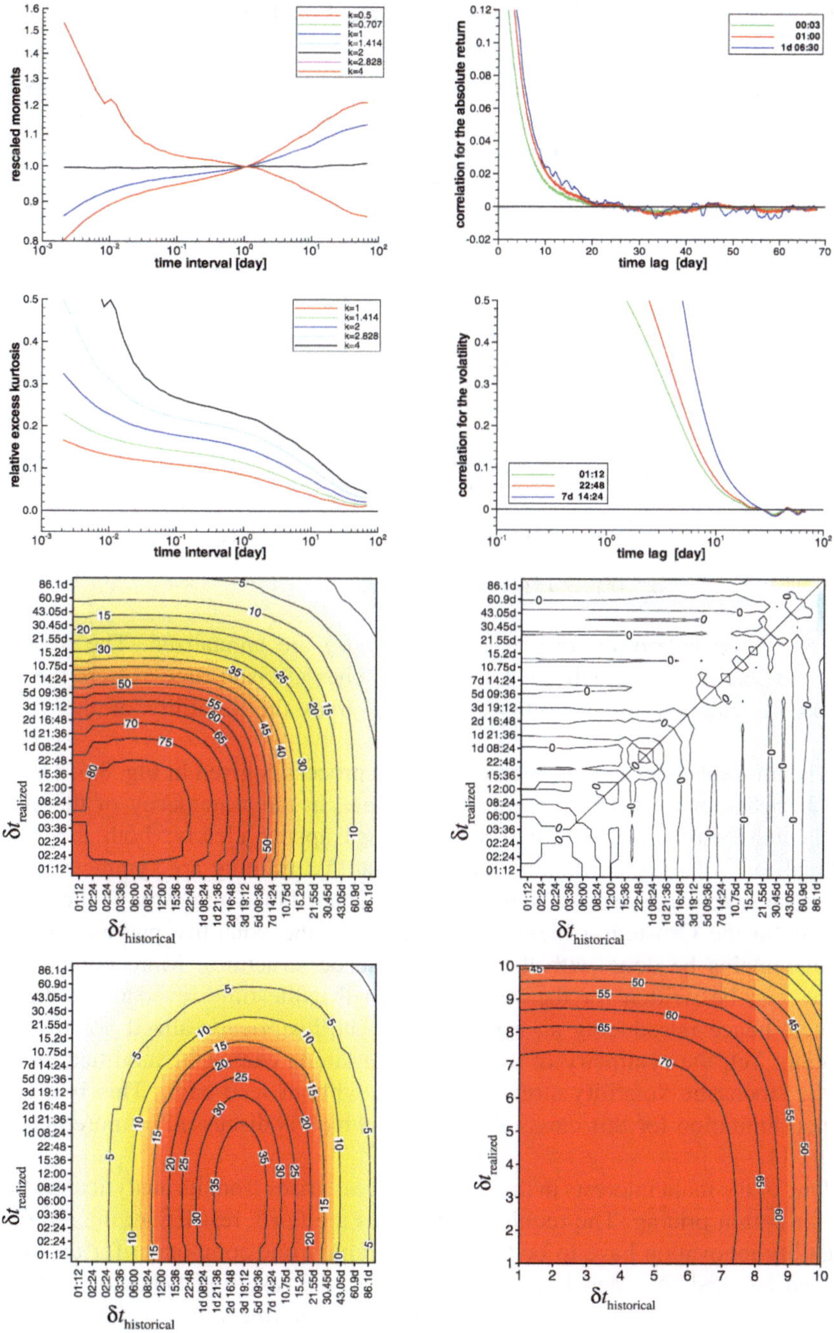

Fig. 8.5 The mug shot for the (one-component) Heston stochastic volatility process, with Student innovations both for the return and volatility

8.3.2 Long-Memory Heston Stochastic Volatility Process

The long-memory Heston process is a minimal modification of the one-component Heston model such as to introduce a volatility cascade from long to short time horizons. The equations for the model are:

$$\tau_k = \rho^{k-1}\tau_1 \tag{8.11a}$$

$$m_k(t-\delta t) = \begin{cases} \sigma_\infty & \text{for} \quad k = k_{\max} \\ \sigma_{k+1}(t-\delta t) & \text{for} \quad k < k_{\max} \end{cases} \tag{8.11b}$$

$$d\sigma_k^2(t) = -\frac{\delta t}{\tau_k}\big(\sigma_k^2(t-\delta t) - m_k^2(t-\delta t)\big) + \gamma\sqrt{\frac{\delta t}{k_{\max}\tau_k}}\sigma_\infty\sigma_k(t-\delta t)\epsilon_\sigma(t) \tag{8.11c}$$

$$\sigma_k^2(t) = \sigma_k^2(t-\delta t) + d\sigma_k^2(t) \qquad k = 1, \dots, k_{\max} \tag{8.11d}$$

$$\sigma_{\text{eff}}^2(t) = \sigma_1^2(t) \tag{8.11e}$$

where all the volatilities σ_∞, $\sigma_{\text{eff}}(t)$, and $\sigma_k(t)$ are annualized. The parameter σ_∞ fixes the mean annualized volatility, the characteristic time τ_k sets the mean reversion scale for the volatility σ_k, and γ fixes the volatility of the volatilities.

The mug shot for the long-memory Heston process is given in Fig. 8.6. The multiscale behavior is clearly present, for example, in the slow decay of the volatility lagged correlation. Yet, two shortcomings appear clearly: both the historical/realized volatility correlation and the volatility graining are symmetric, pointing to an exact time reversal symmetry (see Chap. 11). The origin of the symmetry lies in the Ornstein–Uhlenbeck structure for the volatility process (a formal proof remains be done with the present cascade structure). More complex processes can be written for the volatility, such that the time reversal symmetry is broken, hence introducing an asymmetry in the historical/realized volatility correlation. Yet, the symmetry of the volatility graining cannot be modified, because of the stochastic volatility structure $\sigma \to r$, without feed-back. This is a fundamental limitation for this class of processes, which is discussed in more details in Chap. 11.

One of the main interests in the Heston process arises from its analytical tractability for option pricing. The required conditions are fairly restrictive: the return and volatility innovation have to be Gaussian, and only one component for the volatility can be used. As seen in the mug shot 8.4, the resulting behavior is close to a GARCH$(1,1)$ process. Yet, all the subsequent modifications of the basic structure in order to obtain more realistic stylized facts result in a loss of the analytical tractability for option pricing. Without close formulas, option pricing can only be done using Monte Carlo simulations. But then any process can be used.

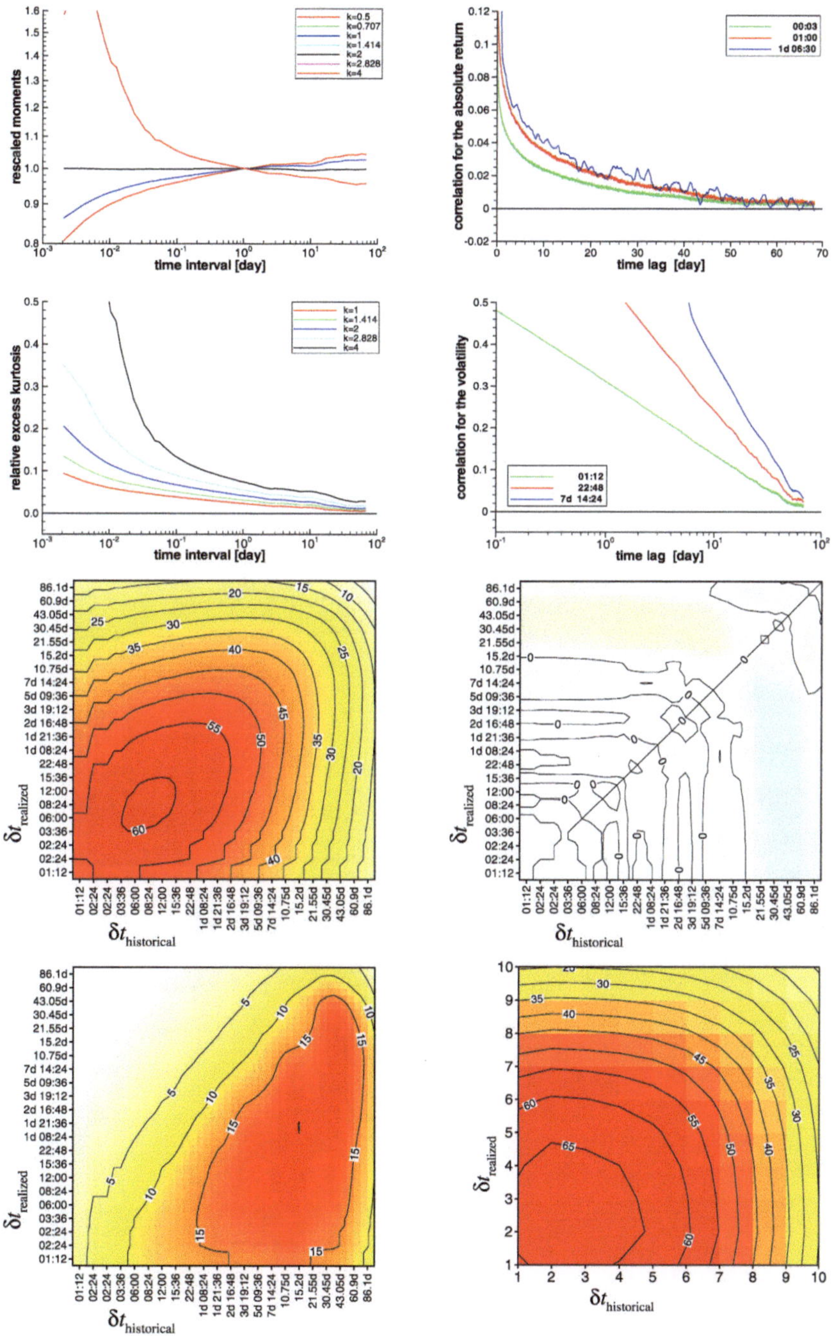

Fig. 8.6 The mug shot for the long-memory Heston stochastic volatility process, with Student innovations both for the return and volatility

Chapter 9
Regime-Switching Process

The idea behind a regime-switching model is that the market can take a few states, for example, like "quiet", "excited", and "shocked". To these states correspond different level of volatility, and the market moves between these states according to a hidden process. In the simplest version of the model, the volatility can take n fixed values $\sigma_1, \ldots, \sigma_n$. The volatility is selected according to a state index $i(t) \in [1, \ldots, n]$, and the dynamics for the index is specified by a Markov chain with the transition probabilities $p(i \to j)$ to jump into state j when in the state i. The probabilities obey the condition $\sum_j p(i \to j) = 1$. In a given state i, the price follows a random walk (with Gaussian or Student residues) with constant volatility σ_i.

This model is structurally close to a stochastic volatility process, as the dynamics of the volatility is independent of the prices. In principle, the specification of the dynamic through the transition probabilities $p(i \to j)$ allows for a broad range of behavior. For a model with n states, the n volatilities and $n(n-1)$ transition probabilities must be given, and the model has a total of n^2 parameters. Clearly, this model is interesting with a few states, in the range $n = 2$ to 4. With more states, it becomes more parsimonious to specify directly the dynamics of the volatility, say with a stochastic volatility process. Although the process has n^2 parameters, the shape of the volatility pdf imposes strong constraints on the parameters in order to have a unimodal distribution with a slower decay on the large volatility side. For example, the probability $p(i \to i)$ to stay in the same state cannot be too large for large volatility, otherwise a peak is created in the volatility pdf. In turn, this bounds the lagged correlation for the volatility which has to decay fairly fast.

Two states in the regime-switching model are not enough to create a realistic pdf for the volatility, as either the pdf shows clearly two peaks, or the pdf decays too fast on the high-volatility side. Three states allow one to get a somewhat realistic volatility pdf, although small peaks still subsist in the distribution. With two or three states, the regime-switching process shows essentially the same behavior as most stochastic volatility processes with one time scale:

- A too fast decay for the lagged correlation and for the historical/realized volatility.
- A fairly symmetric historical/realized volatility.
- A strictly symmetric fine/coarse grained volatility.

G. Zumbach, *Discrete Time Series, Processes, and Applications in Finance*, Springer Finance, DOI 10.1007/978-3-642-31742-2_9,

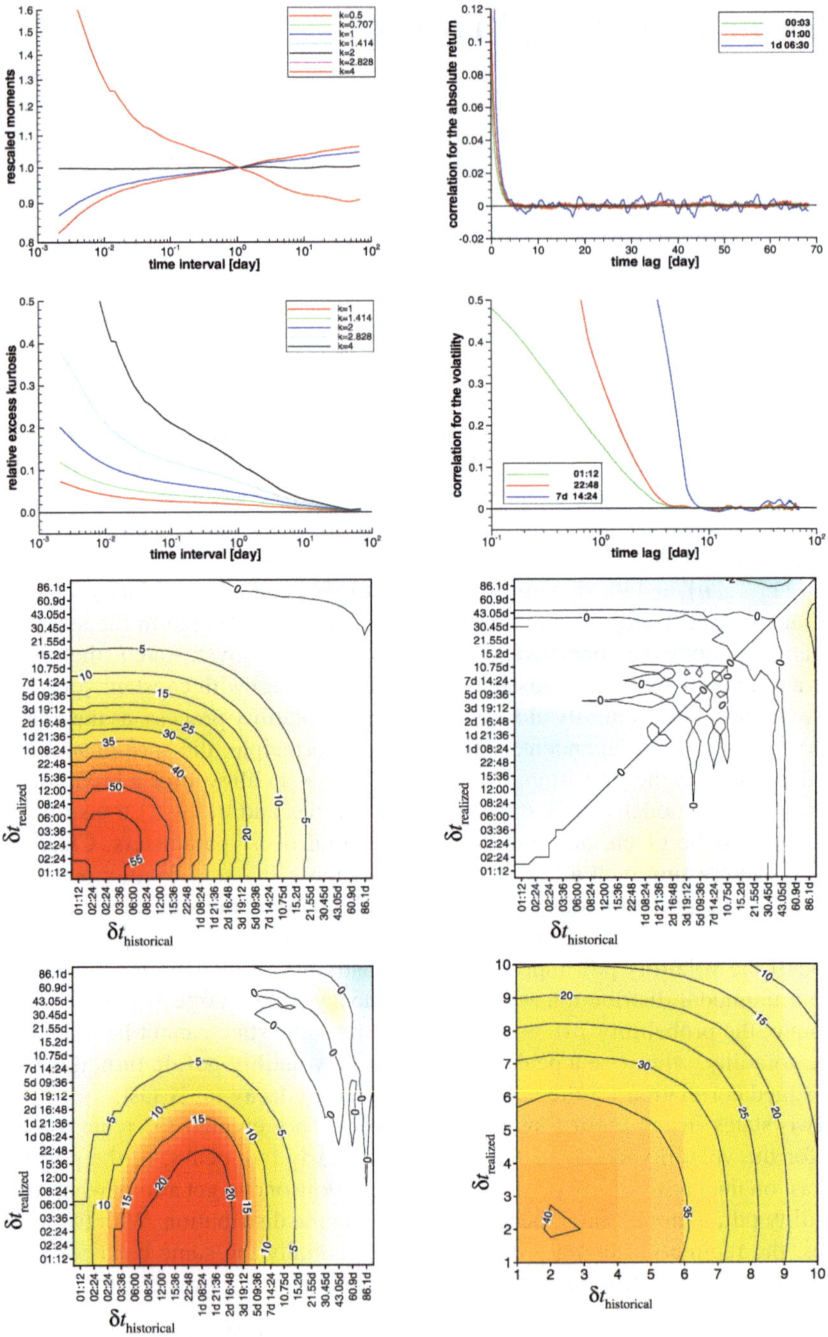

Fig. 9.1 The mug shot for the three-state regime-switching process, with Student innovations

In [123], the author shows that a Markov chain with two states can be rewritten in a form that is manifestly invariant under time reversal. As a regime-switching process is a random walk (which is time invariant) with a volatility depending on the world's state, the time reversal properties follow from the Markov chain. For a regime switching model with two states, all the statistics presented in this paper are compatible with an exact time reversal symmetry, in agreement with [123]. For this reason, the mug shot for this model is presented for three states in Fig. 9.1. Overall, the properties of this process is quite far from the properties of the empirical time series, and adding more states is not improving substantially upon the basic flaws. Therefore, we conclude that this process is not a good candidate as a model for the empirical price time series.

With three states, the parameters for the simulations are:

- $\sigma = [0.09\ 0.15\ 0.30]$
- The probabilities are

$$p(i \to j) = \begin{pmatrix} 0.9996 & 0.00025 & 0.00015 \\ 0.0014 & 0.9980 & 0.0006 \\ 0.0010 & 0.0090 & 0.9900 \end{pmatrix}$$

- $p(\epsilon_r)$: Student with 4.0 degrees of freedom

Chapter 10
Price and Volatility Using High-Frequency Data

10.1 Introduction

A very large fraction of finance is based on daily data. As most investment decisions are medium to long-term based—say with a reference time horizon above two weeks—using daily data is appropriate. For most contracts, the daily close prices are given by the stock exchanges, thus providing for good reference prices. Moreover, using one price per day leads to simple data base and simple computations, without worrying about seasonalities, market opens and closes, different time zones, and inhomogeneous time series. In the short time direction, continuous-time models are used frequently, particularly in the evaluation of options. An opportunity to move closer to the theoretical continuous-time limit is offered by the increasing availability of high-frequency data. The large amount of intra-day information allows one to get much more accurate quantitative measurements. For very-short-term to medium-term decisions and trading, an edge can be gained by using this information, for example, by constructing short-term forecasts. Yet, the path leading to superior information is not that simple.

The first difficulty lays in the complexity of high-frequency data. For equities, the information is provided by the stock exchanges which for most of them are completely electronic. The trading is done through a double auction mechanism for buyers and sellers. The market participants issue market orders (immediate execution at the available market price) or limit orders (possible trade conditional on a specified price). Eventually, they could withdraw previously posted limit orders that are still pending. Essentially, the interactions between market orders and limit orders create the dynamics for the best bid and ask. Depending on the stock exchange, different levels of information are available. The basic data stream contains only information about the best bid and ask positions, typically prices, volumes, and number of orders. More detailed data streams can include the flow of incoming orders and/or the state of the order queues on the bid and ask sides. A large body of literature studies the complex dynamics of the order queue and its statistical properties (see, e.g., [65, 111, 144]). For example, an interesting question is the long term impact on the prices of trades with different volumes.

G. Zumbach, *Discrete Time Series, Processes, and Applications in Finance*,
Springer Finance, DOI 10.1007/978-3-642-31742-2_10,

The foreign exchange market is more complex because decentralized, and there are different competing systems to exchange information and/or to trade. There is no single best bid or ask prices, but instead several competing prices issued by different market makers. Then, a single market maker can issue different prices depending on the required volume or on the counterparty. As the competition for business is intense, the difference between prices is usually tiny, but there is clearly not a single price at a given time. Moreover, the trades are not available, and the orders incoming on the brokers are not available. Compared to equities, the high-frequency information is rather poor and in general limited to the stream of quotes for the best bid and ask prices, as provided by one data provider.

The emerging high-frequency picture is clearly complex, with rich flows of information describing partly or completely the microscopic interactions between the market participants. Yet, our particular needs about this information are limited as we would like to obtain only the equivalent of a single "prevailing" price for each time. The simple solution is to use the best bid and ask prices, and then to compute the logarithmic middle price. But the prices occur at random times, with a finite granularity (i.e., the prices are quoted with a conventional accuracy, for example, of 0.01$). Therefore, there is no convergence to a continuous path. Because of this lack of convergence, the basic diffusion model for the price evolution needs to be revisited for very short time intervals, with important consequences for the volatility.

There are a number of reasons why the continuous idealization of the price path fails at short time intervals. First, the prices are not real numbers but have a finite precision set by the stock exchanges (or by consensus for FX), typically of 1 cent or 5 cents. For stocks, the smallest increment typically depends on a range for the price of the stocks. The detailed rules could be somewhat convoluted, but the end effect is that the prices have a relative precision of the order of 0.1 to 5 BIPS. The rounding effect is not included in the above model, but can be easily added by rounding the price, say to an accuracy of 1 BIPS. The time horizon at which the granularity effect becomes dominant can be estimated by a simple scaling $\sigma\sqrt{\delta t/1\text{ year}} = 1\text{ BIPS} = 10^{-4}$. For a 10 % volatility, the time interval is $\delta t = 30$ seconds. For liquid instruments, quotes and trades occur at a faster pace, and the granularity is therefore a dominant feature of the very short time dynamics. The granularity becomes small for price changes say 10 times larger, leading to a time $\delta t \simeq 1$ hour (using the same scaling). An illustration of the granularity effect is given on Fig. 10.1 for the USD/CHF exchange rate. This illustration is fairly similar to any short-term plot of the prices for most financial instruments.

Second, the market participants disagree on "the" price or want to reach different positions. These differences lead to trades and hence to prices jumping when the best bid or ask prices are modified. In the FX market, the large brokers publish quotes that can differ slightly, depending on their own inventories or opinions. The jitter induced by the quoting activity is also clearly visible on Fig. 10.1. The short time difference in the quotes of the FX market has been called the "incoherent effect" in [46], as the traders need to exchange information through quotes and trades in order to build a consensus about the prevailing price.

Following a famous paper [133], the price jumping is often called "bid-ask bounce". Roll studied transaction prices occurring randomly on the bid or on the

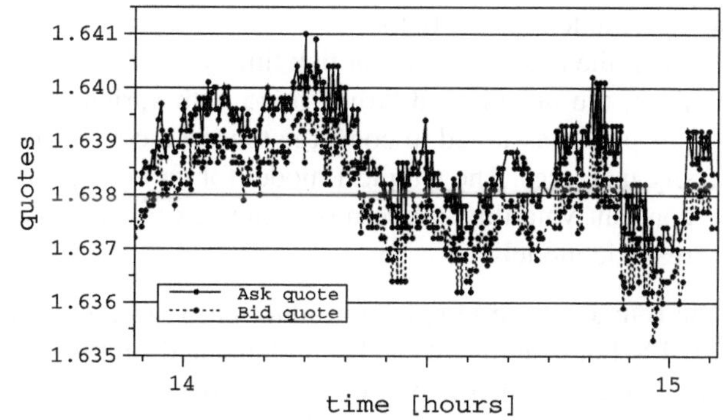

Fig. 10.1 One hour of prices for the USD/CHF exchange rate. The data are from Reuters, the 06.04.2000, with GMT time

ask sides, while the bid and ask prices stayed fixed. Roll's analysis of the short-term bouncing thus applies to time series of transaction prices when it is unknown if the transactions are buy or sell initiated. The end result of Roll analysis is similar to the models below, with a strong negative AR(1) component in the returns. The high-frequency information available today is richer, and such bid/ask ambiguity can be resolved or does not exist when bid and ask prices are available. Therefore, the term "bid-ask bounce" is improper to describe the short-term price moves according to the original paper, but it is often used in the literature.

These basic reasons for the lack of convergence to a continuous price path lead to two basic models for the description at very short time of the prices. Both are based on the distinction between the observed price p and a hypothetical "true" underlying price $\tilde{p}$, which is assumed to have a smooth behavior. The relations between p and $\tilde{p}$ are described in the next section for these two models. A description of general models for $\tilde{p}$ is given in Sect. 10.3, while the focus after is on an i.i.d. random walk. The realized variance and integrated variance are introduced next, with the computation of the upward bias introduced by the difference between p and $\tilde{p}$. Two approaches to correct for the bias in the simplest variance estimators are presented in Sects. 10.5 and 10.6. Another method is introduced in Sect. 10.7 in order to filter the price p for obtaining a good approximation of $\tilde{p}$. Finally, the sampling in tick time is discussed in more details, with the introduction of the volatility per tick and the role of the tick rate.

10.2 The Short-Term Models for the Prices

10.2.1 Prices with Finite Granularity

It is fairly trivial to add a finite precision on the prices. The process in tick time of the observed price can be simply written as

$$p(j) = \lfloor \tilde{p}(j) \rfloor \tag{10.1}$$

where:

- j: the index for the ticks.
- $p(j)$: the observed price at tick time j.
- $\tilde{p}(j)$: the unobserved "true" price with variance σ^2 and no correlation. Often, this part is assumed to converge toward a continuous process when the tick density increases. The simplest model for $\tilde{p}(j)$ is a standard Brownian motion with constant volatility, while more complex processes can be used in order to build realistic models.

The rounding operator $\lfloor \cdot \rfloor$ introduces a finite precision on the prices, typically of 1 BIPS. For example, it can be implemented simply by $\lfloor p \rfloor = 10^{-4}\,\mathrm{int}(10^4 p)$ where $\mathrm{int}(\cdot)$ gives the integer part of the argument. In order to obtain a realistic model for prices, a finite granularity should always be included. Albeit trivial from a numerical point of view, this addition makes the process very difficult to use on the analytical side. For this reason, the next model is more popular in the literature, even though it is less intuitive than the price granularity.

10.2.2 The Incoherent Prices

This second stylized model is frequently used to describe the high-frequency price dynamics as it leads to some easy analytical computations. The process in tick time of the observed price can be simply written as the sum of two terms: the hypothetical process for the true price and the incoherent component modeled as an additive noise

$$p(j) = \tilde{p}(j) + u(j). \tag{10.2}$$

The prices $p(j)$ and $\tilde{p}(j)$ are as previously, and

$$u(j) \sim \mathrm{i.i.d.}\big(0, \eta^2\big) \tag{10.3}$$

represent the incoherent term modeled as an i.i.d. noisy component with a given (stationary) distribution.

Variations of the basic model can be written using the physical time t or the logarithmic middle price $x(j) = \log(p(j))$. Given the model (10.2), the observed k-tick return is

$$r_k(j) = \tilde{r}_k(j) + \frac{u(j) - u(j-k)}{\sqrt{k}} \tag{10.4}$$

and the variance of the observed returns is given by

$$\mathrm{Var}(r_k) = \mathrm{Var}(\tilde{r}_k) + \frac{2\eta^2}{k} = \sigma^2 + \frac{2\eta^2}{k}. \tag{10.5}$$

The observed variance is equal to the "true variance" (the variance of the Brownian motion describing the dynamics of the unobserved price), plus an additional term coming from the incoherent component. Importantly, the two components scale differently with respect to aggregation. The first term scales with the usual diffusion dependency and is constant due to our convention using $1/\sqrt{k}$ in the return definition. The second term scales as $1/k$ and is small for large k. The same model written in physical time leads to the same dependency but with k replaced by δt.

For the model with constant volatility, the autocorrelation of the return at a lag h can be easily computed:

$$\rho(h) = \begin{cases} -\eta^2/(\sigma^2 + 2\eta^2) & \text{for} \quad h = 1 \\ 0 & \text{for} \quad h > 1. \end{cases} \tag{10.6}$$

This implies $-0.5 \leq \rho(1) \leq 0$. The lower bound -0.5 is reached when σ^2 is completely negligible compared to η^2, and the return is the lag one difference of a noise. An empirical autocorrelation around -0.4 is observed for the USD/JPY and USD/CHF [46], which implies $\eta^2/\sigma^2 \simeq 2$. A similar estimation was done by [78] for liquid stocks, by reporting the ratio $\gamma = \eta^2/E[\sigma^2] \simeq 0.1\ \%$, but with σ^2 measured at a daily time scale (i.e., with the normalization used in Sect. 10.4). This ratio depends on the tick rate and therefore is not directly comparable with the FX estimate. Assuming a tick every $10'$ and a market open 8 hours per day, a dimensionless ratio γ' is obtained by the scaling

$$\gamma' = \frac{\eta^2}{\sigma^2} = \gamma \ \frac{8 \cdot 3600}{10} \simeq 3.$$

This indicates that, at tick-by-tick level, the volatility originating from the incoherent component is largely predominant, with the same order of magnitude for FX and for stocks. Therefore this effect should be carefully considered before using data at a very high frequency.

When either k or δt is large enough, the usual long-term volatility σ^2 is obtained, for example, with daily data. When decreasing the time horizon, the variance increases compared to the long-term volatility due to the η^2 term, creating an upward bias in a variance estimator. This has dramatic consequences on measures of volatility computed with high-frequency returns since the noise term can dominate the "naive" volatility estimator. The plot of the variance or of the volatility versus δt has been made popular by [7] under the name of signature plot. A simple solution to reduce the bias consists in taking a time interval for the returns that is long enough so that the bias stays under an acceptable level. This time horizon depends on the time series and is in the range of 5 minutes to 1 hour (in agreement with the rough estimate made in the introduction of this chapter). This solution limits the usefulness of high-frequency data. A better solution is to construct volatility estimators that essentially remove the part in η^2. One way is to add a term like $r_j r_{j+1}$ in the volatility estimator, so that $E[2r_j r_{j+1}] = -2\eta^2$ cancels the bias. Another solution is to measure the volatility at two (or more) time horizons and to combine them so as to remove the term in η^2 that has a different scaling. Finally, another solution is to apply an AR(1) filter on the price, so as to remove the incoherent component before computing the returns. These estimators are presented below.

The above model has no lagged correlation beyond lag one. Stock indices have more structures, with a small and regularly decaying short-term correlation. The model can be easily extended by introducing an autoregressive structure in the return $\tilde{r}(j)$:

$$\tilde{r}_1(j) = \phi \tilde{r}_1(j-1) + \epsilon(j) \tag{10.7}$$

with $\epsilon \simeq \text{i.i.d.}(0, \sigma_\epsilon^2)$. Then the autocovariance structure of the model becomes

$$E[r_l r_{l+h}] = \begin{cases} \sigma^2 + 2\eta^2 & \text{for} \quad h = 0 \\ \phi\sigma^2 - \eta^2 & \text{for} \quad h = 1 \\ \phi^h \sigma^2 & \text{for} \quad h \geq 2 \end{cases} \tag{10.8}$$

with $\sigma^2 = \sigma_\epsilon^2/(1-\phi^2)$. This lagged correlation replicates the empirical data for stock indices as shown in [46], where a small incoherent effect at lag one is present.

10.3 Models for the Underlying Price

A wide range of models can be used for the smooth part of the price process, with the usual trade-off between analytical tractability versus capturing realistic features. The most important stylized facts we would like to include in a model are the fluctuations of the volatility, the seasonalities, and the fat-tailed distributions for the return and innovations. On the mathematical side, a convergence to continuous paths with an increasing tick density together with the micro-structure model (10.2) allows for some powerful analytical results for various estimators of the volatility. But the convergence to continuous paths restricts the choice of the process essentially to Gaussian processes with stochastic volatility. Finally, the processes can be written either using a regular (business) time grid or using a random tick times with a specific process for the arrival of ticks. For practical applications, the behavior of the proposed estimators should also be studied ideally in the presence of outliers and possible gaps in the data. All these possibilities lead to a broad range of models and results studied in numerous articles. A few major topics are introduced in the following sections, and the interested reader should consult the original literature or some reviews for an in-depth treatment. A good review of the realized volatility is given in [104].

10.4 Continuous Price Paths and the Realized Variance

In order to derive rigorous mathematical statements, the unobserved true price $\tilde{p}(t)$ is assumed to be converging to a continuous path when the tick density increases (with some technical conditions to ensure that the tick density increases everywhere). The innovations are assumed to be i.i.d., with a normal distribution. The volatility has an unspecified time dependency; for example, it can be deterministic or stochastic with normal innovations. The micro-structure model is (10.2) and assumes that u is independent from p.

This part of the literature uses other conventions that we will follow in this section. The time indexing is done by separating the time t into a day counter denoted by t and an intra-day index $i = 1, \dots, n_t$, where a number of price ticks n_t occurs during day t. The identification $(t, 0) = (t-1, n_t)$ makes some formulas simpler. The time-dependent variables carry both subscripts, like $p_{t,i}$. The return is a simple price difference without scaling by the time interval, as defined in (6.6)

$$r_{t,i} = p_{t,i} - p_{t,i-1} \tag{10.9a}$$

$$\tilde{r}_{t,i} = \tilde{p}_{t,i} - \tilde{p}_{t,i-1}. \tag{10.9b}$$

Following the description in the previous paragraph, the process for the underlying price $\tilde{p}$ is taken as

$$\tilde{r}_{t,i} = \sigma_{t,i}\epsilon_{t,i} \qquad \text{with} \quad \epsilon \sim N\big(0, n_t^{-1}\big) \tag{10.10}$$

together with (10.9b). The variance $\sigma_{t,i}$ is scaled at a reference time of 1 day.

The main question of interest in this chapter is to evaluate accurately the volatility corresponding to a given day t, using only the intra-day information. The setup is meant for stock time series, with a possible discontinuity between consecutive days occurring during the market closure. This discontinuity is included in the formula below but can be easily removed by modifying the summation boundaries.

The theoretical results are derived in the limit $n_t \to \infty$. The *realized variance* for the day t is defined as

$$\mathrm{RV}_t = \sum_{i=1}^{n_t} r_{t,i}^2. \tag{10.11}$$

Without the microstructure noise, the expected variance over one day is given by

$$E[\mathrm{RV}_t] = E\left[\sum_{i=1}^{n_t} \tilde{r}_{t,i}^2\right] = \frac{1}{n_t}\sum_i \sigma_{t,i}^2. \tag{10.12}$$

The last term is called the integrated variance

$$\mathrm{IV}_t = \frac{1}{n_t}\sum_i \sigma_{t,i}^2. \tag{10.13}$$

It corresponds in the large n_t limit to the integral of the variance over one day (in an option context, beware that IV stands for implied volatility!). A result from the semimartingale processes [8] shows that the volatility is measured with an increasing accuracy when the number of ticks n_t increases, namely $\mathrm{RV}_t \to \mathrm{IV}_t$ in probability in the large n_t limit. Intuitively, the knowledge of the prices on a finer grid allows one to estimate more accurately the (time-dependent) integrated variance, as already emphasized by [107].

In an *ex-post* measurement using the prices $\tilde{p}_{t,i}$, the daily variance can in principle be computed accurately, hence the name *realized variance*. The realized variance computation contrasts with the latent variance estimator, as obtained from a process applied on daily data. The latent variance depends on the process equations and the parameter values, although the dependency is weak in practice. In this sense, the realized variance is essentially model-free, albeit the microstructure noise prevents such a simple evaluation.

Beware that the realized volatility used elsewhere in this book is slightly different from the realized variance in this chapter. The former definition corresponds to the volatility that occurs between t and $t + \Delta T$, with t the current time. The name for that "realized volatility" originates in volatility forecasts, where the information set up to t is used to compute a volatility forecast which is compared against the realized volatility.

Now, the microstructure noise can be added to the underlying continuous price path. The distribution for the noise $u_{t,i}$ is taken as in Sect. 10.2.2, namely is i.i.d. with a variance η^2 and independent from $\epsilon_{t,i}$. Because there is no n_t scaling in the noise, the incoherent term dominates the short-term behavior. With the noise, the expected variance over one day of the (one-step) returns becomes

$$E[\mathrm{RV}_t] = E\left[\sum_{i=1}^{n_t} \tilde{r}_{t,i}^2\right] + n_t 2\eta^2 = \mathrm{IV}_t + n_t 2\eta^2. \tag{10.14}$$

This estimator is studied rigorously by [15], showing that it diverges for $n_t \to \infty$. The last term on the right-hand side increases with n_t, creating the upward bias discussed in the introduction of this chapter and ultimately the divergence for $n_t \to \infty$. When using high-frequency data, the term $n_t 2\eta^2$ gives typically the largest contribution to the left-hand side. Its large size emphasizes the necessity for correcting the high-frequency noise, but also the difficulty of the task as a large term should be removed. Any imperfection in the correction will result in a possibly large error in the estimation of the integrated volatility, in the form of bias or variance.

The return can be computed over kth ticks, which gives with the noise term

$$r_{t,i}^{(k)} = p_{t,i} - p_{t,i-k} = \sum_{j=0}^{k-1} \sigma_{t,i-j}\epsilon_{t,i-j} + u_{t,i} - u_{t,i-k}. \tag{10.15}$$

The expected variance over one day of the k-step return, taken every kth tick, is

$$E\big[\mathrm{RV}_t^{(k)}\big] = E\left[\sum_{i=1}^{n_t/k} \big(r_{t,ik}^{(k)}\big)^2\right] = \mathrm{IV}_t + \frac{n_t}{k} 2\eta^2 \tag{10.16}$$

and the noise term decreases by $1/k$ when increasing the aggregation. With the convention used in this chapter, this is the equivalent of formula (10.5). A similar formula can be written with overlapping returns, introducing some corrections in the boundary terms.

Often, the data are sampled regularly in a physical time scale, with a time increment of δt. With aggregation, the returns are measured over the time interval $\delta t_r = k\delta t$. Up to an annualization factor by 1 day/1 year, the integrated variance corresponds to the variance $\sigma^2[\delta t_\sigma, \delta t_r]$ defined in Chap. 2, with $\delta t_\sigma = 1$ day. A signature plot [7] is a graph of the mean of σ (or σ^2) versus δt_r. Such a plot allows one to find the time interval δt_r such that the bias in the variance is small enough compared to using daily data, for which $\delta t_r = \delta t_\sigma = 1$ day. An optimal balance between bias and variance gives a more rigorous way to select δt_r. Yet, discounting explicitly the microstructure noise is more efficient to evaluate the realized variance.

The mug shots contain a similar plot in the top left figure, with δt_r on the horizontal axis and with δt_σ corresponding to the sample length. All the statistics for the mug shots are done after an AR(1) filter (see Sect. 10.7) is applied to the price time series in order to remove the microstructure noise (up to the limitation of the filter). The mostly flat curve for an exponent 2 shows that the filter removes correctly the short-term noise, while the curves for the other values of the exponent display the nontrivial scaling of the return distributions.

The properties of the price paths with respect to their smoothness at a scale δt_r can also be measured directly. For example, [134] defined the function

$$S[\delta t_r] = \frac{1}{2}\left\{1 + \log_2\left(\frac{E[\sigma^2[\delta t_\sigma, 2\delta t_r](t)]}{E[\sigma^2[\delta t_\sigma, \delta t_r](t)]}\right)\right\} \tag{10.17}$$

which can be related to Besov spaces. Essentially, this formula computes the discrete derivative with respect to δt_r of the sample variance, up to some normalization (the above paper used a different notation and a partition of the time axis on a dyadic grid). Without microstructure noise, the volatility is independent of the time horizon δt_r, and the index has value $1/2$. With noise, the variance becomes larger for smaller return time horizon: the index becomes smaller than $1/2$, indicating rougher paths compared to Gaussian price paths. The previous reference includes an empirical study of the index $S[\delta t_r]$ and of the signature plot, as well as of its behavior for some models. Similar indexes have been introduced by [3] for processes with jumps.

The following sections present different ways to measure the realized variance or to filter the microstructure noise from the price paths.

10.5 Correction with Multiple Time-Scale Estimation

One way to recover the IV_t term from $\mathrm{RV}_t^{(k)}$ consists in using multiple values for k. Because the terms IV and η^2 scale differently with k, they can be inferred from $\mathrm{RV}_t^{(k)}$. With two aggregation values k_1 and k_2, corresponding to Two Time Scales

(TTS), the linear combination

$$\mathrm{RV}_t^{\mathrm{TTS}} = \frac{k_1 \mathrm{RV}_t^{(k_1)} - k_2 \mathrm{RV}_t^{(k_2)}}{k_1 - k_2} \tag{10.18}$$

gives an estimator for the integrated variance IV_t. This is essentially the estimator proposed by [147]. Some variations in the factors are also proposed in order to build better estimators (for example, by using overlapping returns). This family of estimators can be studied rigorously because the variance appears as a linear combination of r^2.

Using more aggregation values, the set of estimators

$$\mathrm{RV}_t^{(k)} \simeq \mathrm{IV}_t + \mathcal{N}_k 2\eta^2 \qquad \text{with} \quad \mathcal{N}_k = n_t/k \tag{10.19}$$

can be fitted by a linear function of $\mathcal{N}_k$. As proposed by [119, 146], the intercept gives a Multiple Time-Scale (MTS) estimator $\mathrm{RV}_t^{\mathrm{MTS}}$ for IV_t (and the slope gives an estimator for η^2). The structure of this estimator is more complex, preventing the derivation of strong analytical results. Intuitively, a multi-scale estimator should have a smaller variance and should be more robust against misspecification of the theoretical model, as it is built with information collected over many time horizons.

10.6 Correction with Lagged Variance

With the return normalization used in this chapter, the lagged covariance between (one step) returns is

$$E[r_{t,i} r_{t,i+h}] = \begin{cases} \sigma_{t,i}^2/n_t + 2\eta^2 & \text{for} \quad h = 0 \\ -\eta^2 & \text{for} \quad h = 1 \\ 0 & \text{for} \quad h \geq 2 \end{cases} \tag{10.20}$$

where the term $1/n_t$ originates in the variance of the price innovations. The combination $r_{t,i}^2 + r_{t,i-1} r_{t,i} + r_{t,i} r_{t,i+1}$ is an estimator for $\sigma_{t,i}^2/n_t$, as proposed by [149]. The properties of this estimator have been studied analytically by [78] and numerically by [46, 162].

Following the same path, the estimator for the realized variance can be improved by using more lags between returns. An estimator for the lagged covariance during the day t is defined by

$$\gamma_h = \frac{n_t}{n_t - h} \sum_{i=1}^{n_t - h} r_{t,i} r_{t,i+h} \tag{10.21}$$

for $h > 0$, and with a similar definition for $h < 0$ (the prefactor is chosen such that all the terms are proportional to n_t, regardless of h). The kernel-based estimator for the realized variance is defined by

$$\mathrm{RV}_t^{\mathrm{kernel}} = \sum_{i=1}^{n_t} r_{t,i}^2 + \sum_{h=1}^{H} k\left(\frac{h-1}{H}\right)(\gamma_h + \gamma_{-h}). \tag{10.22}$$

The weight function $k(x)$ with $0 \le x \le 1$ is such that $k(0) = 1$ and $k(1) = 0$. The definition of [149] is obtained for $H = 1$ (for which $k(x)$ is irrelevant). Some choices for the weight function are $k(x) = 1 - x$ (Bartlett), $k(x) = (1 - x)^2$ (second order), $k(x) = 1 - x^2$ (Epanechnikov), or $k(x) = (1 - \cos(\pi(1 - x)^2))/2$ (Tukey–Hanning). This family of estimators has been studied extensively by [78] and by [16], with some propositions for the choice of $k(x)$ and H. More papers on this topic are summarized in [104].

Another direction is to use simultaneously consecutive returns, as proposed by [47]. Let us consider the vector of M consecutive returns $R_{t,i} = (r_{t,i}, r_{t,i-1}, \ldots, r_{t,i-M+1})'$. For the process with additive noise used above, the covariance matrix for R is given by

$$\Omega = \begin{bmatrix} \sigma^2/n_t + 2\eta^2 & -\eta^2 & & & \\ -\eta^2 & \sigma^2/n_t + 2\eta^2 & -\eta^2 & \ddots & \\ & & \ddots & \ddots & -\eta^2 \\ & & & -\eta^2 & \sigma^2/n_t + 2\eta^2 \end{bmatrix} \tag{10.23}$$

and with zeros at the empty locations. The eigenvalues of the covariance matrix are

$$\lambda_m = \sigma^2/n_t + 4\eta^2 \sin^2\left(\frac{\pi m}{2(M+1)}\right) \qquad m = 1, \ldots, M \tag{10.24}$$

with the corresponding eigenvectors

$$\varphi_m(k) = \sqrt{\frac{2}{M+1}} \sin\left(\frac{\pi m k}{M+1}\right) \qquad k = 1, \ldots, M. \tag{10.25}$$

The projection on this orthonormal basis is known as a discrete sine transform (DST). The eigenvalue with $m = 1$ is the smallest, and approximately $\lambda_1 = \sigma^2/n_t + \eta^2\pi^2/(M+1)^2$, namely the noise term is reduced by $\sim 1/M^2$ (whereas aggregation reduces the noise term by $1/k$). An estimator for the realized variance is given by the intra-day sum of the first eigenvalue, with

$$\mathrm{RV}_t^{\mathrm{DST,min}} = \mathrm{IV}_t + 4\eta^2 n_t \sin^2\left(\frac{\pi}{2(M+1)}\right). \tag{10.26}$$

Although biased, this estimator of the realized variance has a reduced noise term.

The eigenvectors $\varphi_m(k)$ have the interesting property to be independent of the process, namely they are independent from σ_t, n_t, or η. Therefore, the first eigenvector can be used to project the returns so as to reduce the noise term

$$r'_{t,j} = \sum_{k=1}^{M} \varphi_1(k) r_{t,j-k+1} \tag{10.27}$$

and the variance of r' corresponds to the fist eigenvalue. Because of the fixed eigenvectors, the DST variance estimator can be computed easily, without diagonalization of the covariance matrix. As the first eigenvector is the sinus function over half a period, the largest contributions in the sum are at $k \sim M/2$, namely in the middle of the window. Therefore, this filter tends to lag the recent behavior of the returns when removing the noise. This is unimportant for a variance computation, but a drawback for filtering prices or returns.

Similarly to the multi-scale variance estimators presented in the previous section, the bias of the previous variance estimator can be removed by using multiple values for M. The M dependency is summarized as

$$\mathrm{RV}_t^{\mathrm{DST,min}} = \mathrm{IV}_t + \mathcal{N}(M)\eta^2 \qquad \text{with} \quad \mathcal{N}(M) = 4n_t \sin^2\big(\pi/2(M+1)\big). \tag{10.28}$$

A simple linear regression allows one to estimate separately IV_t and η^2, and the intercept is an unbiased estimator $\mathrm{RV}_t^{\mathrm{DST,\,LR}}$ for the realized variance. This estimator has a very small variance, as studied numerically in [47] and in agreement with a theoretical argument based on the Cramer–Rao bound.

10.7 AR(1) Filter on the Prices

Both previous sections focused on measuring the volatility over one day using high-frequency data. For some applications like high-frequency trading or arbitrage, the notion of a prevailing price is relevant. The goal is to obtain a good approximation for $\tilde{p}$ using the available high-frequency noisy prices p, by filtering the short-term noise so as to recover the underlying "true" price.

The return (10.4) computed over one tick for the model with the incoherent term is given by

$$r_i = \sigma_i \epsilon_i + u_i - u_{i-1} \tag{10.29}$$

where the index t has been dropped. This series of returns has a variance $E[r^2] = \sigma^2 + 2\eta^2$, a first lag covariance $E[r_i r_{i-1}] = -\eta^2$, and a zero covariance for larger lags.

An equivalent MA(1) representation is given by the series

$$r_i = w_i - \theta w_{i-1} = (1 - \theta L) w_i \tag{10.30}$$

with L the lag operator and where w_i are i.i.d. random variables. The representation (10.30) can be inverted to give

$$w = (1 - \theta L)^{-1} r. \tag{10.31}$$

The $(1 - \theta L)^{-1}$ operator is related to an exponential moving average (EMA), because $\mathrm{EMA}[\mu] = (1 - \mu)(1 - \mu L)^{-1}$, as seen in Eq. (2.11). Therefore, the white noise term w can be computed by a simple EMA in tick time

$$w = (1 - \theta)^{-1} \mathrm{EMA}[\theta; r] \tag{10.32}$$

with the decay parameter θ. This suggests using an EMA filtering in order to remove the noise term.

The filtered series of returns r' is obtained from the noisy returns by an EMA

$$r'_i = \mathrm{EMA}[\mu; r]_i = (1 - \mu)\big(r_i + \mu r_{i-1} + \mu^2 r_{i-2} + \mu^3 r_{i-3} + \cdots\big). \tag{10.33}$$

The computation of the first two lagged moments of r' gives

$$\begin{aligned} E\big[r'^2\big] &= \frac{1-\mu}{1+\mu}\big(\sigma^2 + 2\eta^2(1-\mu)\big) \\ E\big[r'_i r'_{i-1}\big] &= \frac{1-\mu}{1+\mu}\big(\sigma^2 \mu - \eta^2(1-\mu)^2\big). \end{aligned} \tag{10.34}$$

The goal is to set the EMA parameter μ so that the underlying variance σ^2 is obtained

$$E\big[r'^2\big] = \sigma^2. \tag{10.35}$$

This condition, combined with (10.34), leads to the relation $\eta^2(1 - \mu)^2 = \sigma^2 \mu$. This expression appears on the right-hand side of $E[r'_i r'_{i-1}]$, namely when μ is set so as to obtain the underlying variance, then the first lagged covariance is null. By expanding this last relation, a quadratic equation for μ is obtained:

$$\mu^2 \rho(1) + \mu + \rho(1) = 0 \tag{10.36}$$

with $\rho(1) = -\eta^2/(\sigma^2 + 2\eta^2)$ the correlation at lag one. The solution for μ is

$$\mu = -\frac{1}{2\rho(1)}\Big(1 - \sqrt{1 - 4\rho(1)^2}\Big). \tag{10.37}$$

For $\rho(1) = \eta^2 = 0$ (no noise), the solution is $\mu = 0$, namely no EMA is needed. For the value $\gamma = \eta^2/\sigma^2 = 2$ as measured typically with empirical time series, the first lag correlation is $\rho(1) = -40$ %, leading to $\mu = 1/2$. These values are such that $\mu \ll 1$; they show that the EMA has always a fairly short range (in a tick time scale).

Because the EMA operator is a function of L, filtering the return is equivalent to filtering the price. Therefore, a tick-by-tick price time series, filtered from the incoherent component, can be computed by

$$\tilde{p} = \mathrm{EMA}[\mu; p] \tag{10.38}$$

with the same value for μ.

When prices or returns are needed, the EMA filter must be used. This filter for the prices relies on the validity of the model with the incoherent component and on a good estimation for $\rho(1)$. The parameter $\hat{\mu}$ must be estimated from the first-lag correlation $\hat{\rho}(1)$ of the return, together with Eq. (10.37). The estimation for $\hat{\rho}(1)$ should be done on a long enough sample, so as to reduce the statistical error. For example, this can be done efficiently by using an MA operator on a moving sample (see Sect. 2.5). In practice, both the deviations from the theoretical model and the statistical errors on the estimator lead to an imperfect extraction of the underlying prices.

The variance of the filtered returns r' can also be computed, in order to obtain an estimate for σ^2. Yet, the methods presented in the previous two sections seem to be more robust against a misspecification of the model. More details, mostly from numerical simulations, can be found in [46, 47, 162].

10.8 Tick Time Sampling and the Volatility per Tick

So far, the choice of the sampling scheme for the prices has been left for the most part open. Two choices are natural. A first possibility is to sample the price at regular time intervals according to a chosen time scale, which can be a physical time scale or a business time scale. Essentially, this is an intra-day extension of the familiar daily close price, possibly taking into account the intra-day seasonality. A second possibility consists in using the high-frequency tick stream, by taking a new price either each time the prevailing mid-price changes, or each time a transaction is made, or each time the order queue is modified. This is a fairly natural choice as all the available information is used. The main difference between both time scales is the number of ticks, fixed in the first case, depending on the market condition in the second case.

The realized variance (10.11) for a given day can be rewritten as

$$\mathrm{RV}_t = n_t \cdot \frac{1}{n_t} \sum_{i=1}^{n_t} r_{t,i}^2. \tag{10.39}$$

When using a tick time scale, this formulation separates the variance into the extensive quantity n_t and an intensive measure of the variance per tick. This separation suggests the definition for the *volatility per tick* for a given day t by

$$\left(\sigma_t^{\mathrm{tick}}\right)^2 = \frac{1}{n_t} \sum_{i=1}^{n_t} \left(10^4 r_{t,i}\right)^2 \tag{10.40}$$

and is given in BIPS. The realized variance becomes

$$\mathrm{RV}_t = c n_t \left(\sigma_t^{\mathrm{tick}}\right)^2 \tag{10.41}$$

where c is a constant depending on the chosen reference time horizon for RV. Several remarks can be made.

- Following the setting used in this chapter, the definitions are given for the one-day integrated variance, but the same decomposition can be done for any time interval. For the applications to short time intervals, the separation into extensive and intensive quantities suggests measuring the tick volatility over a longer time interval in order to have a better statistical estimator.
- The volatility per tick measures the typical price change per tick and depends mostly on the details of the double auctions dynamics, like order flows, spread, volume, price granularity, impact function, etc. Its value depends also on the precise definition of "one tick" (modification in the queue, transaction, or change at the best bid and ask prices). Depending on the context, definition (10.40) is called volatility per trade.
- Definition (10.40) is subject to the microscopic noise discussed previously in this chapter. In practice, the estimators discussed in Sects. 10.5 and 10.6 should be used in order to avoid an important upward bias.
- Empirically, a strong positive correlation is observed between the spread and the volatility per tick [30, 151]. This can be understood by an adverse selection mechanism, namely agents posting limit orders become less aggressive when the volatility is high because of fears against more informed agents in the market. The impact of market orders was studied in [30], and a simple model was developed in order to explain the relation between spread and volatility per tick.

  Such a connection relates the microscopic behavior of the trading orders with the volatility per tick. It is a first step toward an explanation of the mesoscopic volatility by the microscopic dynamics and the trades. Clearly, much more work is required in order to construct a clear understanding of the volatility per tick, its dynamics and the relations with other quantities. A point made by [28] for stocks is that the markets operate in a regime of vanishingly small revealed liquidity, namely the volume in the queues and the trades represent a very small fraction of the total number of stocks (of the order of 10^{-5} to 10^{-3}). Such values point to a largely endogenous price formation mechanism, instead of the efficient market hypothesis which is driven by the fundamental economics of the firms and the related news. The very interesting paper [28] discusses extensively this point, see also the discussion about w_∞ in Sect. 7.17.
- The specification of a model in tick time requires to define a process for the arrival time, a process for the underlying price $\tilde{p}$, and the relation between p and $\tilde{p}$. The three processes and their possible dependencies clearly increase the complexity of the problem. For example, the heteroscedasticity can by captured in the tick arrival process, with short (long) time intervals between ticks which are more likely to be followed by short (long) intervals between ticks. Together with a simple constant volatility process for the return, this model already captures the heteroscedasticity on the returns sampled at a fixed time interval (say daily). But part of the heteroscedasticity can also be included in the return process. This joint description has been poorly explored so far, either for the empirical analysis of high-frequency data or for the analytical model using processes.

Chapter 11
Time-Reversal Asymmetry

11.1 Statement of the Problem

Time-reversal invariance (TRI) is a very important concept in science. The idea can be summarized as follows: when a sequence of events is viewed starting from the end, namely with the arrow of time reversed, is it possible to measure a difference compared to the normal time ordering? A rigorous formulation of time reversal invariance is that the transformation $t \rightarrow -t$ is an exact symmetry of the system under consideration. The basic laws of physics are time-reversal invariant (Newton equation for mechanic, Maxwell equations for electromagnetism, Einstein equation for general relativity, Schrödinger and Dirac equations for quantum mechanics, etc.), but the macroscopic world is clearly not time-reversal invariant. This paradox was solved by thermodynamics and the increase of entropy. The same question can be asked about finance, namely if a time series of prices originating in the financial market is time reversed, can we "see" the difference? In order to see the difficulty of the problem, four time series are given on Fig. 11.1. One of the series has been "time reverted", namely the data are presented with the time flowing backward from left to right. Can the reader spot which one? This quiz is clearly difficult because financial time series are dominated by randomness, and the appropriate formulation of this question is whether statistics can show the presence, or absence, of time-reversal invariance.

In a physics context, [123] studied this question for the correlation functions of a system with fluctuation. For a system in a stationary state and for a quantity $x(t)$, all the correlation functions like $E[x(t)x(t+\delta t)]$ are time-reversal invariant. In order to have a quantity sensitive to the direction of time, Pomeau introduces correlation functions of the type

$$E\big[x(t)\big\{x(t+\delta t) - x(t+2\delta t)\big\}x(t+3\delta t)\big]$$

and studied under which conditions they are different from zero.

In an economic framework, [130, 131] have already investigated the presence of time-reversal symmetry. Their idea is to search for differences between up and down

G. Zumbach, *Discrete Time Series, Processes, and Applications in Finance*, Springer Finance, DOI 10.1007/978-3-642-31742-2_11,

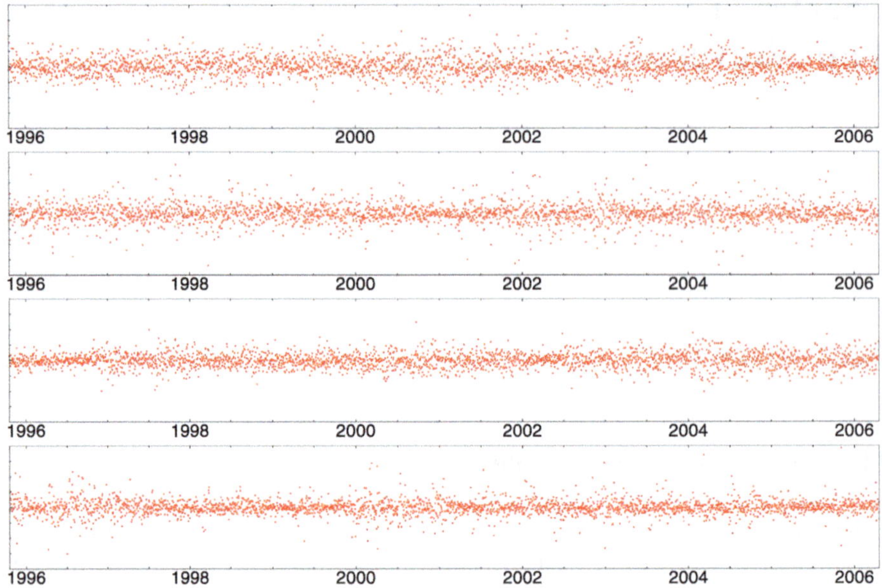

Fig. 11.1 Four daily time series for the annualized daily returns, with one of them time reverted

moves over long time horizons, using yearly economic indicators. This behavior is typically related to business cycles, for example, with long slow rises followed by abrupt decreases. The estimator that is used is given by $E[r^2(t)r(t+k\delta t)]$, where δt is the time increment of the time series, r the return, and $k \neq 0$ an integer index. As the amount of long-term economic data is fairly small, such studies need to rely on a careful analysis of the statistical properties of the indicators [131]. Other indicators have been proposed in [37, 70]. The salient results is that most of the economic time series are not time-reversal invariant, but for a small fraction of them, the null hypothesis of time-reversal invariance cannot be rejected.

In this chapter, adapted from [156], the focus is to study various statistics related to the volatility, for time horizons ranging from 3 minutes to 3 months. For foreign exchange rates, a symmetry between the exchanged and expressed currencies is plausible (at least for major free floating currencies). This symmetry occurs because an FX rate is a conversion factor between two numeraires, and not the price of a security expressed in a numeraire (like for an equity price or of a bond price). Under the exchange of the currencies, an exchange rate is transformed by price $\rightarrow$ 1/price, and the logarithmic returns by $r \rightarrow -r$. Notice that reversing the time induces the same transformation $r \rightarrow -r$ and the reverse ordering of the time series. Therefore, if the exchange of currencies is an exact symmetry, then all statistics that are odd[1] in the returns are zero, like, for example, $E[r^2(t)r(t+k\delta t)]$. The same argument implies that the return probability distribution in even, that is, $p(r) = p(-r)$. We have

[1] An even function is such that $f(-r) = f(r)$, an odd function such that $f(-r) = -f(r)$.

checked the empirical validity of this argument for $p(r)$ and found that it is likely incorrect. Even if ultimately incorrect, the argument points to a small asymmetry for such a term. In order to have a better signature of time irreversibility, a better track for foreign exchange data is to search for estimators that are even in the returns but sensitive to time reversal.

A general definition of TRI is given, for example, in [37]. In particular, they show that the distribution of the returns must be symmetric for a TRI series, and they construct a test based on this property using the characteristic function of the pdf. A more general construction along this line is as follows. Take a quantity σ such that $\sigma \to \sigma$ under the time reversal transformation. For example, σ can be the logarithmic price; in this book, σ is a volatility estimator computed from the returns. Then, the quantity $\Delta\sigma(t) = \sigma(t + \delta t) - \sigma(t)$ is odd under the time reversal transformation: $\Delta\sigma \to -\Delta\sigma$. If the series is TRI, then the distribution of $\Delta\sigma$ must be even. Another idea for testing TRI is based on the covariance or correlation between two quantities x and y. Essentially, the quantity $E[x(t)y(t+\delta t)]$ should be equal to $E[x(t+\delta t)y(t)]$ if the series is TRI, and a test can be constructed on the difference between these two quantities. Because of the symmetry between the arguments of the covariance or the correlation, the two quantities must be different in order to have a nontrivial test. Ramsey and Rothman [130] use r and r^2; we use volatilities σ with different parameters.

The statistics sensitive to TRI are based on various estimators of the volatility (i.e., even in the returns) and are essentially measuring time reversal for volatilities computed with information before t and after t. Three statistical measures are developed in Sect. 11.2, and they all show that empirical data are clearly not time-reversal invariant.

Using the same statistical tool, various processes that should mimic empirical data are investigated. For example, a simple Gaussian random walk is time-reversal invariant, and the three statistics are zero. Much more interesting are the processes from the ARCH family and from the stochastic volatility family. Some of the ARCH-type processes can reproduce the empirical figures, but all the simulated stochastic volatility processes are time-reversal invariant. These results demonstrate that the last family of processes cannot describe some stylized facts of financial time series. It is an important result as TRI statistics allow us to select between models that are structurally very different. The usual method for model selection is to nest the processes and to show that the corresponding parameter is significantly different from zero. But this method is not generally possible for processes that are so widely different as ARCH and stochastic volatility. Besides, the systematic comparison of the results between empirical data and processes allows us to glimpse the origin of time asymmetry.

As these statistics are all based on volatilities, an efficient deseasonalisation procedure is crucial for the present computations. With high-frequency data, the use of a "naive" 3 minutes regular sampling in physical time (after the week-ends have been removed) would mainly show the strong daily pattern related to the opening and closing of the various markets around the world. Such a strong seasonality would hide other interesting stylized facts. Therefore, the use of a proper business time is essential to extract the interesting signals. With daily data, the usual business time (Monday through Friday, no data on the week-end) is enough to eliminate the weekly seasonality. A simple computation using daily time series can be done, with the volatility over n-day computed by a simple sum of squared daily returns for n consecutive days (this is a "microscopic" volatility definition). This simple procedure applied to stocks shows the same asymmetry as observed with high-frequency data, but the resolution at shorter time interval is limited.

11.2 Empirical Time Reversal Statistics

The changes of the volatility are measured by the volatility increment as defined in (2.3)

$$\Delta\sigma = \sigma_r - \sigma_h.$$

With a time-reversal transformation, the volatility increment changes by $\Delta\sigma(t) \to -\Delta\sigma(t)$. The probability density $p(\Delta\sigma)$ of the volatility increment can be estimated, and its asymmetry $a_p(\Delta\sigma) = p(\Delta\sigma) - p(-\Delta\sigma)$ gives a measure of the time irreversibility. This asymmetry is a quantitative measure of the following intuitive perception of the price dynamics. A shock on the market (for example, due to the arrival of an important piece of news) produces a sudden increase of volatility, followed by a slow relaxation toward a normal level of volatility. For the time-reversed series, this corresponds to a slow volatility increase followed by a sudden relaxation to the normal, and the distribution of volatility increments is $p(-\Delta\sigma)$. The asymmetry $a_p(\Delta\sigma)$ measures the asymmetry in the dynamics between the original and time-reversed series. Figure 11.2 shows the probability distribution $p(\Delta\sigma)$.

The pdf appears to be symmetric at first glance. It shows that our intuitive perception as described above is exaggerated. Yet, a detailed examination of $p(\Delta\sigma)$ reveals the expected asymmetry around $\Delta\sigma \simeq 0.02$. Figure 11.3 displays the asymmetry $a_p(\Delta\sigma)$, and a fairly consistent symmetry breaking pattern for various empirical time series is observed. The negative values for $\Delta\sigma \lesssim 0.05$ correspond to the "return to the normal" or to a larger probability for small negative volatility increments. There is a corresponding larger probability for large volatility increments $\Delta\sigma$ (that is, the arrival of news or shocks) that translate into positive values for a_p. Because the volatility is stationary, a simple empirical first moment like $\langle\Delta\sigma\rangle$ converges toward zero with the inverse of the sample size. This shows that more complex statistics should be used to reveal this asymmetry.

The second statistics involves the correlation of volatilities at various time horizons. The correlation between historical and realized volatility, as introduced in (3.6.3), is

$$\rho_\sigma(\delta t_\sigma, \delta t'_\sigma) = \rho\big(\sigma_h[\delta t_\sigma, \delta t_\sigma/24](t), \sigma_r[\delta t'_\sigma, \delta t'_\sigma/24](t)\big)$$

where on the right-hand side $\rho(x, y)$ is the usual linear correlation between two time series x and y. Essentially, this quantity measures the dependency between

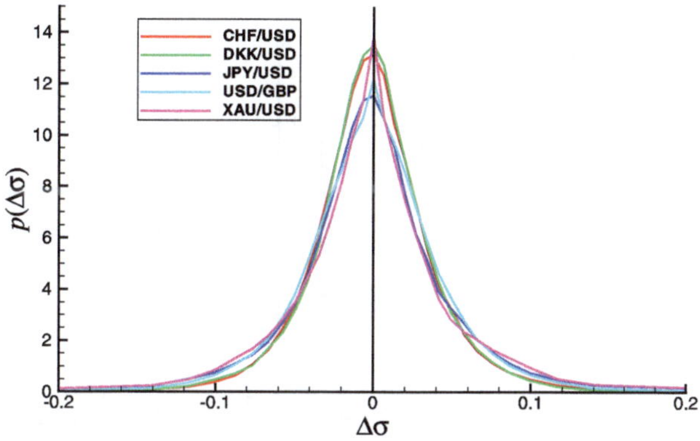

Fig. 11.2 Probability density for the volatility increment $\Delta\sigma$, for the foreign exchanges CHF/USD, DKK/USD, JPY/USD, USD/GBP and for gold XAU/USD. The time horizons are $\delta t_\sigma = 1$ day and $\delta t_r = \delta t_\sigma/24$. Similar figures are obtained for other time horizons

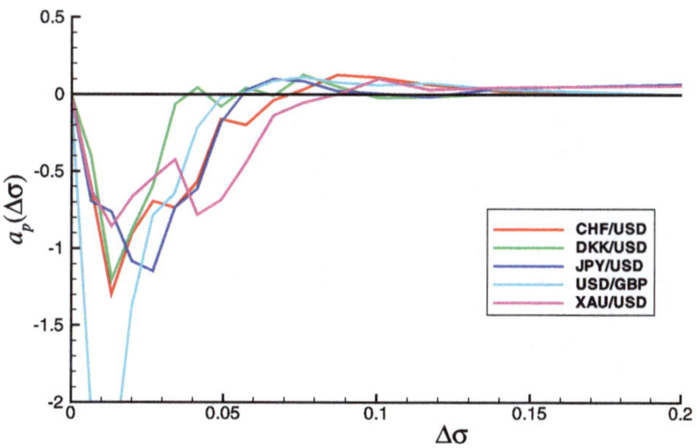

Fig. 11.3 Asymmetry of the probability density for the volatility increment $\Delta\sigma$. The parameters and time series are as for Fig. 11.2

past and future volatilities at the respective time horizons δt_σ and $\delta t'_\sigma$. Overall, this correlation is asymmetric with respect to the exchange of δt_σ and $\delta t'_\sigma$, as can be seen in all the empirical mug shots. As this exchange is directly related to the time-reversal symmetry, a *historical versus realized volatility correlation asymmetry* is defined by

$$a_\sigma(\delta t_\sigma, \delta t'_\sigma) = \rho_\sigma(\delta t_\sigma, \delta t'_\sigma) - \rho_\sigma(\delta t'_\sigma, \delta t_\sigma)$$

and $a_\sigma \simeq 0$ indicates the time-reversal invariance. The natural representation for $a_\sigma(\delta t_\sigma, \delta t'_\sigma)$ is in a two-dimensional plane, and this figure is given in each mug shot on the third row, on the right. The analysis of a_σ for the above empirical time series reveals a fairly consistent pattern, with a maximum of order 6 to 12 % for $\delta t_\sigma \simeq 1$ week and $\delta t'_\sigma \simeq 6$ hours. The overall behavior can be summarized in a one-dimensional cut, as given in Fig. 11.4 for empirical data and processes. The empirical data show a distinct and fairly consistent pattern indicating a clear asymmetry with respect to time reversal.

The third statistics involves correlations of past and future volatilities at the same time horizon δt_σ but different granularities δt_r and $\delta t'_r$, respectively. The volatility correlation sensitive to the granularity has been introduced in (3.9) and is included

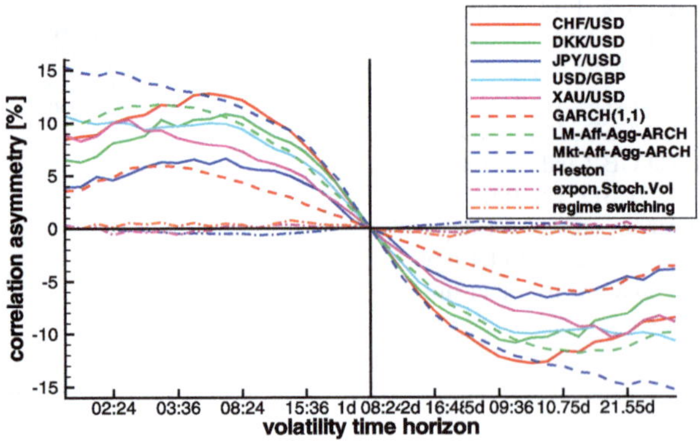

Fig. 11.4 The measure of asymmetry $a_\sigma(\delta t_\sigma, \delta t'_\sigma)$. The parameter $\delta t'_\sigma$ is given on the horizontal axis, the value of δt_σ is the symmetric through the vertical axis (for example, for $\delta t'_\sigma = 8$ h 24 corresponds $\delta t_\sigma = 5$ day 9 h 36)

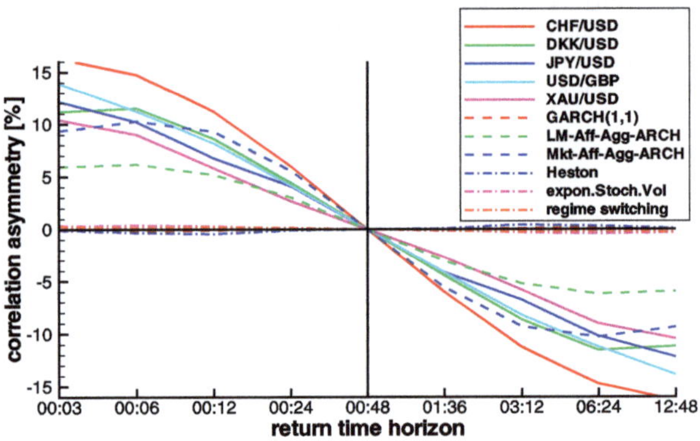

Fig. 11.5 The measure of asymmetry $a_{\text{gr}}(\delta t_r, \delta t'_r)$. The parameters are: $\delta t_\sigma = 2^9 \cdot 3$ minutes $= 1$ days 1 h 36; $\delta t'_r = 2^n \cdot 3$ minutes is given on the horizontal axis; $\delta t_r = 2^{8-n} \cdot 3$ minutes is the symmetric value through the middle point (for example, to $\delta t'_r = 24$ minutes there corresponds $\delta t_r = 1$ h 36)

in each mug shot on the fourth row. For a given δt_σ, the "granularity" dependency was defined by (3.9)

$$\rho_{\text{gr}}(\delta t_r, \delta t'_r) = \rho\big(\sigma_h[\delta t_\sigma, \delta t_r](t), \sigma_r[\delta t_\sigma, \delta t'_r](t)\big)$$

with $\delta t_r \leq \delta t_\sigma$, $\delta t'_r \leq \delta t_\sigma$. As for the second statistics, the exchange of δt_r and $\delta t'_r$ is related to the time-reversal symmetry, and a similar measure of *granularity asymmetry* is defined by $a_{\text{gr}}(\delta t_r, \delta t'_r) = \rho_{\text{gr}}(\delta t_r, \delta t'_r) - \rho_{\text{gr}}(\delta t'_r, \delta t_r)$. The computation of a_{gr} for empirical time series at $\delta t_\sigma \simeq 2$ days shows a consistent and systematic asymmetry, with values in the range of 12 to 18 %. The natural representation is in the two-dimensional space $(\delta t_r, \delta t'_r)$, and a one-dimensional cut is displayed on Fig. 11.5. The asymmetry pattern is clear and consistent between the empirical data. Therefore, the three measures of asymmetry deliver a consistent message, namely foreign exchange time series are not time-reversal invariant, and the asymmetry is quantitatively small.

11.3 TRI in Theoretical Processes

The same statistics can be computed for theoretical processes, using Monte Carlo simulations. The simplest process is a Gaussian random walk that is exactly time reversal invariant (the proof follows from the independence of the increments that allows one to reorder terms under the expectations). The more interesting processes include heteroscedasticity, either with an ARCH form or with a stochastic volatility term. The process parameters are the values that reproduce best the overall figures in the mug shots, as given in the previous chapters, but they have not been optimized further to reproduce the statistics sensitive to time-reversal invariance. A simple exploration around the given process parameters shows that the qualitative features related to TRI are insensitive to the exact parameters values.

Generically, the mug shots for the ARCH processes show that they are not time-reversal invariant. The simplest GARCH(1, 1) process exhibits an asymmetry according to the first measure a_p and the second measure a_σ, but the last measure a_{gr} displays no asymmetry (see the mug shot Fig. 7.1 and Figs. 11.3, 11.4, and 11.5). The multiple time horizon ARCH processes can reproduce quantitatively the above three measures of time irreversibility. As can be seen by comparing the mug shots, a key ingredient to put in these multiscale processes is that the return time horizons δt_r must increase with the volatility time horizon δt_σ. This corresponds to the intuition that short-term intra-day traders use tick-by-tick data, whereas long-term fund managers use daily data. When δt_r is kept at the process time increment, as in the microscopic ARCH process, the time reversal asymmetry is too small, or even zero for the third measure. Taking a volatility granularity of order $\delta t_\sigma = 24 \cdot \delta t_r$ gives roughly the correct quantitative time-reversal asymmetry in these multiple-component processes. On the other extreme, taking $\delta t_\sigma = \delta t_r$ produces too much asymmetry, as for the long-memory pure-aggregated ARCH process (see Sect. 7.10). Clearly, the process parameters can be chosen in order to reach a better numerical agreement with the empirical values (no such optimization has been done so far), although the aggregated ARCH processes are already very good.

The stochastic volatility processes have been introduced in Chap. 8. Two such models with a long-memory volatility cascade are labeled in the graphs and tables as "LM stoch. vol." and "LM Heston". Yet, all these SV processes show time-reversal invariance (up to the statistical errors of the numerical simulation, see Figs. 11.3, 11.4, 11.5 and the corresponding mug shots).

The regime-switching processes are explained in Chap. 9. The transition probabilities $\Pr(i \to j)$ are strongly constrained to give realistic distribution for the volatility. With asymmetric transition probabilities $\Pr(i \to j) \neq \Pr(j \to i)$ and for a number of states larger than two (see Chap. 9), the process is not time-reversal invariant according to the measures a_p and a_σ, as is visible on Fig. 11.6. Yet, the asymmetry is much smaller than the empirical observed values, and it is not possible to modify the transition probabilities to get simultaneously large asymmetry and realistic pdf. Finally, the coarse/fine graining measure of asymmetry a_{gr} for the regime-switching process is compatible with zero, up to Monte Carlo statistical fluctuations. Because of the lack of feed-back of the return on the volatility, one

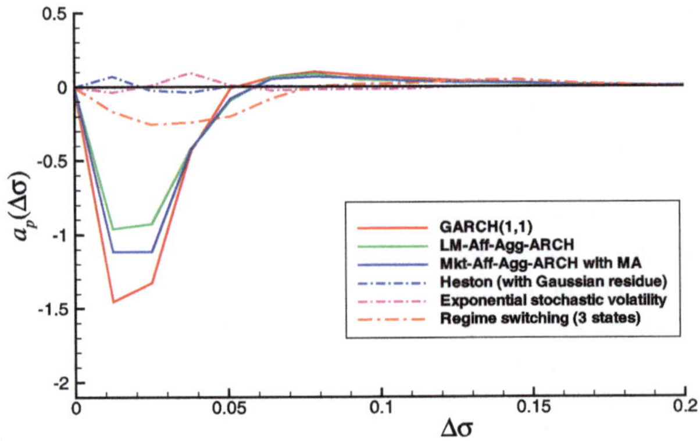

Fig. 11.6 Asymmetry of the probability density for the volatility increment $\Delta\sigma$ for a few processes

can understand intuitively that the stochastic volatility and regime switching cannot include an asymmetry in the third measure of invariance a_{gr} using the volatility granularity.

11.4 Test Statistics

So far, the focus was to construct statistics sensitive to time reversal invariance, to understand what they are measuring and to show the differences between empirical time series and various processes. Yet, definitive conclusions require to use rigorous test statistics. Essentially three methods can be followed in order to obtain distribution information on a statistics: analytical, bootstrap, and Monte Carlo simulations. In order to select the most appropriate method for the asymmetry statistics, one should keep in mind that our statistics are based on volatilities and that volatilities have slowly decaying correlations. The bootstrap method is appropriate only when the data are independent, or possibly when the dependencies decay exponentially fast. Clearly, the slow decay of the correlations rules out a bootstrap approach. Similarly, the analytical approaches are using a convergence toward a limit law, like a Gaussian. The convergence needs to be sufficiently fast, and similar conditions are imposed on the data as for the bootstrap method. Moreover, the asymmetry statistics are not based on simple moments, but on probability distributions and correlations. Both problems makes the analytical approach fairly difficult. Therefore, we have used Monte Carlo simulations to compute numerically the distributions for the asymmetry statistics and the related p-values. The computed p-values are the probability that the statistics is lower than or equal to zero. For a symmetric distribution, we expect the p-value to be around 0.5.

In order to obtain the distribution of the asymmetry statistics for a given process, 200 Monte Carlo simulations are performed. For each simulation, the process is simulated with a time increment of 3 minutes and for a time length identical to the available empirical data, namely 11.5 years. At the end of the simulation, the asymmetry statistics are computed. This is repeated 200 times, and the empirical distributions for the asymmetry statistics are computed. The cumulative probability that a statistic is smaller than or equal to zero can be easily estimated and gives the desired p value. This approach is simple and gives finite sample information, but relies on the process that should reproduce sufficiently well the empirical data.

The distribution statistics can be obtained for given values of the arguments, say, for example, $a_p(\Delta\sigma)$ for a given $\Delta\sigma$. Yet, the sensitivity of the TRI test can be enhanced by simple integration of the above statistics (similarly to a portmanteau statistics for the lagged correlation). The *volatility increment asymmetry* is defined by

$$A_p(\Delta\sigma) = \frac{1}{n} \sum_{0<\Delta\sigma_k<\Delta\sigma} a_p(\Delta\sigma_k) \tag{11.1}$$

where $\Delta\sigma_k$ are the points on the histogram sampling grid and where n is the number of terms in the sum. In the computations below, the upper bound is $\Delta\sigma = 0.06$. For the historical/realized volatility correlation asymmetry, two possibilities are

$$A_{\sigma,\mathrm{tot}} = \int_{\delta t_{\min}}^{\delta t_{\max}} d\delta t_\sigma \int_{\delta t_{\min}}^{\delta t_{\max}} d\delta t'_\sigma \, a_\sigma(\delta t_\sigma, \delta t'_\sigma) \tag{11.2}$$

$$A_{\sigma,\mathrm{cut}} = \int_{\delta t_{\min}}^{\delta t_{\max}} d\delta t_\sigma \, a_\sigma\bigl(\delta t_\sigma, f(\delta t_\sigma)\bigr). \tag{11.3}$$

The first quantity computes the total asymmetry in the historical/realized volatility, the second one the asymmetry along the cut used for Fig. 11.4. In practice, the two statistics give very similar results. Table 11.2 reports the p-values for $A_{\sigma,\mathrm{cut}}$ as the values are related to Fig. 11.4. For the volatility graining asymmetry, the same two integrated measures can be constructed. Again, to stay close to Fig. 11.5, the statistics for

$$A_{\mathrm{gr,cut}} = \int_{\delta t_{\min}}^{\delta t_{\max}} \delta t_r \, a_{\mathrm{gr}}\bigl(\delta t_r, f(\delta t_r)\bigr) \tag{11.4}$$

are reported.

The main statistics are reported in Tables 11.1 to 11.3. For the historical/realized volatility correlation asymmetry, the probability distributions for the processes are given in Fig. 11.7. The p-statistics confirm fully the above picture: the empirical data are compatible only with multicomponent ARCH processes. Notice that the process parameters have not been optimized to reproduce quantitatively the empirical asymmetry. A better agreement can be obtained for the multicomponent ARCH processes as they contain parameters that control the asymmetry. This is not possible for other stochastic volatility processes, say, for example, with respect to the volatility graining asymmetry: regardless of the parameter values, the p-values stay close to $1/2$. Notice also that a simple GARCH(1, 1) process can reproduce the asymmetry for a_p, but the asymmetry for a_σ is too small, and the asymmetry for a_{gr} is not reproduced at all.

Table 11.1 The total measure A_p for the asymmetry of the probability density of $\Delta\sigma$ at a time horizon of $\delta t_\sigma = 1$ day. The columns give respectively the mean, standard deviation, and p-value for A_p

	Mean	stdDev	p-value
CHF/USD	0.64		
DKK/USD	0.39		
JPY/USD	0.65		
USD/GBP	0.95		
XAU/USD	0.63		
GARCH(1, 1)	0.19	0.083	0.014
LM-Aff-Agg-ARCH	0.26	0.082	0.015
Mkt-Aff-Agg-ARCH	0.39	0.1	0.0
exp stoch.vol.	0.062	0.12	0.35
exp LM stoch.vol.	−0.02	0.09	0.48
Heston	0.008	0.075	0.36
LM Heston	0.02	0.1	0.50
Regime switching	0.23	0.11	0.015

Table 11.2 The integrated measure of asymmetry for the historical/realized volatility correlations $A_{\sigma,\text{cut}}$ along a cut in the $(\delta t_\sigma, \delta t'_\sigma)$ plane

	Mean	stdDev	p-value
CHF/USD	0.1		
DKK/USD	0.09		
JPY/USD	0.05		
USD/GBP	0.09		
XAU/USD	0.08		
GARCH(1, 1)	0.007	0.02	0.42
LM-Aff-Agg-ARCH	0.14	0.01	0.0
Mkt-Aff-Agg-ARCH	0.15	0.01	0.0
exp stoch.vol.	0.0009	0.01	0.50
exp LM stoch.vol.	−0.0016	0.01	0.52
Heston	0.0005	0.007	0.49
LM Heston	−0.001	0.01	0.54
Regime switching	0.004	0.01	0.34

Let us summarize the situation. One statistical test applied to one empirical time series gives one number. In order to have an error bound or a confidence interval on this number, additional hypothesis are needed. Most of the time, the additional hypothesis is independent variable, or possibly independent and Gaussian variable (i.e., the sample is made of i.i.d. random (Gaussian) variable). Yet, because of the long-range correlation of the volatility, such simple hypotheses are invalid in our context. Therefore, we turn to a more complex hypothesis, using

Table 11.3 The integrated measure of asymmetry for the volatility graining correlation $A_{\mathrm{gr,cut}}$ along a cut in the $(\delta t_r, \delta t_r')$ plane

	Mean	stdDev	p-value
CHF/USD	0.13		
DKK/USD	0.1		
JPY/USD	0.095		
USD/GBP	0.1		
XAU/USD	0.08		
GARCH(1, 1)	0.0008	0.007	0.48
LM-Aff-Agg-ARCH	0.1	0.01	0.0
Mkt-Aff-Agg-ARCH	0.11	0.01	0.0
exp stoch.vol.	−0.00007	0.008	0.58
exp LM stoch.vol.	−0.0007	0.006	0.54
Heston	0.00007	0.01	0.52
LM Heston	0.0003	0.007	0.59
Regime switching	0.0002	0.01	0.46

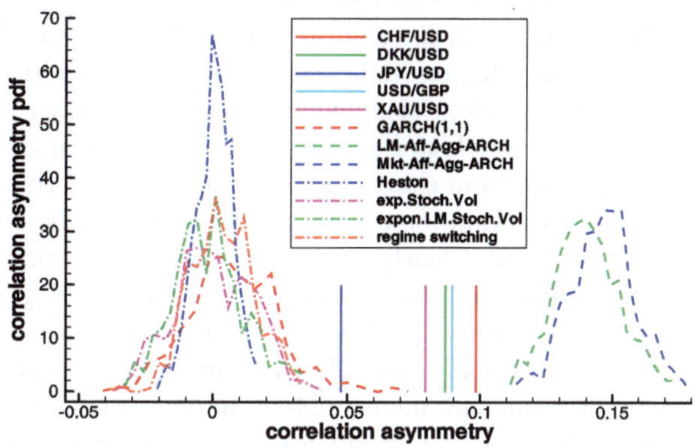

Fig. 11.7 The probability density for the asymmetry measure $A_{\sigma,\mathrm{cut}}$

processes that capture the lagged correlation of the volatility, either approximately (i.e., with an exponential decay) or accurately (i.e., with a long memory). The processes are sampled using Monte Carlo simulation with a sample length equal to the empirical sample size, and the asymmetry statistics are computed. By repeating this procedure, error bounds and p values on the statistics can be obtained for the different alternative hypothesis given by the various processes. As seen on Fig. 11.7, the error bounds on the asymmetry statistics are in the range ± 0.03 to ± 0.04, regardless of the process. As the five empirical values for the $A_{\sigma,\mathrm{cut}}$ statistics are in the range 0.05 to 0.10, *the hypothesis that the empirical time series are time-reversal invariant can be rejected*. Furthermore, the simulated distributions of the statistics are clearly observed to be around zero for some processes or exclude zero for some others. The computed p-values follow directly from the graph, with an almost complete dichotomy between distributions centered around zero and distribution excluding zero in their supports. Therefore, we can also reach strong conclusions about processes that can capture the observed non-TRI of the empirical data and those that are (likely) exactly TRI. In particular, all the stochastic volatility processes cannot reproduce the empirical asymmetry and can therefore be eliminated as possible models for the price data generating process.

11.5 Possible Origins of Time Irreversibility

With the help of both the empirical analysis and the process structures, we can speculate on the origin of the time asymmetry. The systematic study on the processes indicates that the time direction is set by the feed-back loop of the price changes on themselves through the intermediary of the price volatility, as captured by an ARCH process. Moreover, to obtain the measures of asymmetry with the right magnitudes, multiple time scales with aggregation must be used in an ARCH structure. This indicates that the interplay between the different time horizons as well as the feed back loop through the price are enough to create the observed time asymmetry. This argument emphasises the role of the prices as the only vector of information between market participants. By contrast, other "hidden" variables such as an independent stochastic volatility or market states are irrelevant with respect to TRI. This confirms the picture of markets segmented along the time horizons of the market participants. The asymmetry in historical versus realized volatilities points in the same direction, with a cascade from short time horizons to long horizons [92, 169].

Another relevant topic with respect to TRI is the connection between news and market behavior. Clearly, financial markets are open systems, driven by various external sources of information, like exchange of goods, politics, central banks, etc. Even if intuitively clear, the connection between "news" and financial markets is very difficult to establish from a purely empirical point. The two fundamental problems are the economic quantification of a "string of text" and the discounting by market participants of the "expected" portion of a news. With respect to TRI, the important question is whether the irreversibly is of external origin (i.e., from the news) or endogenous (i.e., created by the market participants or the trading rules). As such, this question is very difficult to investigate empirically, but the present study offers some indirect evidence. In all equations for the processes, there is at least one source of randomness, for example, in the return equation $r(t) = \sigma(t)\epsilon(t)$. Possibly, a process can include several sources of randomness, like in a stochastic volatility process. Essentially, the random variables $\epsilon(t)$ capture two different phenomena: the trades of the market participants (for example, through an order queue) and the influence of the external world (i.e., the news). The canonical hypothesis for the random processes ϵ is to assume an i.i.d. distribution (and independence at a given time with the other process variables like price, return, volatility, etc.). A debatable assumption is that of time series independence: it is plausible that the news are serially correlated, as an important piece of information is likely to be followed by more information. But this is difficult to establish directly, again because news are given by texts and not by numbers. In the investigation of processes, we follow the canonical hypothesis of source of randomness that are i.i.d. Within this hypothesis, two processes are particularly interesting with respect to TRI. In a regime-switching process, the "state of the world" $i(t)$ exhibits irreversibility, as controlled by the transition probabilities $\Pr(i \to j)$, and $i(t)$ has nontrivial serial correlations. For this process, a parallel can be established with the serial correlation of the news. Yet, it is not possible with a regime-switching process to reproduce the observed time asymmetry. On the other hand, within the framework of an i.i.d. source of randomness, the multiscale GARCH processes can provide a quantitatively correct description of the time irreversibility observed in the empirical time series. Together, both arguments indicate that the time irreversibility is mostly of endogenous origin, namely created by the interactions between market participants and not from external origin.

In a physical system, a common origin of time irreversibility is in friction or dissipation, and in the increase of entropy. A direct analogy of these quantities in a financial system should be taken with care. For example, the equivalent of friction can be thought as the bid-ask spread, namely the cost for buying, followed by selling directly afterward. However, in a liquid markets, and particularly in modern electronic trading exchanges, the market orders and limit orders play a very symmetric role. Using an argument based on the optimal choice of the market agents between both kind of orders, the authors in [144] show that the profit of a systematic market making trading strategy should be close to zero. Indirectly, this shows that the raw cost of trading represents a very small dissipation in the system. In the same direction, all the simulated processes do not include the cost of transactions, yet all the ARCH processes are not TRI. Both arguments show that the observed asymmetry in time does not originate in the bid-ask spread.

Notice that the above discussion on the origin of time irreversibility presents only indirect arguments, as only an investigation of the market microstructure and of the market participant decision processes can give direct evidence. Yet, such studies are clearly very difficult, and even more on large decentralized markets as foreign exchanges.

All the measures of asymmetry used so far are even in the returns, and essentially related to volatilities. The asymmetry in the return distribution is another measure of time asymmetry, which is odd in the returns. As discussed in the introduction, this symmetry is also related to the exchange of currencies for the foreign exchange time series. The empirical results for the return distribution asymmetry are reported in Table 11.4. All the processes have rigorously symmetric return distributions (because the returns are proportional to the residuals which have symmetric distributions). The results for the processes are in clear agreement with this symmetry, with p-values close to $1/2$. Interestingly, the asymmetry for the empirical data is much larger than for any processes (roughly at 1σ to 2.5σ from 0). This indicates that the argument on the exchange of currencies in an FX rate is likely incorrect. Clearly, more extensive investigations are needed to clarify the empirical facts and to develop processes that can reproduce the observed return asymmetry in the data.

Table 11.4 The total measure of the asymmetry of the probability density of r at a time horizon of 1 day. The columns give respectively the mean, standard deviation and p-value

	Mean	stdDev	p-value
CHF/USD	−0.21		
DKK/USD	−0.23		
JPY/USD	−0.11		
USD/GBP	−0.08		
XAU/USD	0.21		
GARCH(1, 1)	−0.009	0.07	0.54
LM-Aff-Agg-ARCH	−0.005	0.07	0.45
Mkt-Aff-Agg-ARCH	−0.0009	0.09	0.44
exp stoch.vol.	−0.002	0.08	0.42
exp LM stoch.vol.	0.0015	0.08	0.54
Heston	−0.002	0.075	0.50
LM Heston	0.002	0.08	0.55
Regime Switching	−0.005	0.08	0.59

11.6 The Analogy with Hydrodynamic Turbulence

The idea of a volatility cascade was first proposed in [74], using a formal analogy with fully developed turbulence. Using an analogy between hydrodynamic turbulence and price movements from foreign exchange markets over many time scales, the authors present an argument for the presence of turbulence in financial systems. They compare the pdf of the velocity increments in a turbulent fluid with the pdf of returns in foreign exchange markets, and analyze their scaling properties. They discovered a striking similarity between the two systems. From this empirical evidence, they propose that financial markets present a cascade of volatilities, from large to short time scales, similarly to hydrodynamic turbulence that presents cascade of eddies, from large to small sizes.

This conjecture was initially criticized for two reasons [97–99], namely:

1. financial returns have negligible autocorrelation, whereas changes in the velocity of turbulent particles are *anti-correlated*, and
2. Turbulence is characterized by an asymmetric pdf, while the foreign exchange returns that were used in the analogy had a symmetric pdf.

Further evidence of the turbulent cascade was presented in [9]. Using a wavelet decomposition of volatility over many time frames, they found evidence of the cascade from large to small time scales. Yet, this cascade was a smooth process, and the dominant time scales of the market agents have remained undetected.

Staying at the level of a formal analogy, the velocity increments in a turbulent fluid can also be compared to the volatility increment in a financial market. The pdf for the volatility increment exhibits a similar scaling behavior, with an asymmetric pdf. Moreover, they are negatively correlated. Besides, all of these papers discuss the analogy with fully developed turbulence, where the scaling theory developed by Kolmogorov applies [71]. In nature, turbulence generally displays some heterogeneous structures, at Reynolds numbers lower than needed for the onset of fully developed turbulence where the full symmetries of the Navier–Stokes equations are restore in average. In this heterogeneous regime, the energy of the turbulence is unevenly spread along the scales. The direct analogy for the heterogeneity in turbulence is given by the heterogeneity in the financial market, as display in our "mug shot". Therefore, a possibly better analogy is between the velocity increment in heterogeneous turbulence and the volatility increment in a financial market.

Yet, there is a major difference between the two systems in the way that their components interact. Turbulence occurs in 1 time and 3 spatial dimensions. The 3 spatial dimensions allow the formation of eddies that can interact in time. An eddy of a given size can split into two eddies of half the size, giving rise to the energy cascade, as first proposed by L.F. Richardson in 1922. Clearly, this is a cascade from one scale to the next smaller scale. In finance, there is only one time dimension. Therefore, the market participants can only interact in time, and when a trader with any time horizon decides to buy or sell, he will create volatility at the shortest time scales. In other words, the participants interact only when they trade, and it is an "instantaneous" interaction. Because every traders create instantaneous price impacts regardless of their original motivations and reference time horizons, the natural structure in finance is from short to long time scales. Therefore, even though there are some interesting analogies in the two systems, some properties of the cascades are fundamentally different because the two systems have different time-space dimensions. This basic difference induces a key difference in the mechanism underlying the cascade. In turbulence, the eddies cascade occurs from large to small space dimension, whereas in finance the cascade occurs in time, from short to long time horizons.

11.7 Further Thoughts on Time-Reversal Invariance

The systematic study of the time-reversal invariance in finance proves to be a very powerful investigation tool. The empirical time series are clearly not time-reversal invariant, according to three different measures. Although this fact is not obvious at the level of the return time series, this is not completely surprising as markets are driven by humans, who are clearly not time-reversal invariant. In particular, the market participants remember the past, and this memory creates an asymmetry in the time direction. More interesting is the fact that different FX rates show a clear and consistent quantitative pattern. On the modeling side, the ARCH processes can accommodate the empirical finding, using multiscale processes with increasing return time horizons δt_r for increasing volatility time horizons δt_σ. The simplest GARCH(1, 1) process can only reproduce some asymmetry, because it contains only one time scale.

On the other hand, all the stochastic volatility and regime switching processes are essentially time-reversal invariant, at odds with the empirical data. This deficiency is related to the structure of these processes and is therefore a key shortcoming of both classes of models (or at least the ones we have investigated). Yet, only a small sample of the possible processes has been implemented. Therefore, the question is if it is possible to find a stochastic volatility process or a regime-switching process that can also accommodate the empirical non-TRI. Yet, because of their core structure, from volatility to return without feed-back, it is at least impossible to incorporate the granularity asymmetry as measured by a_{gr} and A_{gr}. This is clearly a fundamental objection against both families of models.

Without detailed investigations of the ARCH, stochastic volatility, and regime-switching families of processes, it is quite difficult to differentiate them based only on their ability to incorporate some heteroscedasticity. Just based on the lagged correlation for the squared returns, the three families can produce a quite convincing dynamics for the volatility. In the literature, most articles do not compare the various models, but investigate only variations inside a given class. The book of [140] is an exception, which mentions that "it is very difficult [...] to say which of these two types of models provides the best description of market returns" (introduction, p. xviii) and favor stochastic volatility models (see, e.g., p. 76) based mainly on structural arguments. A tight statistical analysis leads us to a different conclusion.

At this point, the reader may want to give a second look at the teaser given at the beginning of this chapter in Fig. 11.1, before looking at the solution. From top to bottom, the time series are as follows.

1. The first time series is the EUR/USD, presented backward in time.
2. The second time series is a simulation of the exponential long-memory stochastic volatility process. This model is time-reversal invariant.
3. The third time series is the GBP/USD time series.
4. The fourth time series is a simulation of the long-memory aggregated ARCH process. This model reproduces all the observed asymmetry, and most of the stylized facts. The detailed statistics shows that the time asymmetry is too strong with these parameter values.

In fact, the figures have been ordered with respect to the time asymmetry, but clearly, they are difficult to differentiate without the appropriate statistics.

This focus for the time reversal study is on properties of the volatility, and we do not claim to have a "final" model for financial time series. Some known features are missing, for example, the correlation between past return and volatility (called the leverage effect) that is present for stocks and stock indexes. Still other terms should be added in a process to capture (even) statistical dependencies on the returns, but they will not change the main conclusion because adding more effects on the return (odd part) will essentially not influence the statistical properties of the volatility (even part).

Finally, even though the various theoretical processes have very different structures and in particular are not nested, TRI provides us with a strong selection criterion: a large number of processes *cannot* reproduce the stylized facts related to the time irreversibility observed in empirical time series. It seems very difficult to reach similar conclusion with the standard econometric approach using maximum likelihood and information criterion. The complex properties of a given process cannot be summarized into one number, and a range of statistics and graphs should be used in order to get a complete picture of the process properties.

Chapter 12
Characterizing Heteroscedasticity

12.1 Introduction

References [95] and [64] are the first to mention the conditional behavior of the return magnitude. The quantitative study of the heteroscedasticity of financial time series started in the seminal paper [59], soon followed by more theoretical processes [25, 61] and an estimation methodology for processes. The dynamics of the volatility is now better understood and is used extensively in risk evaluation or option pricing, to name just two broad areas of applications. But despite the acceptance of heteroscedasticity as a major stylized fact of financial time series, the existence of long memory is a lingering topic since 20 years, and a simple and broad characterization of the volatility memory is still lacking. Our goal for this chapter is to find a simple stylized description of the volatility lagged correlation that gives a good description of its shape for a broad set of time series, for lags ranging from a few days to a few years. Essentially, the volatility lagged correlation measures the dependency from the past volatility in a time series, and this information should be used to build processes and forecasts. Our approach in this chapter consists in staying strictly at the level of the statistical description of the lagged volatility correlation because of the many possible processes that can be built to capture the same or similar volatility structures.

Despite the practical importance of a simple characterization of the available information, few studies have been published on this subject. The large statistical errors on the estimated correlations for lags above a few months constitute one of the main difficulties. This is due to the limited sample lengths, typically of the order of 10 to 20 years. As the length of the empirical time series cannot be increased by an order of magnitude, the characterization of the lagged correlation in the range of one month to a few years stays difficult. Another issue is the absence of rigorous statistical comparison of two different decay shapes estimated on the empirical lagged correlation, say, for example, an exponential and a power law, both estimated with a χ^2 figure of merit. Since these shapes are not nested, there is no statistical testing procedure to accept or reject one shape against the other. These difficulties are reflected by the lack of literature: the author has indeed not been able to find

G. Zumbach, *Discrete Time Series, Processes, and Applications in Finance*, Springer Finance, DOI 10.1007/978-3-642-31742-2_12,

any article solely dedicated to the characterization of the volatility lagged correlation. Similarly, the book by [125] devoted completely to volatility forecasts does not contain any section on the empirical volatility correlation (or more generally on the lagged dependencies for the volatility), despite that the mere existence of a dependency is of core importance for its subject matter.

The traditional finance literature deals with the characterization of the lagged correlations mostly indirectly through process estimations, such as a comparison between the likelihoods for FIGARCH and GARCH. Through the process structure, these studies assume a given shape for the lagged correlation, which is for the vast majority a hyperbolic decay for long-memory processes [14, 53]. Moreover, log-likelihood estimations are weakly sensitive to long-term dependency because they are set at the shortest time interval, typically one day. Some of the econophysics literature also favor power law [90, 126]. Following the cascades and multifractal model initiated by [94], another thread in the literature favors a log-normal decay [9, 13, 112]. For these processes, the log decay shape follows from the assumed volatility cascade on nested scales generated by the tree-like structure. On the empirical side, [154] observed on a large data set that a log-decay shape provides a better description of the lagged correlations than a power law. Clearly, there is no consensus in the literature about a simple stylized fact characterizing the decay of the lagged correlation of the volatility.

A closely coupled topic is the optimal choice of the volatility and correlation estimators. The volatility σ is mainly given by the estimator $\sigma^2 = 1/n \sum r^2$ where n is the number of terms in the sum. For large n, the volatility estimator is better, but the sample size for the lagged correlation is reduced as a nonoverlapping volatility measure is obtained every n steps. The lag range is also reduced for increasing n, as the shortest lag without overlap is n. These trade-offs for the choice of n are fairly complex, and the choice of the optimal volatility estimator to be used as input for the lagged correlation is a nontrivial question. The correlation can also be computed by various estimators: the usual Pearson estimator is the most common, but robust estimators have their merits in particular with fat-tailed data. On the analytical side, the usual variance computation for the volatility and correlation estimators are based on Gaussian independent random variables, whereas financial time series have fat-tailed distributions and a long memory. As the analytical tools break down in a realistic setup, the question of the optimal estimators can only be investigated using Monte Carlo simulations, as presented in Sect. 12.5. The analysis is done both on sample sizes comparable to the available data lengths and on a sample long enough to be considered as infinite. This Monte Carlo study allows us to understand the optimal choice for n and the variance of the lagged correlation estimator, in a setup comparable to the empirical data.

Equipped with the optimal estimators, empirical time series are investigated in Sect. 12.6. A large set of 225 daily financial series is systematically estimated for four analytical shapes, and the cost functions are compared. This cross-sectional study allows us to get a broad view of the typical shape for the volatility lagged correlation. Another study is conducted on the DJIA time series, with data spanning one century (Sect. 12.7). The much larger sample size improves the correlation estimates, and the different analytical shapes can be compared. The implications of this analysis, in particular for processes, are presented in the last section. This chapter is based on the paper [159], where more details can be found.

12.2 Volatility and Correlation Estimators

12.2.1 The Volatility Estimator

The volatility $\sigma(t)$ at time t, measured over a span of n days, is given by

$$\sigma^2(t) = \frac{1}{n} \sum_{t-n\,\delta t<t'\leq t} r^2(t') \tag{12.1}$$

where r is the daily return. This definition is straightforward, and the only open issue is the optimal choice for the number of days n. For a sample of size T, the trade-off is between a better volatility estimator for large n (with a variance that decreases as $1/\sqrt{n}$) and a larger sample for the lagged correlation for small n (a nonoverlapping—but dependent—volatility estimator is obtained when moving by n steps, leading to an effective sample size of the order of $(T-n)/n$). Assuming the independence of the successive volatility values, the variance of the correlation estimator is of the order of $1/\sqrt{n(T-n)/n} \simeq 1/\sqrt{T}$. As the leading n dependency disappears, the optimal value for n results from the next order volatility dependency. Because of the complex analytical implications of this parameter, the issue is investigated with the help of Monte Carlo simulations.

On one hand, a good measure of volatility should be homogeneous of degree one with the returns because the market cares about the magnitude of the price changes. The traders and quants follow this argument, using mostly $\sigma^2 \simeq r^2$. On the other hand, the volatility has a probability density which is well approximated by a log-normal distribution. This can be used as an argument to justify that the "good" variable should be $\log(\sigma)$, as this transformed variable has a normal distribution. Some theoretical processes take $\log(\sigma)$ as the basic variable [96, 114]. This is a convenient way to enforce the positivity of the volatility in a stochastic volatility process and to get a distribution closer to the observed one (for normal innovations). All the computations and figures below have been done for σ and $\log(\sigma)$. For both volatility measures, the figures and results are essentially unchanged, and the same conclusions are drawn. Such empirical study therefore does not allow us to settle the question of which variable is better suited to capture the dynamics of the volatility. The figures below are given only for the lagged correlations of σ.

12.2.2 The Lagged Correlation Estimators

By default, the lagged correlation is evaluated by the usual Pearson product of standardized variables. Yet, for random variables with fat-tailed distributions, this estimator has the disadvantage of being very sensitive to extreme events. To mitigate this effect, we also use another estimator of dependency which provides for a good compromise between robustness and computational efficiency.

The usual correlation estimator is the Pearson lagged correlation

$$\rho(l) = \rho\big(\sigma(t), \sigma(t+l)\big) = \frac{1}{T-l} \frac{\sum_t (\sigma(t) - \bar{\sigma})(\sigma(t+l) - \bar{\sigma})}{\mathrm{Var}(\sigma)} \tag{12.2}$$

with T the sample length, $\bar{\sigma} = 1/T \sum_t \sigma(t)$ the mean volatility, and $\mathrm{Var}(\sigma)$ the variance of the volatility,

$$\mathrm{Var}(\sigma) = \frac{1}{T} \sum_t \big(\sigma^2(t) - \bar{\sigma}\big)^2. \tag{12.3}$$

This Pearson estimator has the drawback to be fragile against outliers. A robust correlation measure is given by the Kendall tau estimator τ (as described, for example, in the Numerical Recipes [129]). In [89], the author proves that for elliptical distributions, the relation between the linear correlation and the Kendall tau is $\tau = 2/\pi \sin^{-1} \rho$. Therefore, the Kendall tau correlation estimator is $\rho_\tau = \sin(\pi\ \tau/2)$. This estimator is robust, but the evaluation time grows as $\sim T^2$. For large scale computations, this is a clear drawback even with today computers.

[76] have introduced another robust estimator for dependency. In a finance context, it has been used, for example, by [89] to study correlations in elliptical distributions with fat-tailed marginals. The estimator is based on standardized sums and differences and is called ρ_{SSD}. For a time series x, the standardized time series $\tilde{x}$ is

$$\tilde{x} = \frac{x - \bar{x}}{\sqrt{\mathrm{Var}(x)}} \tag{12.4}$$

with $\bar{x}$ the mean of x and $\mathrm{Var}(x)$ the variance of x defined in (12.3). Then, the usual (Pearson) correlation can be written as

$$\rho(x, y) = \frac{1}{T} \sum_t \tilde{x}(t)\tilde{y}(t) = \frac{\mathrm{Var}(\tilde{x} + \tilde{y}) - \mathrm{Var}(\tilde{x} - \tilde{y})}{\mathrm{Var}(\tilde{x} + \tilde{y}) + \mathrm{Var}(\tilde{x} - \tilde{y})}. \tag{12.5}$$

A robust estimator for the correlation is obtained by replacing the usual variance estimator (12.3) by a robust one. In this work, we have used the mean absolute deviation (MAD) as a robust estimator:

$$\mathrm{MAD}(x) = \frac{1}{T} \sum_{t=1}^{T} \big|x(t) - \mathrm{median}(x)\big|. \tag{12.6}$$

The τ_{SSD} dependency measure is defined by

$$\tau_{\mathrm{SSD}}(x, y) = \frac{\mathrm{MAD}(\tilde{x} + \tilde{y}) - \mathrm{MAD}(\tilde{x} - \tilde{y})}{\mathrm{MAD}(\tilde{x} + \tilde{y}) + \mathrm{MAD}(\tilde{x} - \tilde{y})}. \tag{12.7}$$

Empirically, this estimator is close to the Kendall tau; therefore, a similar relationship with the usual linear correlation seems plausible. We define a robust measure of the lagged correlation based on the standardized sum and difference as

$$\rho_{\mathrm{SSD}}(l) = \sin\left(\frac{\pi \tau_{\mathrm{SSD}}(\sigma(t), \sigma(t+l))}{2}\right). \tag{12.8}$$

This estimator is robust (but less robust than the Kendall tau), and the computational time grows as $\sim T$ (the median evaluation is done in $\mathrm{O}(T)$ time as a complete ordering is not needed [129]). As 15 years of data is clearly large $T \simeq 3900$, this estimator proves to be substantially faster than the Kendall tau.

12.3 Lagged Correlation Shapes

The volatility lagged correlation can be approximated by simple functions. The goal is to have different decreasing shapes, depending on a few parameters. Four "popular" functions are explored in this chapter, all depending on two parameters.

12.3.1 Exponential Decay

The exponential function is given by

$$\rho(l) = \rho_0 \exp\left(-\frac{l - l_{\mathrm{ref}}}{\tau}\right) \tag{12.9}$$

with two parameters ρ_0 and τ. The reference lag l_{ref} can be chosen freely and induces only a change of value for $\rho_0 = \rho_0(l_{\mathrm{ref}})$ given by

$$\rho_0\left(l'_{\mathrm{ref}}\right) = \rho_0(l_{\mathrm{ref}}) \exp\left(\frac{l_{\mathrm{ref}} - l'_{\mathrm{ref}}}{\tau}\right). \tag{12.10}$$

The value for l_{ref} has been set to 10 days because this value results in a small correlation between the statistical estimates for ρ_0 and τ for a given time series and because the cross-section correlation between ρ_0 and τ for different time series is small.

Most of the simplest processes used to model financial time series lead to an exponential decay for the lagged correlation, like GARCH(1, 1), the exponential stochastic volatility, or the Heston process. The core reason is a multiplicative structure in the equations of the type $\sigma^2(t + \delta t) = \mu \sigma^2(t) + \cdots$. When iterated, such structures lead to products of μ and hence to exponentials.

12.3.2 Two Exponentials with Fixed Ranges

In [92], a detailed study of the volatility dependency at different time ranges shows structures at one day, one week, and one month, related to the natural human activity cycles. For large lags, the longest human cycles are influencing the dependency, and it is fairly natural to assume a one month and one year time ranges. This observation leads to the "two exponentials with fixed range" shape

$$\rho(l) = \rho_1 \exp\left(\frac{l - l_{\text{ref}}}{\tau_1}\right) + \rho_2 \exp\left(\frac{l - l_{\text{ref}}}{\tau_2}\right). \tag{12.11}$$

The time ranges τ_1 and τ_2 are fixed to 21 and 260 days, and the parameters ρ_1 and ρ_2 are estimated for each time series.

12.3.3 Power Law Decay

The "power law" lagged correlation, also called hyperbolic decay, is given by

$$\rho(l) = \rho_0 \left(\frac{l_{\text{ref}}}{l}\right)^{\nu} \tag{12.12}$$

with two parameters ρ_0 and ν. Most of the processes built to capture the long memory assume a hyperbolic decay for the lagged correlation. Power laws are also popular among physicists, in particular because they are scale-free. This particularity calls for universal mechanisms in order to create power laws, and therefore could explain naturally the similarity observed between widely different time series.

The basic idea of a "long memory" is to have a lagged correlation function that decays at least more slowly than an exponential for large lags. There is no complete agreement about the definition for long memory, but in general people refer to a hyperbolic decay for the lag correlation. In this sense, the power law shape is the only one in our set that is compatible with a long memory for the lagged correlation, as the three other shapes decay at least exponentially fast for sufficiently large lags. Yet, these considerations are theoretical as the accessible range of lags is bounded above by one to a few years. Clearly, we can make statements for lags in the analyzed range, but the behavior at larger lag values can differ.

12.3.4 Logarithmic Decay

While developing a new risk methodology, [154] observed that the lag correlation for the volatility is well described by a logarithmic decay

$$\rho(l) = \rho_0 - s \log\left(\frac{l}{l_{\text{ref}}}\right) = \rho_0 \left(1 - \frac{\log(l/l_{\text{ref}})}{\log(\tau/l_{\text{ref}})}\right) \qquad l \le l_{\text{cut-off}} \tag{12.13}$$

with two parameters ρ_0 and s, or alternatively ρ_0 and $\tau = l_{\text{ref}} \exp(\rho_0/s)$.

The first parameterization is more convenient as s is weakly dependent on the sample size, whereas ρ_0 and τ are strongly dependent on the sample size T. The reference lag l_{ref} can be chosen freely and leads to a change of the base correlation $\rho_0 = \rho_0(l_{\text{ref}})$ given by

$$\rho_0\left(l'_{\text{ref}}\right) = \rho_0(l_{\text{ref}}) + s\log\left(\frac{l_{\text{ref}}}{l'_{\text{ref}}}\right). \tag{12.14}$$

We choose $l_{\text{ref}} = 65$ days, because this value gives a small dependency between l_{ref} and s, both in the cost function below and in a cross sectional study.

The raw lagged correlations, the human behavior, and the multicomponent ARCH processes all suggest that an upper cut-off exists, above which the decay is different from a log-decay. For this purpose, an upper cut-off $l_{\text{cut-off}}$ is introduced, above which an exponential decay is assumed:

$$\rho(l) = \rho_{0,\exp}\exp\left(-\frac{l}{\tau_{\exp}}\right) \qquad l \geq l_{\text{cut-off}}. \tag{12.15}$$

The parameters $\rho_{0,\exp}$ and $\tau_{\exp}$ are computed at $l_{\text{cut-off}}$ so that the functions (12.13) and (12.15) have the same values and slopes. Finally, the correlation is zero in the case $\tau \leq l < l_{\text{cut-off}}$.

The computations presented below have been done for a few cut-off $l_{\text{cut-off}}$: 6 months, 1, 2, and 4 years. The results are in favor of a cut-off between 6 months and 2 years, yet it is not a critical parameter. All the figures below are presented for $l_{\text{cut-off}} = 1$ year. In [49], a more detailed and complex HARCH process is estimated on high-frequency data using a log-likelihood estimator (the estimated parameters for this model are compatible with a long memory for the volatility). The above paper uses high-frequency data and is therefore weakly sensitive to time horizons on the order of several months. Moreover, a likelihood estimator focuses on the innovations (at the highest frequency), whereas the upper cut-off lies at the other end of the spectrum. A process estimation with log-likelihood is therefore quite different from the present lagged correlation which is directly sensitive to the time scale of interests. Despite these significant differences in the methodologies, all the results point to a cut-off of the order of one year.

12.4 The Cost Function and Parameter Estimation

The cost function is given by a standard χ^2 figure of merit, modified so as to include an "integration measure" on the lag axis. The idea is to weight the lag l so that the sum is equivalent in the continuum limit to the integration

$$\int d\ln(l) = \int \frac{dl}{l}.$$

This measure avoids an excessive importance of the large lags. Intuitively, the measure makes the formula behave like a "fit-by-eyes" in a graph for the lagged correlation versus the logarithm of the lag. With this modification, the normalized χ^2 cost function is

$$\chi^2 = \max_\theta \left[\sum_{l=n}^{l_{\max}} \mu(l) \left(\frac{\rho_{\text{emp}}(l) - \rho(l;\theta)}{\sigma(\rho_{\text{MC}}(l))} \right)^2 \right] \tag{12.16}$$

with $\rho_{\text{emp}}(l)$ the empirical (Pearson or robust) correlation, and $\rho(l;\theta)$ the theoretical form for the lagged correlation with parameters θ. The standard deviation for the correlation $\sigma(\rho_{\text{MC}}(l))$ is derived from Monte Carlo simulations, as discussed at the end of Sect. 12.5. The measure $\mu(l) = C/l$ is normalized so as to be a unit measure, namely $\sum_l \mu(l) = 1$.

In formula (12.16), the magnitude of the correlation differences is measured in the natural unit of the standard deviations. For the sake of the argument, let us assume now that the shape given by $\rho(l;\theta)$ and the standard deviation $\sigma(\rho_{\text{MC}}(l))$ both correspond to the empirical correlation $\rho_{\text{emp}}(l)$, up to statistical fluctuations (with the parameter vector θ that maximizes the χ^2 function). This statement means that

$$\rho_{\text{emp}}(l) = \rho(l;\theta) + \sigma\big(\rho_{\text{MC}}(l)\big)e(l) \tag{12.17}$$

where $e(l)$ are random variables with zero mean and unit variance. Because $\mu(l)$ is a unit measure and because $E[e^2(l)] = 1$, the magnitude of χ is given by $\chi^2 \simeq 1$. In practice, smaller values are obtained, even in the controlled environment of the Monte Carlo simulations where relation (12.17) can be checked. The discrepancy is due to the correlation of the errors $e(l)$ for neighboring values of l. This correlation can be partly absorbed into the fitted shape $\rho(l;\theta)$ by adjusting θ, leading to a reduction of the fluctuations $e(l)$. Even if the magnitudes provided by $\sigma(\rho_{\text{MC}})$ are right, the maximization in (12.16) makes that $E[e^2] < 1$. Therefore, the deviation of the magnitude of χ^2 from one measures the dependency across lags of the correlations, together with the fitness of the proposed shape $\rho(l;\theta)$. Because of this dependency, it is difficult to make a theoretical statement about the distribution of χ^2, and Monte Carlo simulations are used instead.

12.5 Monte Carlo Simulations

For Monte Carlo simulations, a realistic process has to be chosen. In particular, the process should capture with a reasonable accuracy the long memory and fat-tailed distributions observed in the empirical data. For this purpose, the long-memory affine aggregated ARCH process is used (see Sect. 7.9), with the weights χ_k for the volatility component σ_k decreasing as a logarithm with a characteristic time of 1300 days, corresponding to a slope $s = 0.1395$. The weight for the mean volatility

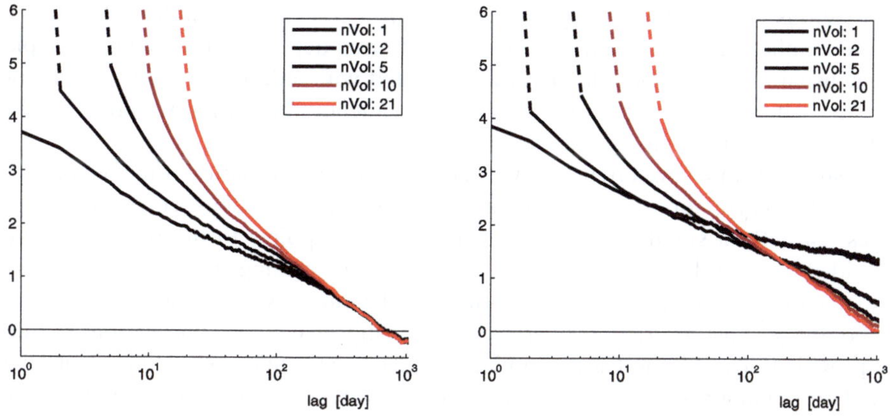

Fig. 12.1 The signal-to-noise ratio versus the lag l (*horizontal axis*), for a fixed sample length $T = 4000$ and for increasing volatility sizes n (*color curves* with the volatility size given in the caption). The *left* (*right*) panel is for the Pearson (robust) correlation estimator

is $w_\infty = 0.05$, and the distribution of the innovations is a Student with 5 degrees of freedom.

The Monte Carlo simulations aim at studying the influence of the volatility estimator parameter n, for the Pearson and robust correlation estimators. For this purpose, the long-memory ARCH process is used to generate data samples of lengths $T = 2000$, 4000, 8000, and 100,000. The empirical time series have lengths ranging from 2500 to 6000 days, the bulk of them with lengths around 4000 days. Both lagged correlations $\rho(l)$ are estimated on the samples. The procedure is repeated 1000 times, and the mean and standard deviation of the lagged correlations are computed.

The mean of the lagged correlation for the Pearson and robust correlation estimators are very close, with slightly smaller values for the robust estimator. The interesting difference appears at large lags, the most difficult region: the robust correlation estimator with small n picks up small correlation values and a finite sample analysis shows that these are more accurate values. The signal-to-noise ratio is defined as the ratio between the mean correlation and its standard deviation. Figure 12.1 shows the signal-to-noise ratio. On the left panel, the Pearson estimator with $n = 5$ for the volatility is a good compromise as it gives a large signal for small lags. Yet, the Pearson correlation for all n values is unable to pick a good signal for large lags. On the right panel, the robust estimator offers a different picture as the value $n = 1$ clearly gives the largest signal for large lags. As we are mostly interested in the region between one month and a few years, this estimator provides for the best overall signal in the targeted domain. The underlying reason for this difference is likely to be found in the dependency between normal levels of volatility. The robust estimator picks this signal, whereas the large volatility events due to the fat-tailed distribution of the innovations destroy the dependency as measured by the Pearson correlation. In order to measure this effect on empirical data, the lagged correlations are measured using both couples ($n = 5$, Pearson) and ($n = 1$, robust). Let us emphasize that they measure slightly differing dependencies. The first couple is the usual linear correlation measure associated with a decent volatility estimator ($n = 5$). The second couple measures the dependency for normal volatility levels, with the largest possible sample ($n = 1$).

A study of finite size effects shows that the Pearson correlation is unable to pick the positive correlations present for large lags, whereas the robust correlation measures correctly a small dependency. This difference points to the better measure provided by the robust estimator. The analysis for increasing sample size T shows also that the dependency on T is important for the intensity ρ_0 but not for the slope.

For the χ^2 cost function, the variance of the empirical lagged correlations should be provided. The Monte Carlo simulations give a convenient way to evaluate the standard deviation σ of the estimated lagged correlations in finite samples. The results have been fitted with the form

$$\sigma = \sigma\big(\rho_{\mathrm{MC}}(l)\big) = \big(a - b + p\log(l)\big)\big(\exp(-l/\tau) + b\big) \tag{12.18}$$

with a, b, p, and τ simple functions of $\log(n)$ and $\log(T)$. The coefficients of the functions are adjusted to all the curves with varying n and T, and separately for the Pearson and robust correlation estimators. Then, the form for σ is used in the Monte Carlo simulations for computing χ^2 and the optimal parameters θ for the various decay shapes. As a further check, a simpler form for σ without dependency on l gives consistent results for the estimated parameters.

12.6 Lagged Correlations for Empirical Time Series

The empirical data set for this chapter contains 225 daily time series, with 41 foreign exchanges, 48 stock indexes, 49 stocks, 68 interest rates, and 19 commodities. It is a subset of the ICM data set described in Sect. 17.3. In the scatter plots, the symbols are as follows: FX: blue circle; stock indexes: green square; stock: black triangle pointing down; IR: red with various symbols according to the maturities (1 day: '×'; 1 month: '+'; 1 year: five points star; 10 years: six points star); commodity: black triangle pointing right. The lengths of the time series are from 10 years to 24 years, with the majority between 12 and 18 years. The χ^2 figures of merit are computed up to lag $l_{\max} = 260$, 520, and 1040, corresponding to 1, 2, and 4 years. The overall results are similar for different value of $l_{\max}$ and are presented below only for $l_{\max} = 1040$. The χ^2 minimization for the lagged correlation is done for each time series, for each shape, and for both selected combinations of n and correlation estimators.

The cross sectional comparison of the cost function between decay shapes is presented in Fig. 12.2 for the ($n = 1$, robust) correlation estimator. The results for the ($n = 5$, Pearson) correlation estimator is consistent with this figure. None of the decay shapes dominates the other ones for all time series, but some decay shapes

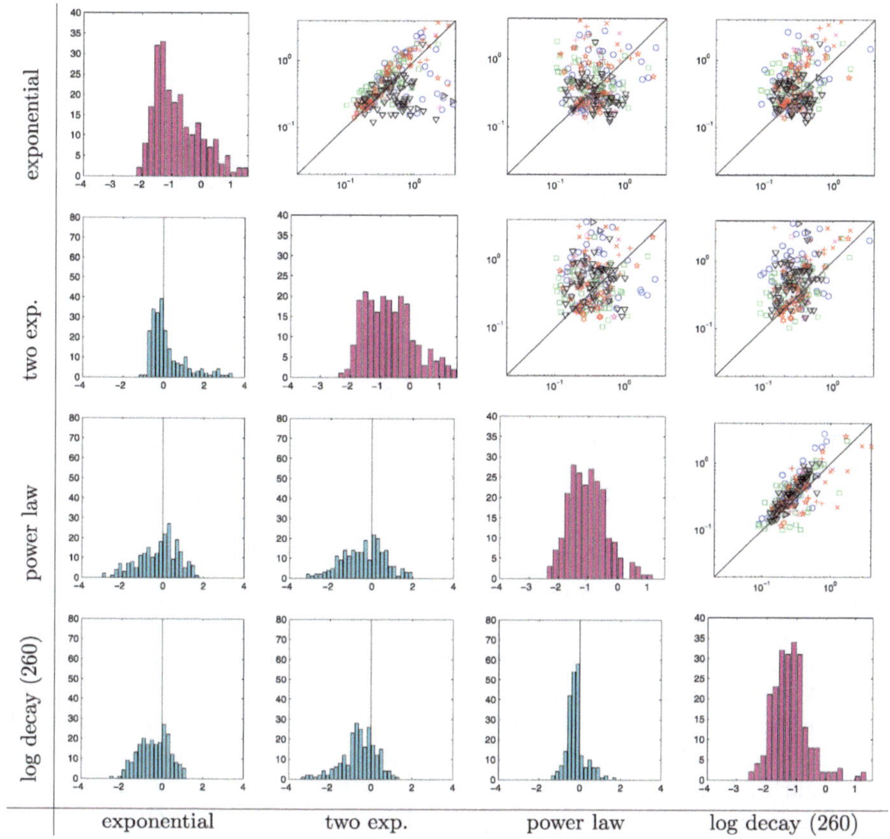

Fig. 12.2 The scaled χ^2 figures of merit, for the various decay shapes, for $n = 1$ and the robust correlation estimator. The estimation is done up to $l_{max} = 1040$. The meaning of the figure at the intersection of the X column and Y line is the following. The diagonal displays the distribution for $\log(\chi^2)$: the model is better when the distribution is shifted toward the smaller value. A *plot above* the diagonal displays a cross-section plot for χ^2 with the model $X(Y)$ on the *horizontal* (*vertical*) *axis*. The decay shape X is better than Y when most of the points are above the diagonal (i.e., when $\chi^2(X) \leq \chi^2(Y)$). A *plot below* the diagonal displays a histogram for the difference $\log(\chi^2(Y)) - \log(\chi^2(X))$. The decay shape X is better than Y when most of the distribution lies on the positive side. The layout is such that to a cross-section graph X–Y above the diagonal corresponds a histogram graph Y–X (*below the diagonal*) of the distance to the diagonal of the first graph

are better overall. For both lagged correlation estimators, the log-decay shape is clearly better than the alternative shapes for most time series. Then, the one- and two-exponentials follow, while the power law is consistently the worst overall description of the lagged correlation. For the log-decay versus power law comparison, the difference is clearly in favor of the first form. These conclusions are generic and robust against changes in the parameters of the computation, like l_{max}, n, or the use of the lagged correlation for σ or $\log(\sigma)$.

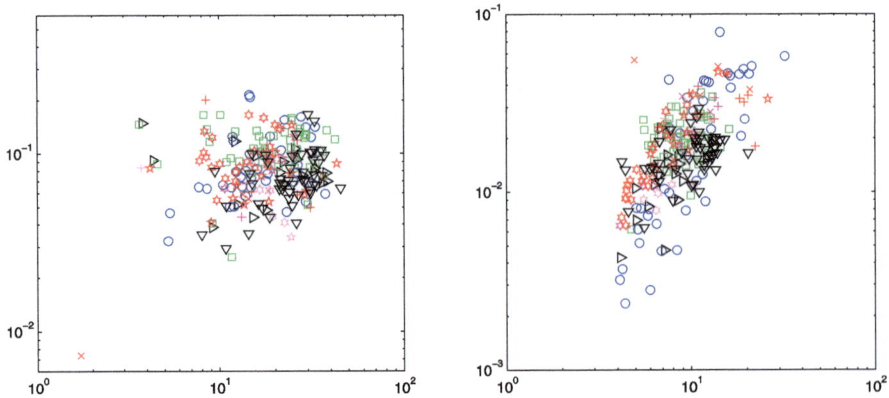

Fig. 12.3 The cross-section of the parameters estimates for a log decay, with cut-off at 260. The *horizontal axis* is ρ_0 in % for a reference length l_{ref} of 65 days, the *vertical axis* is s. The estimation is done up to $l_{\max} = 1040$. The *left* (*right*) *panel* is for $n = 5$ ($n = 1$) and the Pearson (robust) correlation estimator

The estimated parameters for the log-decay are given in Fig. 12.3 for a reference lag l_{ref} of 65 days. Interestingly, the bulk of values for ρ_0 and s are similar for each time series. At the chosen reference lag, the correlation is of the order of 10 %. A small segregation by asset classes can be seen on the graph, particularly on the right panel. For the FX rates (blue circles, all against USD), two clouds of points can be observed: the cloud with the largest decays s refers to emerging countries (Russia, China, Brazil, Thailand, Philippine, Ukraine, India, etc.), while the smallest decays correspond to developed countries (EUR, GBP, Australia, Poland, Japan, etc.). Most of the stock indexes have very similar values, with a few points with larger decay (Romania, Russian, Pakistan, Peru, etc.) and smaller decay (SP500, Poland, Korea, etc.). For 10 years interest rates (red 6 points stars), the decay coefficients are clustered with a few points with higher decay (India, Singapore, South Africa, Hong Kong government debt). The US stocks (black triangle pointing down) are clustered, with large decays all corresponding to the financial industry (Pmi group (mortgages, debts, and structured products based on them), AIG, American express, JP Morgan, Bank of America, etc.), while small decays correspond to industrial group (ChevronTexaco, 3M, Medtronic, Hewlett-Packard, Intel, etc.). Let us emphasize that the sample ends on 28 April 2008 and therefore does no include the credit crunch between financial institutions during Fall 2008. Across the whole data set, the emerging overall picture is fairly consistent, with larger decays associated to more risky time series inside a given asset class.

To summarize the parameter values at the overall level, all time series behave similarly with respect to volatility clustering, with similar slopes. At a finer level, the time series with a more eventful history have a higher slope (i.e., shorter ranges). Yet, there is no obvious relationship between slope and volatility, as can be seen on Fig. 12.4. The volatility separates roughly the time series by types (horizontal

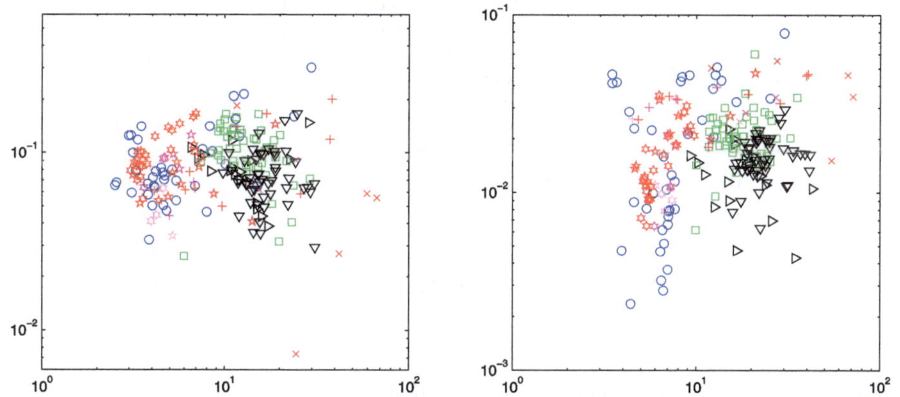

Fig. 12.4 The cross-section of the slope parameters estimates for a log decay (*vertical axis*), versus the volatility, in % (*horizontal axis*). The estimation is done up to $l_{\max} = 1040$. The *left* (*right*) *panel* is for $n = 5$ ($n = 1$) and the Pearson (robust) correlation estimator

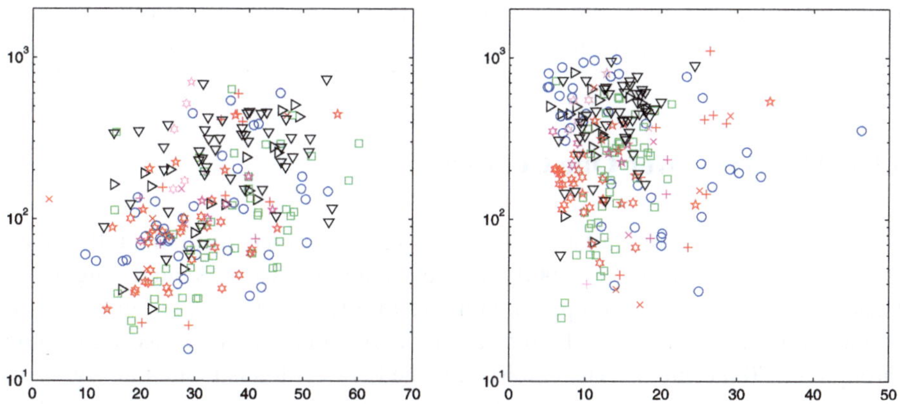

Fig. 12.5 The cross-section of the parameters estimates for an exponential decay. The *horizontal axis* is ρ_0 in % for a reference length l_{ref} of 10 days, the *vertical axis* is τ measured in days. The estimation is done up to $l_{\max} = 1040$. The *left* (*right*) *panel* is for $n = 5$ ($n = 1$) and the Pearson (robust) correlation estimator

axis), for example, with growing volatility between long-term interest rates, stock indexes, and single stocks. Nevertheless, the slopes (vertical axis) show no obvious relationship with the volatility but give another characterization of the time series.

The optimal parameters for an exponential shape are reported in Fig. 12.5. As can be seen on Fig. 12.6, this shape is not good for small lags, but the range parameter τ gives a good indication for an equivalent range for the lagged correlation. The ranges depend on the lagged correlation estimators but are essentially between 3 months to 4 years for all time series. A deeper analysis reveals the same fine differentiation for the range τ as observed for a log-decay slope s. This graph shows that similar parameter for the decay can be used for all time series.

Table 12.1 The detailed results for the DJIA over one century for the ($n = 5$, Pearson) estimator. The parameters (θ_1, θ_2) for each shape are (ρ_0, τ) exponential, (ρ_1, ρ_2) two exponential, (ρ_0, ν) power law, and (ρ_0, s) log-decay. The last two columns give the p-values in percent, estimated by Monte Carlo simulations or with a log-normal distribution

	χ^2	θ_1	θ_2	p-val MC	p-val log-normal
Exponential	0.92	0.51 ± 0.08	350 ± 272	0	0.015
Two exp	0.75	0.15 ± 0.29	0.47 ± 0.16	0	0.05
Power law	0.47	0.57 ± 0.09	0.26 ± 0.12	1	0.41
Log-decay	0.087	0.37 ± 0.07	0.09 ± 0.04	39	43

Table 12.2 As for Table 12.1, but for the ($n = 1$, robust) estimator

	χ^2	θ_1	θ_2	p-val MC	p-value log-normal
Exponential	0.32	0.16 ± 0.03	685 ± 850	0.3	0.096
Two exp	0.31	0.08 ± 0.16	0.20 ± 0.12	0.5	0.11
Power law	0.52	0.21 ± 0.05	0.19 ± 0.12	0	0.005
Log-decay	0.10	0.15 ± 0.05	0.035 ± 0.02	9	10.8

12.7 The DJIA over One Century

Very few time series are available with a long history; among them, the DJIA stock index is recorded since before 1900. This very long time series is interesting as the lagged correlation can be computed with lags of a few years with reasonable error bounds (but this assumes implicitly that the correlation structure has been stable through the 20th century). The empirical lagged correlation is plotted in Fig. 12.6, along with the best estimates for the four decay shapes. The χ^2 and corresponding parameters are given in Tables 12.1 and 12.2. With this amount of data, the results are clear cut: the log-decay shape gives the best fit of the empirical data, while the other shapes look very unlikely.

In order to quantify this last statement, Monte Carlo simulations are run over a length of 30,000 days using the long-memory process. The corresponding distribution of χ^2 values is computed for the various shapes. The Monte Carlo distribution of $\log(\chi^2)$ for the log-decay shape is very well approximated by a normal distribution, with the mean and variance computed with the usual moment estimators from the Monte Carlo sample. The cumulative distribution is used to compute the p-value of a χ^2 at least as large as the value obtained with the DJIA. The p-values can be estimated either by using the Monte Carlo χ^2 distribution, or by using the estimated normal distribution. Both values are reported in the tables for every shape. Clearly, at the usual 1 or 5 % threshold, the log-decay is accepted while the other shapes are rejected.

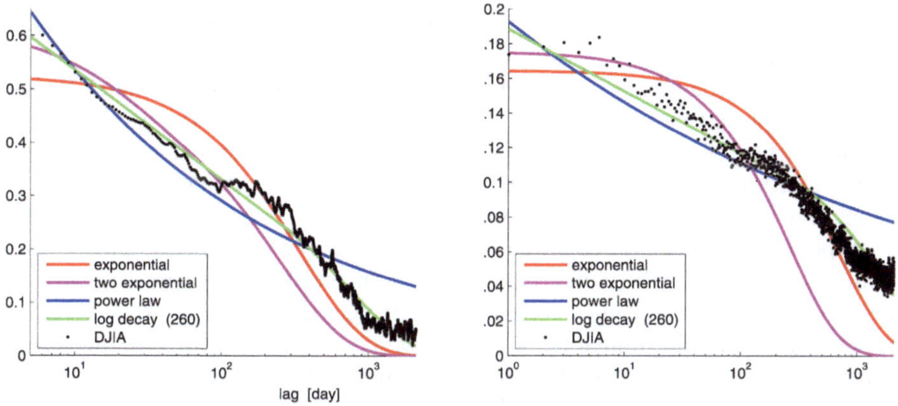

Fig. 12.6 The empirical and estimated lagged correlations for the DJIA. The estimation is done up to $l_{\max} = 2080$, with the empirical correlation computed on a sample from 2.1.1900 to 31.12.2008 corresponding to 29,691 days. The *left* (*right*) *panel* is for $n = 5$ ($n = 1$) and the Pearson (robust) correlation estimator

12.8 Summarizing the Findings About the Heteroscedasticity

The Monte Carlo simulations for a realistic process lead us to select two pairs for the volatility and correlation estimators, namely ($n = 5$, Pearson) and ($n = 1$, robust). They measure slightly different dependencies, with the Pearson correlation weighting equally all the volatilities, while the robust estimator puts emphasis on typical volatilities. The optimal slope parameters s show that the dependencies extend further for the typical volatilities, and conversely that large volatility events reduce the lagged dependency. This points to a complex dependency structure that is not completely captured by the usual linear measure of dependency. A possible explanation is the difference between external large and disruptive news versus the normal dynamics of the market associated with the steady flow of "normal" news. The usual Pearson linear correlation is equally sensitive to every input value, and the disruptive news create large volatility peaks with a reduced dependency. The "normal" events are best measured by the robust correlation estimator and show a longer memory. A similar finding is reported in [79]: large events are less correlated than the normal dynamics.

Concerning the best overall characterization of the lagged correlation decay, the empirical study shows that the best shape is given by a log-decay shape, with a cut-off of the order of one year. On the DJIA long sample, the agreement is even stunning. The slopes are depending mainly on the volatility estimator n and to a lesser extend from the time series. In the other direction, a power law can be excluded. This finding eliminates some processes or structures that have been proposed for capturing long memory with a power law decay. An example is the FIGARCH process, where the decay shape is fixed by the fractional difference operator, and therefore cannot be modified. A single exponential shape gives also a decent characterization of the empirical data in the large lag region, but this shape does not capture well the whole range of lags. This finding eliminates many popular processes with a single time scale, like GARCH(1, 1), Heston, or SABR.

For the log-decay shape estimated over an empirical data set containing 225 time series, the optimal parameter values show interesting properties. The parameter values are fairly well clustered, and no separation by asset classes can be found. This points to a universal economic mechanism generating the log-decay dependency, a mechanism that remains to be found.

A possible explanation could be found in the multiple time scale ARCH processes (see Chap. 7), where the weights given at increasing scales can be set so as to reproduce the observed lagged correlations. The lagged correlation for the volatility is not particularly discriminating, and many processes can be built with the correct decay. As discussed in Chap. 11, only the multiscale ARCH processes are able to reproduce the observed asymmetry with respect to time reversal. The structure of these processes points to the economic source for the long memory, namely the presence of market participants actives at very different time horizons. Let us emphasize that all these processes are able to *reproduce* the observed lagged correlations, but that they do not *explain why* such a structure emerges from the markets.

Another possible explanation might be found in the multifractal model for financial time series [9, 13, 112]. These processes are built on a cascade structure for the (logarithm of the) volatilities over a set of geometric time scales. This construction leads naturally to the lagged correlation of the logarithm of the volatility to decay as a logarithm of the lag, as in the log-decay shape (12.13). The lagged correlations have been evaluated both for σ and for $\log(\sigma)$, with very similar results. Therefore, the present chapter supports multifractal models. Yet, these are stochastic volatility processes that are time-reversal invariant, and therefore they cannot reproduce the observed empirical statistics developed in Chap. 11.

At a finer level, the decay is faster for time series with an "eventful" history, but no direct relationship with the volatility can be found. For example, emerging countries have systematically faster decays. In a study of the generalized Hurst exponent on a set of 89 time series, [103] find that this exponent is related to the degree of development of a market. The intuitive argument used to explain this relation is that developed markets have a heterogeneous structure, with market participants working at different natural time horizons. In an emerging market, some groups of participants may be much smaller or missing, particularly the groups working at long time horizons like pension funds or life insurance companies. This varying market composition leads to a shorter memory, hence to a smaller Hurst exponent, in agreement with the present results. Consequently, the current findings add support to the heterogeneous market hypothesis which explains the multiscale structure of the lagged correlation as due to the range of natural time horizons of the market participants (see also [92]). What remains to be explained is why the weights for the increasing time horizons have similar values for various financial contracts across the world. In an econometric perspective, it would be interesting to study in more details the relationship between the decay parameter s, the Hurst exponent $H(q)$, and other properties or characterizations of the financial time series.

Chapter 13
The Innovation Distributions

13.1 Empirical Probability Distribution for the Returns

An important concern for many financial applications is the probability distribution for large price changes, namely the tails behavior of $p(r)$. It is nowadays widely acknowledged that this distributions are fat-tailed, regardless of the time series. Because the mean volatility of time series can be very different, it is easier to take as the variable to study $r' = r/\,\mathrm{stdDev}(r)$ with

$$\mathrm{stdDev}^2(r) = \frac{1}{n}\sum_t r^2(t) \tag{13.1}$$

and where n is the number of terms in the sum. Notice that this standardization of the returns uses the full sample information and that the mean return $\langle r\rangle$ has been neglected. The tail behavior is conveniently studied by the empirical cumulative probability density cdf(r'). For negative residuals, the tail corresponds to the convergence of the cdf toward zero, whereas for positive residuals, the tail is given by the convergence toward one. In order to have similar figures for both tails, we plot the cdf versus $-r'$ for the negative tail and $1 - \mathrm{cdf}$ versus r' for the positive tail.

The resulting return cumulative distributions are given in Figs. 13.1 and 13.2 for the negative and positive tails, respectively. The data set used in this computation is a subset corresponding to the G10 countries of the larger data set used in the previous chapter. Clearly, the distribution at one day is well described by a distribution with fat tails, but badly by a Gaussian. The graph at 1 day shows that regardless of the asset, the return distribution is well described by a Student distribution with a number of degrees of freedom between 3 to 5. With increasing time horizons, we can observe a slow convergence toward a Gaussian. Concurrently, the sizes of the data samples diminish, and it becomes increasingly difficult to make clear assertions about the tail behavior for $p(r)$. What is more surprising is the universality of the return distributions in the space of assets and asset classes: essentially, commodities, stocks, foreign exchange rates, and interest rates have very similar distributions.

G. Zumbach, *Discrete Time Series, Processes, and Applications in Finance*,
Springer Finance, DOI 10.1007/978-3-642-31742-2_13,

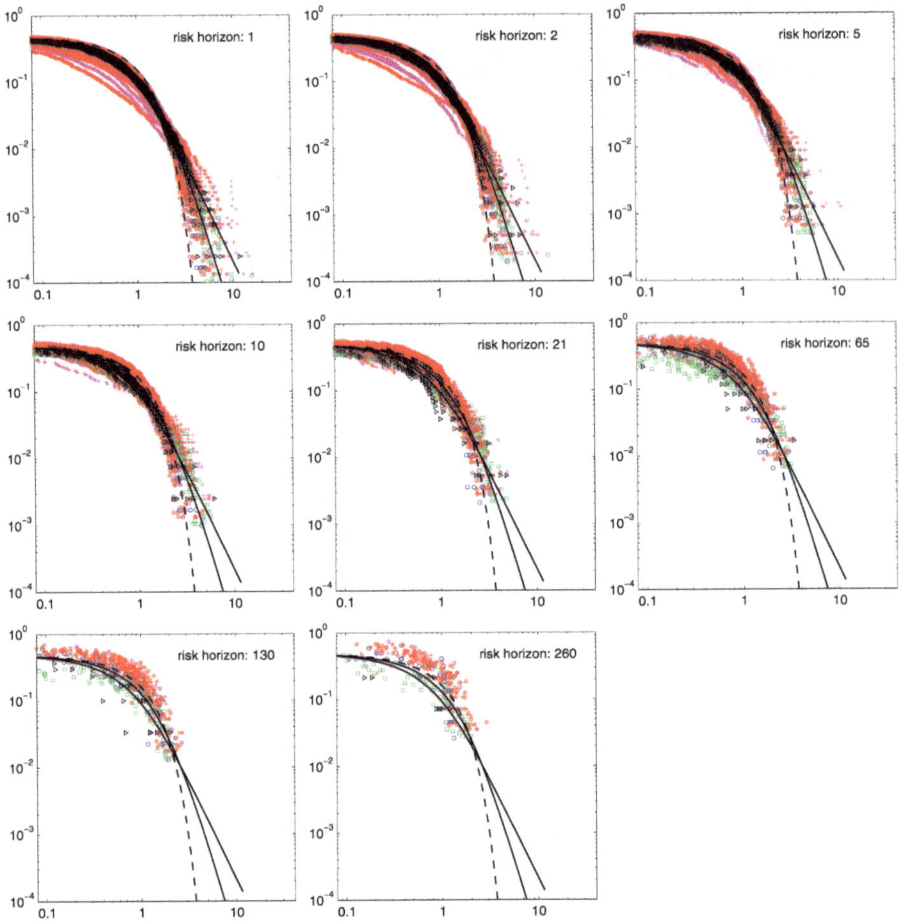

Fig. 13.1 The cumulative probability density for the returns $\text{cdf}(r')$ versus $-r' = -r/\text{stdDev}(r)$ for the negative tails. The time series are the G10 data set. The *solid line* corresponds to a Student distribution with 3 and 5 degrees of freedom (rescaled to have a unit variance); the *dashed line* corresponds to a standard normal distribution

13.2 The Mapping Between Returns and Innovations

For a process, the innovation distribution $p(\epsilon)$ must be studied. The relation between return r and innovations ϵ has been introduced in (5.2), which is reminded here:

$$r(t+\delta t) = \mu(t) + \sigma_{\text{eff}}(t)\epsilon(t+\delta t) \tag{13.2}$$

where the return $r = r[\delta t]$ is computed at the process increment δt. In this equation, μ and σ_{eff} are known at time t (i.e., they are measurable at t) and correspond to the one-step mean and standard deviation of the process. It is also interesting to do the same study at a longer time horizon $\Delta T \geq \delta t$, in particular for risk evaluation

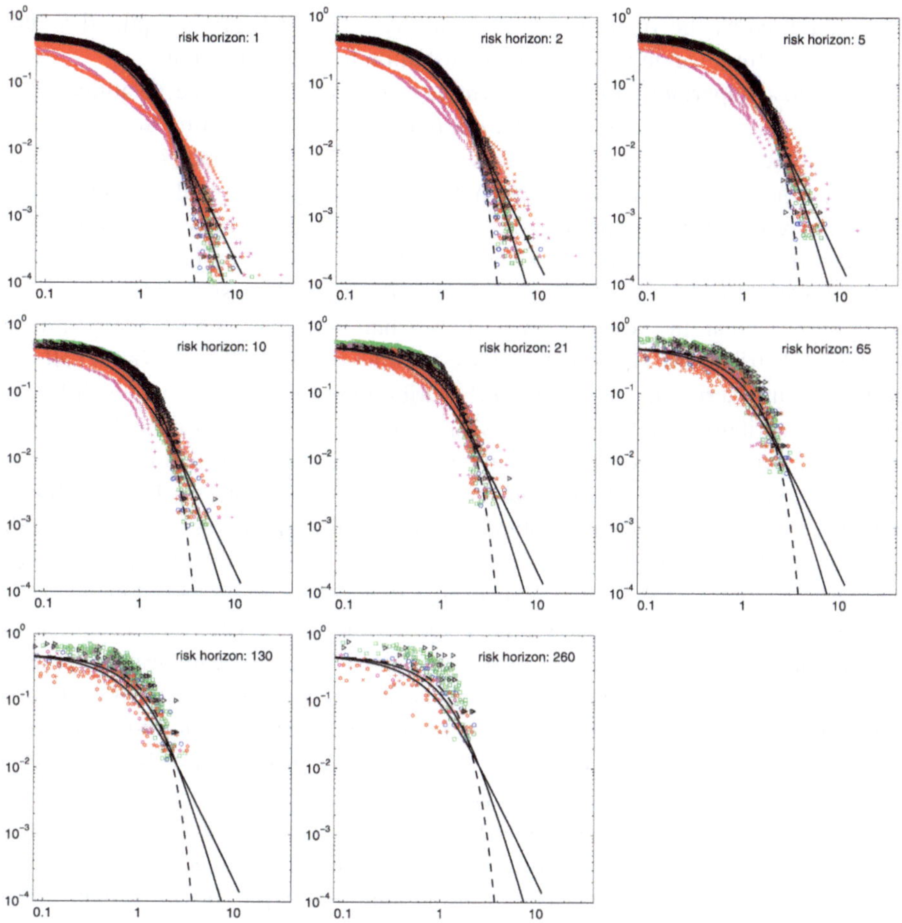

Fig. 13.2 The cumulative probability density $1 - \text{cdf}(r')$ versus r' for the positive tails

(see the next chapter). In this case, the time interval δt corresponds to the increment of the available time series (say daily for risk applications), and ΔT corresponds to the chosen risk horizon. In this setting, the forecasts for the return $\widetilde{r}[\Delta T](t)$ and volatility $\widetilde{\sigma^2}[\Delta T](t)$ are computed at t using the available information. Then, the relation between returns and innovations is given by

$$r[\Delta T](t + \Delta T) = \widetilde{r}[\Delta T](t) + \widetilde{\sigma}[\Delta T](t)\epsilon(t + \Delta T) \tag{13.3}$$

with the shorthand for the volatility forecast $\widetilde{\sigma} = \sqrt{\widetilde{\sigma^2}}$. Both settings use the same mapping between innovation ϵ and return r, differing by the way the mean return and standard deviation are computed. In these formulas, the forecasted return and volatility act respectively as time-dependent location and size factors (see Sect. 15.2). In both approaches, the idea is that the time-dependent heteroscedasticity is captured by the volatility forecast and the expected return is captured by the forecasted return, leaving a stationary distribution for the innovations.

The most important part in this formula is the forecast for the volatility $\widetilde{\sigma}(t)$. This term should capture the heteroscedasticity of the financial time series, so that the pdf of the innovations is stationary. This is expressed by the fact that $p_{\Delta T}(\epsilon)$ depends on the time increment ΔT, but not on t, and only the volatility forecast will make the returns larger in periods of high volatility. The volatility forecast is itself a time series, with a non trivial distribution. All volatility forecasts are based on the (magnitude of the) past returns, and therefore, they will share some properties with $p(r)$. In particular, the volatility forecast distribution has also a fat tail (say similar to the $|r|$ distribution). Because the above formula is essentially $r = \widetilde{\sigma}\epsilon$ and all the terms are time series, the *unconditional* distribution of the returns $p(r)$ is different from the distribution of the innovations $p_{\Delta T}(\epsilon)$. Intuitively, because the distribution of the volatility forecasts captures some of the fat tails, the distribution of the innovations should have less fat tails than the distribution of the returns. The question is whether the volatility forecast captures all the fat tails and leaves innovations with a Gaussian distribution, or whether a part of the fat tails for the returns originates in the innovation distribution. The answer is dictated by empirical findings and is investigated in this chapter using a large set of time series. The result is similar to the various ARCH processes investigated in Chap. 7, namely the innovations always display fat-tailed distributions. The outcome is insensitive to the details of the return and volatility forecasts. In particular, the innovations have a similar fat-tailed distribution regardless of the process used to build the volatility forecast (I-GARCH(1), GARCH(1, 1), long memory ARCH, etc.).

Let us emphasize again the important difference between the conditional and unconditional distribution of the returns. The distribution specified in a process and a risk methodology is the distribution of the innovations $p_{\Delta T}(\epsilon)$. It is identical to the conditional distribution of the returns (up to shift and magnitude factors) but not to the unconditional distribution of the returns (because of $\widetilde{\sigma}(t)$).

Formula (13.2) or (13.3) is used to evaluate the random forthcoming return at $t + \Delta T$ from a draw of the innovation $\epsilon(t + \Delta T)$. In order to establish and test this formula, historical data for r is used, and the formula is solved for ϵ, leading to

$$\epsilon[\Delta T](t + \Delta T) = \frac{r[\Delta T](t + \Delta T) - \widetilde{r}[\Delta T](t)}{\widetilde{\sigma}[\Delta T](t)}. \tag{13.4}$$

On the right-hand side, the return $r[\Delta T](t + \Delta T)$ corresponds to the historical return at the time $t + \Delta T$, or to the realized return at t. This formula is similar in structure to the "standardization" of a time series by removing its sample mean and dividing by the sample standard deviation; the key difference is that the respective forecasts at t are used. The motivation is to discount the realized return by the available information at t in order to build a random variable ϵ that is well behaved, namely i.i.d.

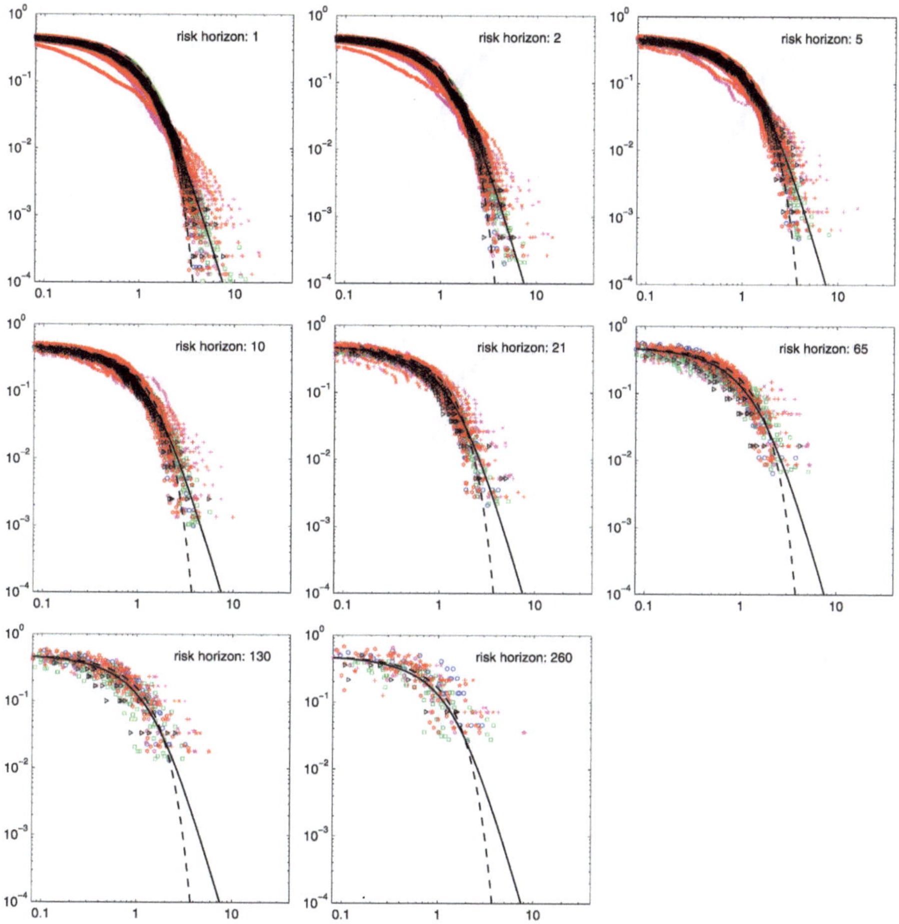

Fig. 13.3 The cumulative probability density cdf(ϵ) versus $-\epsilon$ for the negative tails. The time series in the G10 data set are used. The *solid line* corresponds to a Student distribution with 5 degrees of freedom (rescaled to have a unit variance); the *dashed line* corresponds to a standard normal distribution

13.3 Empirical Probability Distributions for the Innovations

The pdf $p_{\Delta T}(\epsilon[\Delta T])$ can be computed empirically and compared to different models for $p_{\Delta T}$. The resulting figures for $p_{\Delta T}(\epsilon)$ for the G10 data set are given in Figs. 13.3 and 13.4 respectively for the negative and positive tails. The return and volatility forecasts are given by a long-memory ARCH process as used in the RM2006 risk methodology (see the next chapter). The salient feature of these graphs is a very good data collapse. This shows that one analytical form for the innovation pdf should be enough for all assets and all risk horizons.

At a one day risk horizon, the Gaussian distribution can be easily excluded as it clearly misses the fat tails of the empirical data. On the other hand, a Student

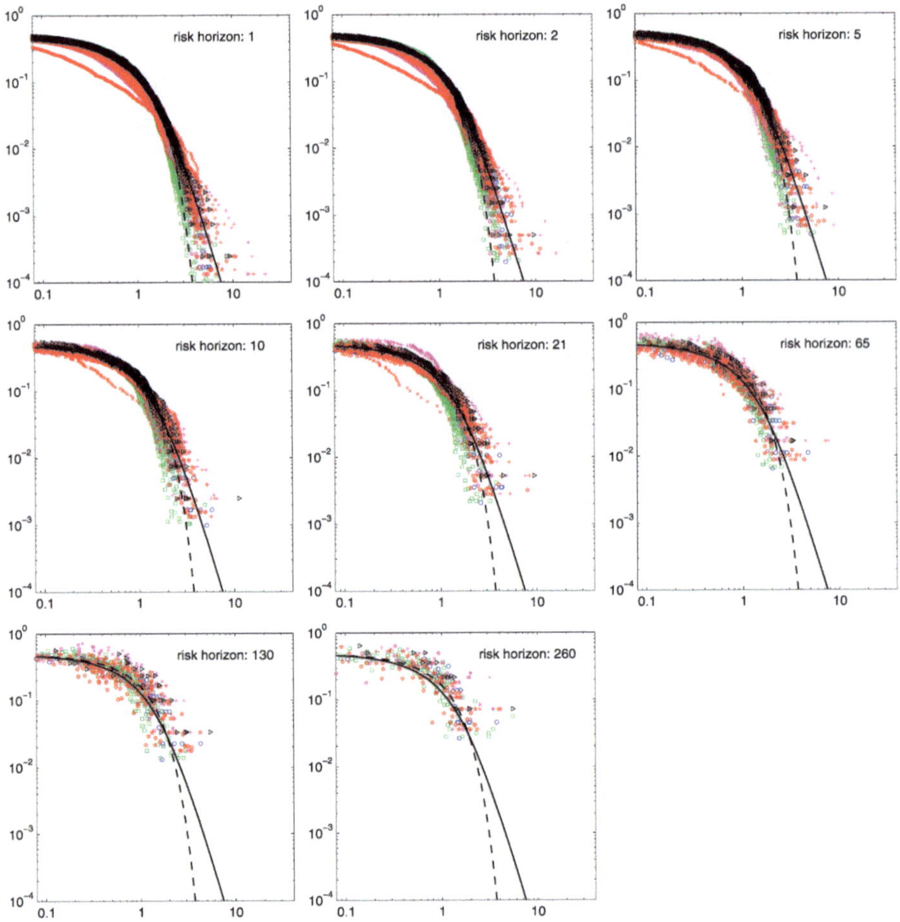

Fig. 13.4 The cumulative probability density $1 - \text{cdf}(\epsilon)$ versus ϵ for the positive tails

distribution with 5 degrees of freedom provides for a good global description of the cdf, including the tails. Changing the number of degrees of freedom ν for the Student distribution shows that $\nu = 3$ gives clearly too much tails, whereas $\nu = 8$ does not give enough tails. This range provides for good bounds on the possible values for ν.

With longer risk horizons, the amounts of data in the tails diminish, and it becomes increasingly difficult to make a clear distinction between different analytical descriptions. Up to one month, the Student distribution with a fixed number of degrees of freedom $\nu = 5$ seems to provide for a good description of the innovation pdf, but at 3 months and above, it is not possible to differentiate between the Student with $\nu = 5$ and a Gaussian.

Notice that we cannot use an analytical hint for $\nu(\Delta T)$, as there is no aggregation property for the innovations. In [154], a Monte Carlo simulations of a Market-

ARCH process is used to generate a long time series, which is subsequently analyzed for the innovation distribution. The conclusion is that a fixed ν is good, or possibly a weak increases of ν with ΔT can be used. But even with a simulation length of $\sim$1500 years, the issue is not clear-cut. Such a study shows that the innovation distribution does not converge toward a Gaussian at a pace compatible with a central limit theorem, and having a good tail estimate for long horizons remains a difficult problem.

Chapter 14
Leverage Effect

14.1 Introduction

Stock and stock index time series show a specific correlation called "the leverage effect". This stylized fact designates the negative correlation between historical return and realized volatility. Essentially, when a stock price goes down, the subsequent volatility is higher that when the stock price goes up. Intuitively, a decreasing stock price triggers sale orders which increase the volatility in the forthcoming time period. In the other direction, an increasing stock price induces the market participants to keep their positions, reducing the forthcoming volatility. Because a stock index is a weighted sum of selected stocks, the same effect is observed for indexes. The first evidence of the effect goes back to [21].

As the return and volatility are depending on their respective time intervals, a multiscale analysis of the effect should be done. In this chapter, we restrict the analysis to the case $\delta t_r = \delta t_\sigma$ for daily data. This effect is important for time interval on the order of months, and therefore using daily data is adequate. Moreover, the investigation for a large number of stocks is interesting with respect to the universality and size of the effect. Using daily data allows one to cover a large number of time series, without seasonality issues.

14.2 Empirical Statistics

The data set used for the empirical analysis contains daily close prices for the stocks of three indexes: SPI, DJ EuroStoxx, and S&P 500. The stocks are filtered for too short histories or too low liquidity. The final empirical sample contains 777 stocks, with a length between 2000 and 7000 days up to February 2010. The historical return and realized volatility are computed for the time horizons $\Delta T = 1, 2, 5, 10, 21, 42, 65, 130$, and 260 days. The empirical volatility is given by the simple microscopic volatility of the daily returns. This data set is analyzed extensively in [167] using a comprehensive set of statistics, whereas this chapter presents exclusively a detailed analysis of the leverage effect.

G. Zumbach, *Discrete Time Series, Processes, and Applications in Finance*,
Springer Finance, DOI 10.1007/978-3-642-31742-2_14,

Fig. 14.1 Scaling box plot for the heteroscedasticity, measured by the historical–realized volatility correlation $\rho(\sigma_h[\Delta T], \sigma_r[\Delta T])$

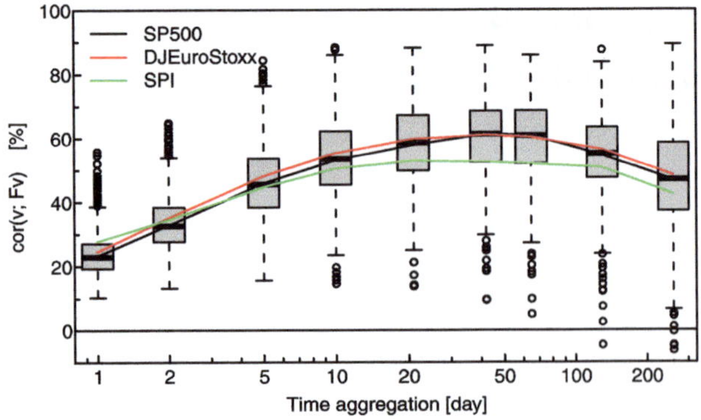

Fig. 14.2 Scaling box plot for the leverage effect, measured by the correlation between the historical return and the realized volatility $\rho(r_h[\Delta T], \sigma_r[\Delta T])$

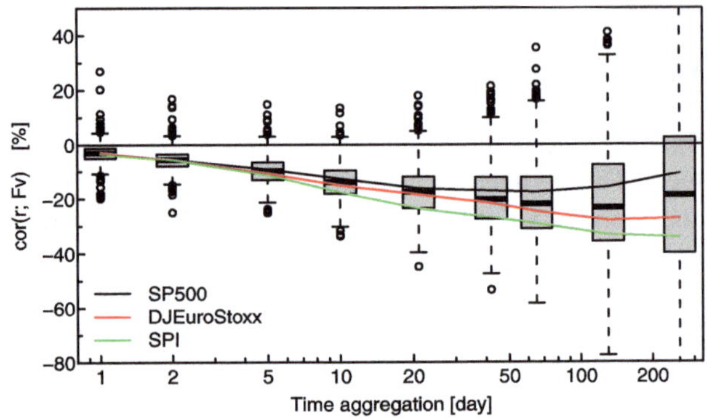

As with other financial time series, the largest correlation is between historical and realized volatility $\rho(\sigma_h[\delta t_\sigma], \sigma_r[\delta t_\sigma])$, as induced by the heteroscedasticity. Figure 14.1 presents a scaling plot of the cross-section of the volatility–volatility correlation. More precisely, the correlations $\rho(\sigma_h[\Delta T], \sigma_r[\Delta T])$ for all the stocks are displayed as one box plot for each ΔT, with all time series gathered together. The center of the box is given by the (cross-sectional) median, the box extends from the first to the third quartile, and the whiskers extend to 1.5 times the interquartile range or to the most extreme data point (whichever is closer to the median). The colored lines give the mean values for the stocks included in the three indexes. This representation gives us a good view of the cross-section properties for characteristic times ΔT ranging from 1 day to 1 year. The historical–realized volatility correlation is the largest correlation in financial time series, with typical values ranging from 20 to 60 %. The maxima of these correlations occur for time ranges between 1 to 3 months. For shorter time intervals, the volatility is not well estimated as the number of daily returns entering into the formula for computing $\sigma(t)$ is small. As the volatility for small ΔT contains a large statistical amount of noise, subsequent correlations are lower. For time ranges longer than a few months, the historical–realized volatility correlation decreases due to the decreasing memory on the volatility. Both effects lead to the observed hump shape.

Figure 14.2 displays the leverage effect using the same box plot scaling representation. The magnitude of the correlation $\rho(r_h[\Delta T], \sigma_r[\Delta T])$ is increasing with the time interval ΔT, with a maximum around 6 months. In the range between 1 month to 6 months, the mean correlation is of the order of -20 %, with most values between -10 to -30 %. Therefore, this is also a quantitatively large effect that should be included in a process when modeling stock time series.

14.3 Processes with Leverage and Monte Carlo Simulations

An important advantage of the ARCH family of processes is its flexibility to accommodate new stylized facts. In the present case, a dependency of the volatility on past returns should be added into the equations. For the GARCH(1, 1) process, extensions have been proposed by [75] and [57]. In a multiscale process, this is easily done by modifying the measure of volatility in (7.32), as explicited here for an aggregated definition:

$$r_k(t) = r[\Delta T_k](t) \tag{14.1a}$$

$$\sigma_k^2(t) = \mu_k \sigma_k^2(t - \delta t) + (1 - \mu_k)\left\{1 - \lambda_{\text{leverage}} \tanh\left(\frac{r_k}{\lambda_{\text{range}}\sigma_\infty}\right)\right\} r_k^2(t). \tag{14.1b}$$

The return r_k is a shorthand for the (aggregated) return measure at the time horizon $\Delta T_k = l_k \delta t$. The historical variance σ_k^2 is an EMA of the past squared returns, and the term in braces introduces the leverage effect. The intensity for the leverage is given by the parameter $\lambda_{\text{leverage}}$ with the constraint $-1 \leq \lambda_{\text{leverage}} \leq 1$. The sign in the equation is chosen so that a positive coupling corresponds to a negative correlation. The tanh function provides for a smooth leverage effect for small returns. The range for the smoothing is given by the parameter λ_{range} in units of the mean volatility. Intuitively, the term in braces models the behavior of market agents active at the corresponding time horizon, with a different response depending on the observed return.

The other process equations are unchanged. With only one component, the defined process is very similar to the GARCH(1, 1) process with leverage introduced by [57, 75], but the equations are not identical. The specification for the leverage in these articles is introduced with a $\text{sign}(r)$ term, corresponding to taking a very small value for λ_{range} in the above equation.

In order to capture stock time series, furthers effects should be included in the process. A first one is the finite precision for the prices, usually given as 1/100 of the monetary unit. The rules set by the stock exchange can be fairly complicated and depend on the stock price. Beyond the details of these rules, the precision is typically of the order of 1 BIPS or 0.1 BIPS (i.e., between 10^{-4} to 10^{-5}). The second effect is the inclusion of Holidays. Several methods can be used to capture Holidays, depending on the setting. In today's global world and cross-border trading, a simple method is to treat Holidays as randomly occurring day with zero return.

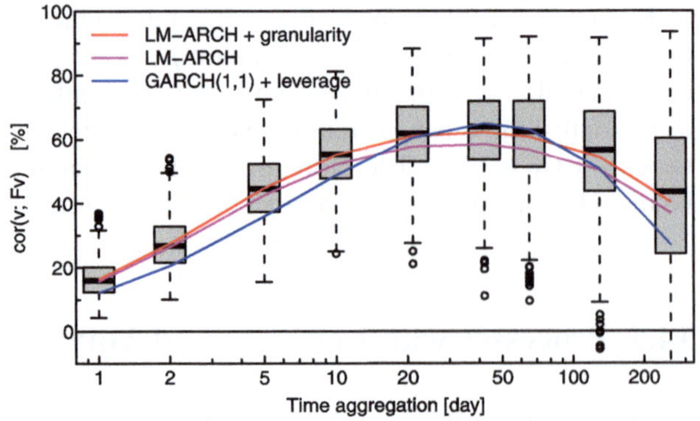

Fig. 14.3 The heteroscedasticity for the LM-ARCH process with granularity (*box plot*) and the means for a few processes (*colored curves*)

The inclusion of both effects in an ARCH process is straightforward; the details are given in [167].

With a long-memory aggregated ARCH process, the process parameters have been optimized so as to reproduce a broad set of statistics over the above time intervals. The leverage effect for stocks is well reproduced by the coupling constant $\lambda_{\text{leverage}} = 0.5$, while the parameter $\lambda_{\text{range}} = 0.5$ is less important. A similar study on the S&P 500 index shows a larger empirical correlation between historical return and realized volatility, of the order of -45 % for time intervals between 1 month to 6 months. The larger values for the empirical leverage effect are reproduced by a coupling constant $\lambda_{\text{leverage}} = 0.95$ in a multiscale ARCH process.

The processes are used to perform Monte Carlo simulations, so as to replicate the empirical data set in terms of the number of time series and their respective lengths. One set of parameter values is used, and the fluctuations in the statistics computed from the simulations are due only to the different realizations of the random price paths. The results for two simpler processes are included for comparison purposes. These processes are a GARCH(1, 1) process with Student innovations and leverage, and the long-memory ARCH process with leverage and Student innovations but without granularity and Holidays. The parameters are set according to the same global screening for the statistics. For the GARCH(1, 1) process, the parameters are $\tau = 32$ days, $w_\infty = 0.115$, and $\lambda_{\text{leverage}} = 0.6$. The setup for the scaling graphs are similar to the empirical ones, with the boxes computed from the LM-ARCH process with granularity and Holidays, and the curves corresponding to the mean for the various processes.

The heteroscedasticity for the processes is show in Fig. 14.3. The too narrow range provided by a GARCH(1, 1) process is clearly visualized, similarly to the mug shot for the foreign exchange rates. Let us emphasize that the location and height of the maximum can be adjusted by changing parameters, but the shape cannot be made broader. The heteroscedasticity with the long-memory process is essentially accurate, however too small in the short ΔT region. This is due to the simulations based on a daily increment: the historical volatilities σ_k are fairly noisy, and this noise limits the induced correlation.

The simulated leverage effect is shown in Fig. 14.4. Interestingly, the shape for the one-component GARCH process is quite similar to the empirical one. As for the

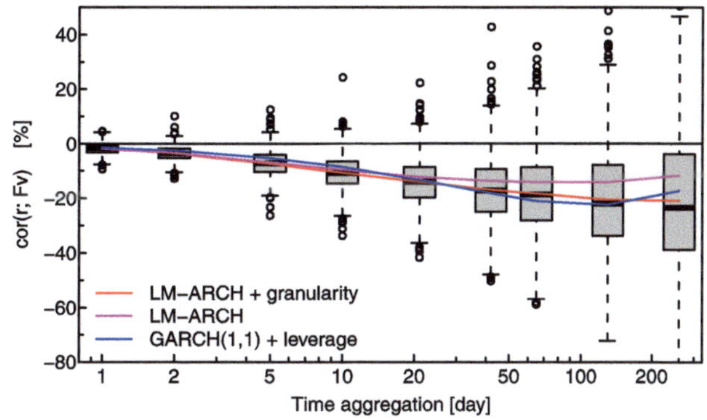

Fig. 14.4 The leverage effect for the LM-ARCH process with granularity (*box plot*) and the means for a few processes (*colored curves*)

heteroscedasticity, the shape is slightly too narrow. The multiscale ARCH process is better at reproducing the empirical statistics. As visualized by the box plot, the dispersion of the empirical statistical values across the time series is also very similar to the dispersion in the simulations. This similarity between empirical stock time series and simulations done for a single set of parameter values is observed for all statistical estimators (see [167] for details). This finding points to a fairly universal description of financial data, with most parameters fixed at default values.

Summarizing this multiscale analysis of the leverage effect in stock time series and for selected processes, the magnitudes of the empirical correlations and of the coupling constant $\lambda_{\mathrm{leverage}}$ show clearly that the leverage effect is quantitatively important for stock and stock indexes. A simple modification of the basic multiscale ARCH processes can capture correctly the leverage for increasing time horizons. A good description is already obtained by a one-component process with leverage (similar to an asymmetric GARCH(1, 1)).

Chapter 15
Processes and Market Risk Evaluation

15.1 Background on Risk Evaluation

Financial risks are many, but the two important ones are credit risks and market risks. Credit risks assess the potential losses due to the default of the counterparty in a contract, while market risks quantify the potential losses due to adverse price fluctuations. Market risks are mainly due to loss of the market value of a security, for example, an equity in a foreign currency can be affected by the depreciations of the equity and of the foreign exchange rate. But market risks also include credit risks, for example, a bond emitted by a country or a legal entity includes in its pricing the information related to the credit worthiness of the bond's emitter, namely the risk of default before the bond mature. To deal with such mixed cases, in this particular example mixing interest rates with credit information, the pricing is decomposed into a pure market part, e.g., risk-free interest rates, and a pure credit part considered as a spread over the former part. Such decompositions allow one to separate the specificity of each part and to use various models and parameters most appropriate for them. The premises used to compute market and credit risks are very different. Market risks are estimated by using information from the past, measuring volatilities and correlations. This approach makes the strong implicit hypothesis that the past contains enough information to construct meaningful forecasts. On the other hand, for a given legal entity, default has never materialized before, and therefore the past credit history is of limited use. Because of their strong relations with processes, we discuss only pure market risks in this book. The reader interested to credit risk and the interplay with market risks can see, for example, [83].

A risk measure is a statement about the future, namely an estimation of the possible large losses between the current time t and some future time $t + \Delta T$. The core object underlying all risk measures is a forecast for the probability distribution $\tilde{p}_{\Delta T}(r[\Delta T])$ of the returns $r[\Delta T]$ over the corresponding interval ΔT. From this probability distribution, risk measures can be computed easily, like Value at Risk (VaR) and Expected Shortfall (ES). These risk measures focus on the tail of the loss distribution and quantify the severity of infrequent adverse price moves. This choice of risk measure is different from the standard deviation, which measures the

G. Zumbach, *Discrete Time Series, Processes, and Applications in Finance*, Springer Finance, DOI 10.1007/978-3-642-31742-2_15,

size of the usual price fluctuations. If the probability distribution of profit and losses is fixed, say a Gaussian or a Student distribution, the differences between all these measures are simple scaling factors. In such a case, it is enough to compute the expected standard deviation and to scale the result appropriately, depending in the assumed probability distribution and risk measures. Such an approach is perfectly valid for the simplest linear instruments, like equities and FX (provided that the assumed distribution captures correctly the empirical one). Yet, many securities have nonlinearities, for example, all bonds and options, and the actual return distribution can be quite different from a simple symmetric Gaussian or Student distribution. For these instruments, the risk measures have to be estimated from the tails of the return distributions (see point 4 below).

The VaR and ES quantify how bad are large losses that occur with a small probability, typically 1 % or 5 % of the time. The complement of this probability is denoted often by the letter α and can be chosen freely by the user. Essentially, the closer is α to one, the more extreme the events that are considered.

The value at risk (VaR) is defined by

$$\mathrm{VaR} = \arg\min_{l} \left\{ \int_{-\infty}^{l} dl' \, \tilde{p}_{\Delta T}(l') \geq \alpha \right\} \tag{15.1}$$

where $l = -r$ is the loss, defined simply as minus the profit. This simple change of sign is a mere language convenience, as ones speaks about large (positive) losses, and not about very negative profits. Consequently, the interest focuses on the upper tail of the loss distribution and therefore to a cumulative probability close to 1. For continuous probability distributions, the above definition can be rewritten in the more intuitive form of an implicit equation for the quantile corresponding to VaR:

$$\alpha = \int_{-\infty}^{\mathrm{VaR}} dl \, \tilde{p}_{\Delta T}(l). \tag{15.2}$$

In these definitions, the VaR value separates the mild losses (and profits) occurring with frequency α from the large losses occurring with frequency $1-\alpha$. It should be emphasized that VaR is the *smallest* of the large losses, and *losses larger than VaR should occur with frequency* $1-\alpha$. These events are often called "VaR breaches". If the VaR breaches do not occur with the chosen probability $1-\alpha$, the evaluation of the probability distribution $\tilde{p}(r)$ is likely incorrect. It should be emphasized that observing less VaR breaches than expected is not a sign of a careful management or a wise financial expertise, but denotes only an inaccurate computation of the risks.

The expected shortfall (ES), also called expected tail loss (ETL) or conditional VaR (CVaR), is defined by

$$\mathrm{ES} = \frac{1}{1-\alpha} \int_{\mathrm{VaR}}^{\infty} dl \, \tilde{p}_{\Delta T}(l) l. \tag{15.3}$$

Essentially, the ES measures the average loss for losses larger than VaR. Using the inequality $l > \mathrm{VaR}$ under the integral, the inequality

$$\mathrm{VaR} < \mathrm{ES} \tag{15.4}$$

between both risk measures is obtained. Similarly to the expected shortfall, higher moments of the tail distribution can be defined, but they are less intuitive.

Both risk measures have advantages and drawbacks.

1. Both risk measures are monetary amounts, either in the position currency or in the portfolio currency. For example, on a given portfolio, the values can be VaR = 1M USD and ES = 1.4M USD. These numbers give a direct sense about how large the losses can be and can be compared against the corresponding provisions.
2. VaR is easier to understand (the minimal losses occurring with a given frequency) than ES (the expected loss given that losses larger than VaR occur).
3. VaR is not concave, whereas ES is. Concavity in the present context means that the ES of a sum of two portfolios is less than or equal to the sum of the expected shortfalls of each subportfolio. Essentially, concavity says that diversification should lower the risk. In two important papers, [10, 11] studied the concavity properties of risk measures. In particular, VaR does not have this desirable property. From a conceptual point of view, the nonconcavity of VaR is a strong argument in favor of using ES. Yet, carefully crafted positions should be constructed in order to have an actual example of a nonconcave portfolio. Therefore, even if the theoretical objection is important, in practice this is not a crucial criterion, and VaR is widely used because it is easier to understand.
4. VaR is mildly sensitive to the tail (it is a quantile), whereas ES measures the weight of the distribution up to "infinity". The case in point are instruments that have a profit and loss profile that are very asymmetric, similar to a life insurance. These instruments show frequent small profits (the incomes corresponding to the life insurance premium) and infrequent large payouts (the insured amount in case of death). Indeed, very asymmetric instruments can be found mostly in the credit area, like bonds, mortgages or CDS (credit default swaps). This asymmetry was at the core of the difficulties of AIG (American Insurance Group) when many large firms, underlying in CDS contracts, defaulted during the financial crisis in Fall 2008. For such instruments, measuring risk with the ES provides for a better quantification of the tails.
5. Other coherent measures of risks can be introduced. The idea is to start from the axiomatic properties such measures should have, like homogeneity or concavity, and then to construct all measures that satisfy the properties. In general, they are given by an integral of the probability p over the interval [0, 1] of the inverse cumulative distribution function, with a decreasing weight depending on p. The original contributions are [1, 2], and the book [108] contains an excellent summary of this approach. In practice, these general measures are not much used because not very intuitive, except for the expected shortfall.

At the end, both VaR and ES should be reported, and it is to the users to decide about their favorite risk measures. Beyond VaR and ES, it is also important to compute the partial risk contributions in a portfolio, as well as the risk sensitivities to a change in the weights in a portfolio [109]. In order to modify a portfolio to mitigate the risks, a more in-depth analysis in a portfolio allows one to gain a proactive view, using the partial risk contributions along subportfolios, positions, and risk types.

Given the forecasted probability distribution for the returns $\tilde{p}(r)$, the computation of the actual risk measures is straightforward. Therefore, the core problem of market risk evaluation is to *compute* a good forecast for $\tilde{p}(r)$ and to *validate* this forecast. A *risk methodology* is essentially an algorithm to compute the desired probability distribution forecasts, while *backtesting* measures the performances of the forecasts on historical data. Sects. 15.2 to 15.5 discuss the core ingredients used to build risk methodologies and their relationships with processes. This chapter then concludes with a discussion of backtesting in Sect. 15.6, with an illustration using the DJIA over the last century. More detailed results can be found in [154, 155] and [168].

15.2 "Return-Based" Versus "Innovation-Based" Risk Methodologies

Broadly speaking, the risk methodologies can be based on two variables, the returns r or the innovations (or residuals) ϵ. The assumptions and algorithms behind both families of risk methodologies are fairly different. This section discusses the basic ideas underlying both approaches, while the innovation-based risk methodologies are discussed in more details in the subsequent sections.

1. The return-based risk methodologies are the simplest from a conceptual point of view. The historical methodology is the key representative in this family and is widely used. The idea is to take a sample of historical daily returns, typically one or two years, and to resample the historical data. For longer risk horizons, a simple random walk scaling with $\sqrt{\Delta T}$ is used. In terms of probability distributions, the forecast $\tilde{p}(r)$ is spanned by the historical daily returns contained in the sample of chosen length, scaled by $\sqrt{\Delta T}$.

   The strong appeal for this methodology comes from its simplicity. It sounds hypothesis free, as it is equivalent to use only the past returns, without any treatment. This simplicity is deceiving, and the methodology misses some important stylized facts. The forecasted volatility is given by the equal-weight historical sample volatility, and this misses the point that recent returns carry more information than distant ones. Fat tails for the return distribution are included, but the historical sample should be long enough to include enough extreme events. A long sample would be needed to better capture the tails, but then the heteroscedasticity is missed. With this methodology, a nonoptimal trade-off must be done between a sample long enough to capture the fat tails and short enough to capture the volatility dynamics. Then, the square root scaling is fine for the variance, but not for the whole probability distribution, and in particular for its tails. This simple scaling does not take into account the change of the shape of the return probability distribution with the forecast horizon, as would be implied by the central limit theorem (with the caveat that the convergence is much slower for financial time series). Essentially, the historical methodology is correct for stationary processes, but the actual returns are heteroscedastic.

An attempt to improve the historical methodology is to weight the past returns, for example, with an exponential decaying function. Then, the risk is evaluated by drawing at random the past returns with a probability given by the weights. This procedure emphasizes the recent past, in agreement with the observed heteroscedasticity. But then, it will miss fat tails. The probability distribution for the forthcoming return implied by this algorithm is spanned by the historical returns with the chosen weights. With exponential decreasing weights, the distribution is fairly singular as based effectively on a small sample. Such a solution improves on the heteroscedasticity side but makes the situation worse with respect to extreme events.

2. The methodology based on the innovations corresponds to a /location/size/shape/ decomposition of a probability distribution. For any probability distributions (with support over $\mathbb{R}$) with a given shape, the location (mean) and size (variance or standard deviation) are two parameters of the distribution, while other parameter(s) might be present. For financial time series, the /location/size/shape/ decomposition corresponds to the /mean return/standard deviation/innovations distribution/ decomposition

$$\tilde{p}(r) = \tilde{r} + \tilde{\sigma}\ p(\varepsilon). \tag{15.5}$$

In this form, the heteroscedasticity is captured by $\tilde{\sigma}$, leaving (in principle) a stationary distribution $p(\varepsilon)$ for the innovations. In this equation, the analogy with processes is clear, the only difference being that $r[\delta t]$ is used in a process (for risk, δt would be one day) while $r[\Delta T]$ is used for risk evaluations. Because of this difference, the innovation distribution $p_{\Delta T}(\varepsilon)$ is depending on the risk horizon ΔT. To summarize the innovation-based approach, the goal is to obtain an i.i.d. random variable ϵ for a given risk horizon ΔT, after removing the predictable part in the mean and the variance of the price changes.

It is fairly easy to build many risk methodologies based on different forecasts for the mean return, volatility, and probability distribution of the innovations. The issue is then to find a constructive method to build efficient risk methodologies, and the relationship with processes helps to select the various components.

In essence, the return-based risk methodologies are deceivingly simple (i.e., resample the past), but they fail to capture accurately the most important stylized properties of financial time series, namely heteroscedasticity and fat tails. On the other hand, the innovation-based risk methodologies offer a natural structure to capture the specificity of the financial time series. When this freedom is used appropriately, they provide for more accurate return distribution forecasts. Extensive backtesting just confirms this simple structural analysis of the return- versus innovation-based risk methodologies. For these reasons, only the presentation of the innovation-based risk methodologies is pursued in the next sections.

15.3 General Considerations on Innovation-Based Risk Methodologies

The first innovation-based risk methodology was established in a seminal paper [110] by researchers at JP Morgan (see also [109]). This methodology was used for the internal risk control inside the bank, and its public disclosures raised a strong interest in the financial industry. The demand for risk control systems leads to the spin-off of the risk control group and subsequently to the creation of a whole industry around risk management for various financial institutions. We call this methodology *RM1994*. Some key advantages of this methodology is its simplicity, conceptually, and computationally. This also points to its limitations, and several improvements have been proposed since.

For the methodologies based on the innovations, there is a strong relation between risks, processes, and forecasts. Clearly, the possible behaviors of the risk factors are directly related to the risk in the following period, as large moves can lead to important losses. A process is needed to model mathematically the dynamics of the time series, and through conditional expectations, a process also provides for the return and volatility forecasts. If the process is well chosen, for example, a quadratic ARCH structure, the conditional expectations can be evaluated analytically, giving simple explicit equations for the desired volatility forecast. The volatility forecast for $r[\Delta T]$ corresponds to the size of the return probability distribution $\tilde{p}(r[\Delta T])$. Therefore, a process can be associated with most risk methodology, and the performance of a risk methodology follows from the ability of the process to reproduce some key stylized facts across a large spectrum of time series. For a quadratic volatility process, the variances forecasts $\widetilde{\sigma^2}$ for increasing risk horizons ΔT can be evaluated. We are using the convenient shorthand for the volatility forecast $\tilde{\sigma} = (\widetilde{\sigma^2})^{1/2}$.

Using a process to infer the volatility forecasts has another key advantage: the same process and parameters can be used for all forecast horizons ΔT. In this way, one gains a strong consistency between risk horizons. Thanks to this consistency, long term horizons can also be dealt with, even though the amount of independent data could be insufficient to statistically validate a "brute force" volatility forecast. In this way, meaningful risk figures can be obtained up to a one year horizon. This is of critical interest for pension funds or insurance firms that have a basic business cycle of one year. Yet, the whole financial industry benefits from being able to assess market risks for horizons longer than a few days. For example, regulatory capital for banks as set by the Basel agreement would be much less procyclical if based on a long-term criterion.

The core raw information used by all market risk methodologies is the historical returns. This supposes that there is information in the history relevant for market risks in the forthcoming period ΔT. In a Bachelier world, the volatility is a constant for each time series, and the largest possible sample for each time series should be used to compute the related historical volatility (in a multivariate setting, the largest sample should be used to compute the historical covariance matrix). These historical volatilities (or covariance) correspond to the desired volatility forecast. In the actual financial world, the need for a more sophisticated volatility forecast in the risk computations is rooted in the heteroscedasticity of the financial time series. Because the volatility is time varying and with a (long) memory, the recent past carries more information than the distant past, and the decay of the information content with the time lag suggests to weight with decreasing importance the historical volatility information. All risk methodologies are associated with a weighting scheme applied to the historical squared returns.

The volatility dynamic is nowadays better understood, and the broad characterization of the heteroscedasticity has been presented in Chaps. 3 and 12. In particular, the long memory of the volatility is observed in almost all time series. It is therefore a very natural step to build a risk methodology based on a long-memory process like LM-ARCH, as done by [153, 154] in the RM2006 methodology. The analytical computations related to quadratic ARCH processes have already been presented in Sect. 7.7 and are quite straightforward. Yet, two specific issues need to be addressed to build a risk methodology: the evaluation of the process parameters and the distribution of the innovations.

Today, portfolios are large, ranging from 100 to 100,000 contracts, and the investment universe is global, ranging from liquid high-quality bonds to stocks, bonds, and currencies from emerging countries. Consequently, a risk methodology has to be robust. Very robust. And simplicity is a good mean to reach the desired robustness. The robustness requirement points to a minimal number of parameters and eliminates all schemes where the parameters depend on individual time series. In particular, the mean volatility parameter is time series dependent. The distinction between linear and affine processes has already been presented in Sect. 7.4. The simplicity and robustness criterion eliminates all affine processes for risk estimation because of the mean volatility parameter, as, for example, GARCH(1, 1). Instead, linear processes should be used, like I-GARCH(1) or the long-memory linear ARCH process.

The second point that should be addressed is the probability distribution of the innovations. In a risk methodology, the relevant distribution concerns the conditional distribution, as captured by the innovations $\epsilon = (r - \tilde{\mu})/\tilde{\sigma}$. As shown in Chap. 13, the distributions for the innovations are observed to also have fat tails, albeit with a slightly larger exponent than the unconditional returns. The Gaussian assumption used in the standard risk methodologies is at odd with these empirical observations. Several authors (see, e.g., [27, 117]) have criticized this aspect of the standard methodologies and have proposed to use fat-tailed distributions instead of a Gaussian. Yet, it should be clear that the probability distribution for the innovations can be chosen as needed: a Gaussian choice is a good first-order approximation, while a Student distribution captures more accurately the empirical distributions. As risk estimations are concerned mainly with large events, it is important to capture correctly the tails of the innovation distribution, and a Student distribution is clearly better.

Finally, an inconsistency in the scale appears when a volatility forecast extracted from a process is used. The inconsistency can be understood by starting from the base equation (15.5) expressed for the random variables r and ε:

$$r[\Delta T] = \tilde{r}[\Delta T] + \widetilde{\sigma}[\Delta T]\varepsilon. \qquad (15.6)$$

To simplify the discussion, let us take $\tilde{r} = 0$. The empirical variance for the historical returns is

$$\langle r^2[\Delta T]\rangle = \langle \widetilde{\sigma^2}[\Delta T]\rangle \langle \varepsilon^2\rangle. \tag{15.7}$$

The scale for the innovations is fixed according to the condition $E[\epsilon^2] = 1$. Using empirical data for a sample long enough, the condition becomes $\langle \varepsilon^2\rangle \simeq 1$. The scale for the returns is fixed by the actual data, and the scale of the volatility forecast is fixed by the historical data and by the process with its parameters. But there is no reason that $\langle r^2[\Delta T]\rangle \simeq \langle \widetilde{\sigma^2}[\Delta T]\rangle$. In order to be able to respect the condition $\langle \epsilon^2\rangle \simeq 1$ when computed from empirical data, a "volatility scale factor" $\gamma[\Delta T]$ depending on ΔT has to be introduced. This term is incorporated in the volatility forecast as a simple multiplicative factor, and when not mentioned, the default value is $\gamma[\Delta T] = 1$. This correction is slightly larger than one for the RM2006 methodology (see Eq. (15.10))

15.4 The Main Innovation-Based Risk Methodologies

15.4.1 The RM1994 Risk Methodology

The RM1994 risk methodology is built on the following basic ingredients:

1. a zero return forecast,
2. an EWMA for the volatility forecast, and
3. a Gaussian distribution for the innovations.

In the RM1994 methodology, the volatilities and correlations are estimated by an exponential moving average (EWMA) depending on one parameter, namely the decay factor μ of the exponential. This corresponds to weights decaying exponentially fast with a characteristic time $\tau = -\delta t / \ln(\mu)$. For assessing risk at a one day horizon, empirical studies show that the decay factor $\mu = 0.94$ provides for good volatility forecast for all assets. The usual values for this parameter are $\mu = 0.94$ or 0.97, corresponding respectively to an exponential characteristic time of 16 or 32 business days. Having only one parameter for all time series contributes to the simplicity of the RM1994 methodology and to its wide acceptances. The volatility scale factor is the default $\gamma[\Delta T] = 1$.

The exponential moving average produces a simple and effective volatility forecast depending on only one parameter. This scheme corresponds to the I-GARCH(1) process with the same value for μ. Yet, the I-GARCH(1) is not able to reproduce correctly the multiscale aspect of the volatility dynamics. In order to produce good volatility forecasts and risk measures at the desired risk horizon ΔT, the optimal decay factor must be adjusted with ΔT. This inefficiency occurs because a long-term forecast must use more information from the distant past than a short-term forecast, but because the I-GARCH(1) process has only one exponential time scale, the volatility forecast is a constant independent of the risk horizon (up to the trivial

$\sqrt{\Delta T}$ scaling). A long-memory process is better as it captures the multiple time scale dynamics of the volatility.

15.4.2 The RM2006 Risk Methodology

The RM2006 risk methodology is built on the following basic ingredients:

1. a return forecast based on an autoregressive model for the returns,
2. a long-memory microscopic ARCH process in order to build the volatility forecast,
3. a Student distribution with 5 degree of freedoms for the innovations.

Some general features are presented below; the details are given in the original article [154].

As computed explicitly in Sect. 7.6, the volatility forecast without an AR term can be expressed in the simple form

$$\widetilde{\sigma^2}[\Delta T](t) = \frac{\Delta T}{\delta t} \sum_{i \geq 0} \lambda(n, i) r^2(t - i\delta t) \tag{15.8}$$

with $\Delta T = n\delta t$. The coefficients λ obey $\sum_{i \geq 0} \lambda(n, i) = 1$ for all n and can be computed by recursion equations. This form for the forecast shows the strong similarity between the one-step forecast, as used in the process definition in (7.37), and the n-step forecast. The leading term in (7.38) is given by $\widetilde{\sigma} \simeq \sqrt{\Delta T/\delta t}$, namely by a "square root law" of the risk horizon. This leading term originates in the diffusive nature of the process, namely the (logarithmic) price process follows a random walk. Then, the n-dependency in the coefficient $\lambda(n, i)$ brings in the corrections due to the particular memory structure of the process.

The RM1994 risk methodology with EWMA weighting corresponds to an I-GARCH process, for which the weights become

$$\lambda(n, i) = \lambda(i) = \frac{1 - \mu}{1 - \mu^{i_{\max}}} \mu^i \tag{15.9}$$

namely there is no n-dependency in the forecast. In other words, the volatility forecasts for all time horizons are given by the volatility as computed with the single EWMA, scaled by $\sqrt{\Delta T/\delta t}$.

Figure 15.1 shows the weights $\lambda(n, i)$ for an I-GARCH process and for a long-memory process, in semi-logarithmic scales. The black and blue straight lines correspond to the I-GARCH process, with no dependency on the forecast horizon. The curved lines correspond to the long-memory process, where the black line gives $\lambda(i) = \lambda(1, i)$ (i.e., the one day forecast), and the colored curves correspond to the forecast horizons $n = 5$, 21, 65, and 260 days. We see clearly that with increasing forecast horizons, the weights decrease on the recent past and increase on the distant past. This behavior is fairly intuitive, as a long-term forecast should use more

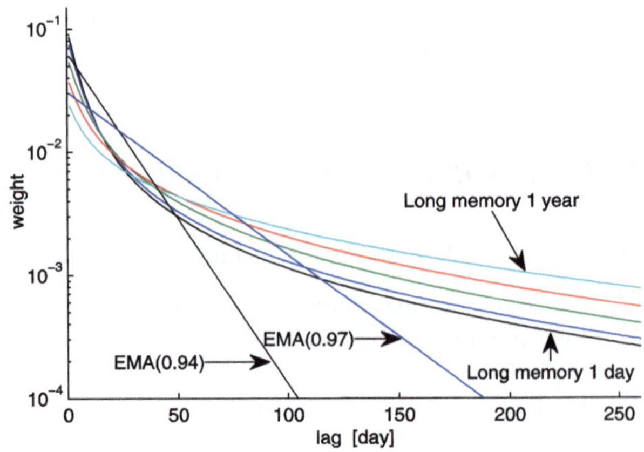

Fig. 15.1 The weight $\lambda(n, i)$ versus the lag i for the I-GARCH and long-memory processes. The curves with labels "long memory 1 day" and "long memory 1 year" correspond to the long-memory process, for a forecast horizon of one day and one year, respectively

information from the distant past, whereas a short-term forecast is dominated by the recent events. Let us emphasize that the dependency of the forecasting weights on the forecast horizon follows from the process equations.

For the RM2006 risk methodology, the volatility scale factor $\gamma[\Delta T]$ was estimated empirically:

$$\gamma[\Delta T] = 1.06 + 0.008\big(\ln(\Delta T/\delta t)\big)^2. \tag{15.10}$$

Its values are slightly larger than 1, showing that the volatility forecasts as computed from the LM-ARCH process equations are not large enough to account for the actual return variances. The origin is likely to be found in a slight misspecification of the process.

Detailed empirical studies show that an autoregressive term μ_{eff} must be included in order to construct a process that captures well the properties of the financial time series. The theoretical approach is to consider μ_{eff} as small and treat it in perturbations. This introduces corrections for the volatility forecast, as well as a nonzero forecast $\tilde{r}[\Delta T]$ for the return at the risk horizon ΔT. The analytical computations related to μ_{eff} are more complex and given in [154]. The key point is that, because the process includes only linear and quadratic terms, we are able to evaluate analytically the conditional expectations needed to compute return and volatility forecasts.

15.5 Market Risk Evaluations in Practice

A risk methodology should be rooted in the empirical properties of the financial time series. A thorough understanding of the "stylized facts" is needed in order to select the appropriate model and parameters. Because a risk methodology is applied to so many time series, it is essential that the model be "universal", namely that it captures the generic properties of the data. To some extend, building a risk methodology is an engineering problem: one has to pick the important parts and neglect smaller or particular effects. For this purpose, large-scale empirical studies should be done, the generic features described, and possibly approximations made. The aim is not to be rigorous (say in the sense of hypothesis testing), but to be "overall quantitatively correct". This approach contrasts with academic studies, where often a rigorous answer to a well-posed question is sought after.

This chapter discusses the core of the risk methodologies, namely it explains for one "simple" asset how to compute risk estimates and the relation with processes. But the application of these ideas to large portfolios at an industrial scale requires several supplementary level of complexities.

The first level of complexity is related to the derivative contracts. A "simple" base time series is called a risk factor. They correspond essentially to linear assets, like equities or foreign exchange currencies. The derivatives should be priced as functions of the underlying risk factors, and these contracts can be fairly sophisticated. A simple example of "derivative" are bonds: the underlying simple risk factor is the interest rate curve as a function of the maturity. A bond with its coupons occurring at various time points should be priced from the underlying interest rate curve. More complex derivatives can be included in a portfolio, say, for example, swaptions. Moreover, some trading strategies involve long and short positions in derivatives and the underlying assets. These kinds of strategies are often used by option market makers and in hedge funds, and they can expose the discrepancies between the option pricing used in the market and in the risk evaluation. Therefore, the option pricing algorithms used in risk evaluation has to follow closely the best practices used in the market.

A second level of complexity is related to the actual details of the financial contracts. For example, a simple bond pays coupons according to a given schedule. In order to price this bond from the interest rate curve, the actual coupon schedule must be known. Therefore, it is not enough to have long time series for the values of the risk factors, the terms and conditions of the actual contracts should also be available.

A third level of complexity is related to the detailed analysis of the risks inside a portfolio: beyond the global risk figure, one also would like to understand where the risks originate from. For example, an analysis according to the currencies or industrial sectors can give a diagnostic that will help reducing the risk exposure. Insights about global changes can be obtained by scenario analysis, or "stress test" analysis (what if oil prices rise? What if the USD interest rates decrease? etc.). Possibly, hedges can be added in a portfolio following such an analysis.

At the end, the techniques presented in this chapter give the foundations of risk evaluations, yet they represent only a tiny fraction of the software size in actual applications developed for risk management. A large part of such softwares corresponds to the first and third points above, whereas the second point requires maintaining large data bases feeded by many data sources. Even though the core risk methodologies can be more or less complex, it is just a small part in a much larger machinery.

This chapter uses only a univariate analysis, but a very large number (say of the order of 10.000!) of time series can be needed for the computations in an actual portfolio. The extension of the current analytical framework to multivariate processes is natural, as the equations involve only linear and quadratic terms. For example, the scalar term r^2 appearing in the scalar variance is replaced by the matrix $r_\alpha r_\beta$ for the covariance matrix, with α and β indexing the risk factors. Such a natural extension is used for actual portfolios, under the implicit hypothesis that the minimal multivariate extension is optimal. Yet, the multivariate empirical performances should be investigated. A risk evaluation on a portfolio raises many new questions, related to the multivariate aspect of the volatility forecast, to the joint distribution of the innovations, and to the multivariate backtesting. This is still an active research area, and some aspects are presented in Chaps. 17 and 18.

15.6 Backtesting

15.6.1 Stating the Problem

As we mentioned above, the key quantity in risk evaluation is the forecast $\tilde{p}_{\Delta T}(r)$ for the return probability distribution. This forecast is time dependent, and therefore we have actually a time series $\tilde{p}_{t,\Delta T}$. Corresponding to the chosen risk horizon, the time series of realized returns $r[\Delta T](t + \Delta T)$ is computed from the historical data sample. Essentially, we have a sequence of distributions and a sequence of single draws from the corresponding distribution. The goal of backtesting is to assess if the sequence of realized returns could have been drawn from the sequence $\tilde{p}(r)$.

Because of the time dependency of the probability distributions, the problem is substantially more difficult than assessing if a dice is fair. The difficulty is that because $\tilde{p}$ is changing with time, the returns cannot be directly compared. There are two routes to tackle this problem. A first solution is to study exceedances over a VaR threshold. This corresponds to mapping the sequence of returns into a sequence of 0s and 1s depending on whether the return values are below or above the selected VaR level. The idea is simple but has a number of drawbacks: (1) it leads to more complex statistics; (2) for large VaR thresholds, most of the data are discarded (i.e., mapped to 0); (3) the quantitative information is lost (i.e., the excesses can be large or small, they are mapped to 1). A better solution is to map the returns to the corresponding probtiles, as presented in the next subsections. A comparison of both approaches can be found in [35].

15.6.2 Mapping to the Probtiles

The quite peculiar setup of the backtesting problem can be addressed elegantly using probtiles, following an idea that likely goes back to Paul Lévy and is used in [135]. Using the forecast $\tilde{p}$ for the return probability distribution, we compute the corresponding cumulative density function (cdf). The cdf is used to map the realized return to the corresponding (realized) probtile $z(t)$:

$$z[\Delta T](t) = \widetilde{\mathrm{cdf}}_{t,\Delta T}\big(r[\Delta T](t+\Delta T)\big) \qquad z \in [0,1]. \tag{15.11}$$

The realized return specifies a *quantile* of the distribution, and z is the cumulative *probability* for the quantile. The contraction *probtile*—standing for the probability of a quantile—gives a convenient name to z. Let us emphasize that the mapping given by $\widetilde{\mathrm{cdf}}_t$ is time dependent. If the realized returns are drawn from $\tilde{p}$, then the probtiles should be uniformly distributed in [0, 1]. Because the probtiles should have the same (stationary) distribution, we can compare them and in particular test for uniformity using the whole sample of probtiles.

This backtesting framework uses the output of any risk methodology, namely the forecasts for the return probability distribution and the realized returns. Therefore, it can be used for all methodologies, either return-based or innovation-based. This can also be used for complex portfolio, including nonlinear instruments like derivatives.

For risk methodologies based on residuals, and for simple time series corresponding to risk factors, the backtesting framework can be formulated in a simpler way than Eq. (15.11). The returns and residuals distributions are related by Eq. (15.5). Similarly, the realized returns can be mapped to the corresponding (realized) residuals using Eq. (13.4), and the residuals should have a fixed distribution defined by the risk methodology. Therefore, we can map the residuals to the corresponding probtiles, and Eq. (15.11) becomes

$$z[\Delta T](t) = \mathrm{cdf}_{\mathrm{th}}[\Delta T]\big(\epsilon[\Delta T](t)\big) \qquad z \in [0,1]. \tag{15.12}$$

In this case, the $\mathrm{cdf}_{\mathrm{th}}$ is given by the risk methodology (e.g., corresponds to the cdf of a Gaussian or Student distribution). Similarly, for the historical methodology, Eq. (15.11) can be written in the more convenient form

$$z[\Delta T](t) = \widetilde{\mathrm{cdf}}_t\left(\sqrt{\frac{\delta t}{\Delta T}}\, r[\Delta T](t+\Delta T)\right) \qquad z \in [0,1]. \tag{15.13}$$

In this case, the cdf corresponds to the cumulative density function of the daily returns, evaluated using 1 year of historical data.

Another desirable property of the probtiles is to be independent. The underlying idea is that a given exceedance at time t carries no information about the future exceedances. The conditions on the uniform distribution and absence of serial dependencies can be summarized as

$$z(t) \overset{\text{i.i.d.}}{\sim} U(0,1). \tag{15.14}$$

As it is known that returns exhibit heteroscedasticity (i.e., cluster of large events), a good risk methodology should discount for the changing volatility. In backtesting, an improper volatility discounting results in cluster of probtiles, and we want to test for possible serial correlations. Because the mapping between residuals to probtiles is fixed for innovation-based methodologies, the independence condition

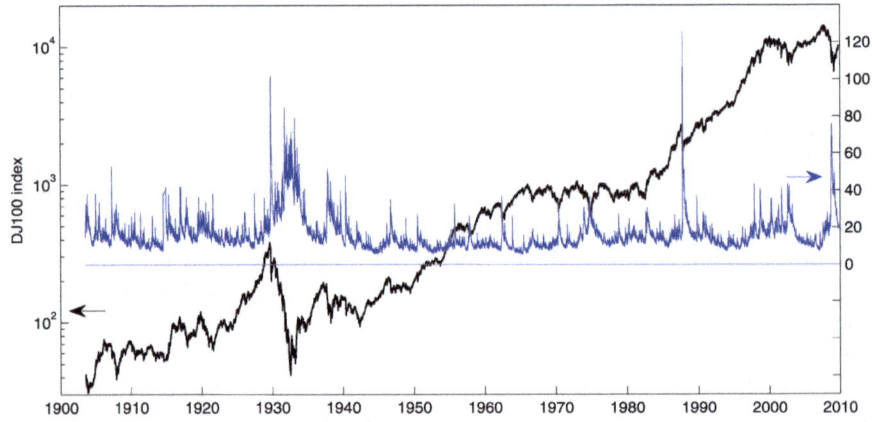

Fig. 15.2 The level of the DJIA (*black, left logarithmic scale*) and the volatility (*blue, right linear scale*). The volatility is computed with the long-memory weights used in RM2006

on the probtiles corresponds to the independence of the residuals (because if two random variables are independent, any functions of them are also independent).

The mapping to probtiles is used extensively in [155] in order to test the performances of several risk methodologies for risk horizons ranging from 1 day to 1 year. The main conclusions are in line with the arguments developed in this book, mainly:

1. Fat-tailed distributions for the innovations should be used.
2. Discounting for the heteroscedasticity is best achieved by a volatility evaluated with a long-memory kernel.
3. The historical methodology is not very good (and cannot be improved).

15.6.3 An Example: Backtesting the DJIA over One Century

Very few time series are available with a long history; among them the Dow Jones Industrial Average (DJIA) stock index is known since 1900. This index is important as it gives a simple and accurate measure of the overall health of the US economy. With the recent market turmoil, this very long series offers an interesting perspective, in particular, when compared to the history of the whole 20th century. The goal of this section is to apply the above probtile backtest to one century of DJIA index data.

The price and volatility are displayed on Fig. 15.2 for a time period extending from 7 July 1903 to 23 October 2009. The major crises can be seen clearly, for example, the first world war, the 1929 market crash and the subsequent great depression, the second world war, the stock market crash of 1987, the .com bubble, and the recent mortgage and credit crises. With the quantitative perspective of one

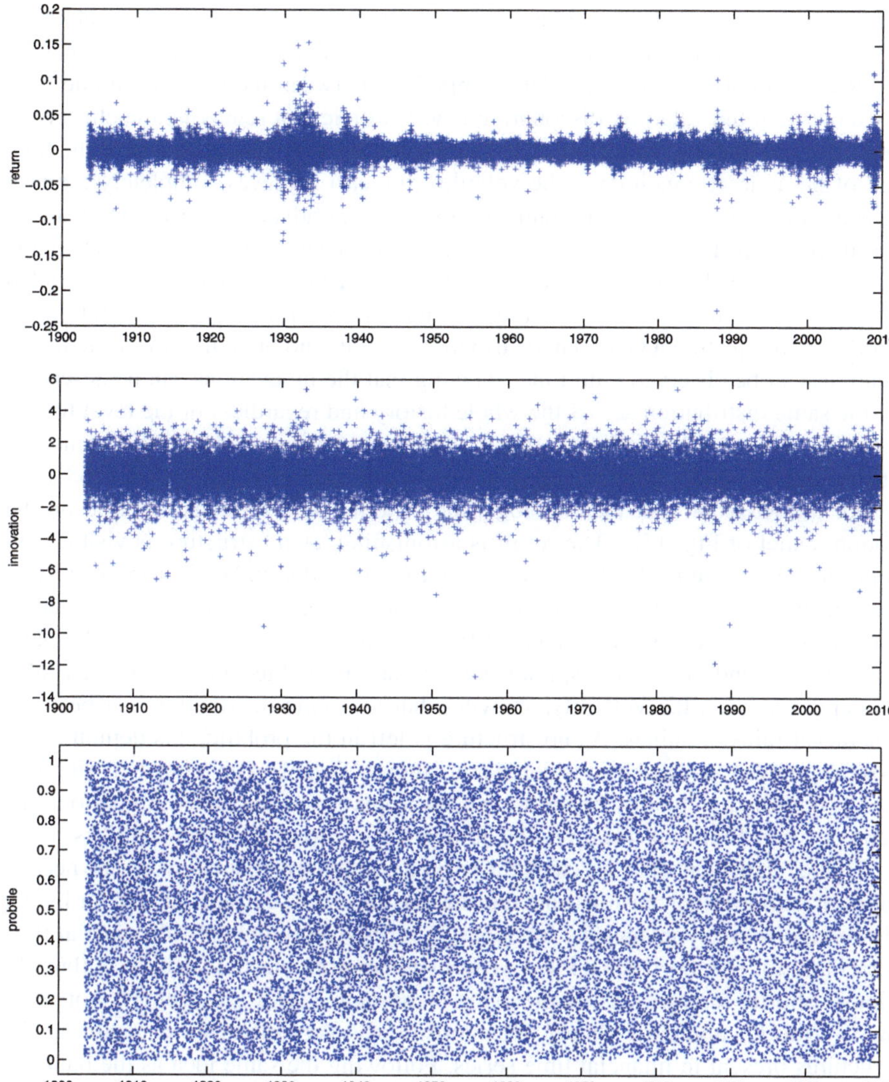

Fig. 15.3 The time series of the returns (*top*), innovations (*middle*), and probtiles (*bottom*). The number of data points of each graph is of the order of 25,000!

century, the recent crises are certainly large, but still significantly smaller than the great depression, and the claim that both events are comparable is unjustified.

Figure 15.3 shows different views of the price changes history. The most straightforward is the daily returns r, given on the top panel. The clusters of volatility are clearly visible, with the periods of large or small price changes. This figure also shows why the return-based risk methodologies have problems adapting to changing market conditions, as essentially enough returns in a new regime should be included in the historical sample before a meaningful risk forecast can be achieved. The time series of the innovations $\varepsilon = (r - \tilde{r})/\tilde{\sigma}$ is plotted in the middle panel for the RM2006 risk methodology. The comparison between the top and middle panels shows that the volatility discounting is very efficient at removing the clusters of volatility and that the innovations have a distribution that is independent of the time and of the market conditions. The volatility forecast achieves completely the task of adapting to a changing environment, a significant advantage of the innovation-based approach. The innovations are defined so as to have a unit variance. The large number of points below -4 and above $+4$ is a clear visual signature of a fat-tailed distribution for the innovations, validating the choice of a Student distribution (or similar) and rejecting a Gaussian distribution. Finally, no obvious slow scale modifications can be observed with time, showing that the innovations can be described by the same distribution across the whole history and regardless of the level for the volatility. This stability indicates that the statistical behavior of the DJIA remained unchanged (beyond a small decrease in the overall level of volatility, see [168]).

For the RM2006 risk methodology, the probtile time series is reported on the bottom panel of Fig. 15.3. The result is astonishing, as no structure is visible (except for the closure of the New York stock exchange due to WWI from 31 July to 13 December 1914). This shows that the basic properties of this stock exchange time series have remained unchanged through the 20th century, despite dramatic changes in economics, politics, financial regulations, and technology. There is also no dependency on the volatility, showing that the same distribution can be used, whether it rains or shines. As no structure is left in the probtile, this demonstrates that the innovations can be considered as a random variable with a fixed distribution. The extraordinary stability of the statistical properties of the DJIA throughout one century likely founds its origin in the human attitude toward profits and risks.

Tighter statistical analyses (with the related confidence bounds) would be needed to reach definitive conclusions, and the analysis should be carried out also for longer risk horizons. Such an analysis was done in [155] over a large set of financial time series and for risk horizons up to 1 year, using appropriate statistical tests (but obviously shorter samples). The conclusion is the same even for longer risk horizons, namely the /location/size/shape/ decomposition is very effective at capturing the structures present in financial time series. Following the same idea as the bank reporting to their supervisors, a very simple test is to count for the number of yearly exceedances. This simple path is followed in [168] on the DJIA very long sample and for the historical, RM1994 and RM2006 risk methodologies. Because of the large sample size, there is no statistical surprise: the broad properties of various risk methodologies just follow from the ingredients used to build them, with their respective advantages and inconveniences.

By capturing more accurately the known properties of financial time series, more accurate risk evaluations can be achieved. The broad criticisms about market risk evaluation found in the press in the wake of the 2007/2008 crisis mostly refer to the return-based approach. As we have seen, this methodology has some fundamental limitations, but it is wrong to generalize its particular problems to all risk evaluations. If the return-based risk methodology is fairly constrained and cannot be modified, the innovation-based risk methodologies can be modified to better capture the empirical properties of the financial time series. Even though spaces for improvements are left, short-term market risk evaluation is a success story that developed over the last 20 years and is used now by most financial institutions through the world.

15.7 Measuring Accurately Shocks

So far in this chapter, the goal was to measure large potential losses on large portfolio, using daily data. Large portfolio correspond to the holdings of financial institutions, large losses correspond to their financial risk, and daily data are widely available and with an almost universal coverage. Yet, the developed method is not accurate in quantifying particular shocks and their timing. For this purpose, one should focus on a single time series, using intra-day data. This section presents such an approach in order to build a "scale of market shocks", similar to the Richter scale used to measure the magnitude of earthquakes. The presentation is kept at a general level, more details can be found in [163, 164].

When a market experiences important turbulences, the media commonly talk about "crashes" or "crises". In order to go further than heuristic analogies, it is necessary to develop objective ways to measure the state of financial markets and to provide a quantitative assessment of market conditions. The size of price movements over a daily interval and the daily innovations used in the previous sections both are subject to high measurement noise and fail to pick up essential intra-day information. Similarly, quantities like daily historical volatility or implied volatility from options only allow one to distinguish poorly between normal and turbulent behavior, particularly at short time horizons. Moreover, all these quantities do not offer a direct measure of the probability of an event to occur. A "scale of market shocks" (SMS) has been developed in order to quantify the importance of market movements. In some other fields, there are widely used scales to quantify event or shock strengths. A familiar example is the Richter scale in geophysics [132] that measures the essentially the (logarithm of the) energy released in earthquakes. In sailing, the Beaufort scale measures the (logarithm of the) speed of winds and the state of the sea.

To improve the identification of market shocks, high-frequency market data should be used. In financial markets, extreme movements are initiated by external events such as political and economic news announcements. These external events translate into quantifiable price changes due to the fast internal dynamics of the markets. Intra-day information should be used in order to compute a short-term market scale measuring the (un)likelihood of events. The advantage of a well-accepted scale is that shocks can be compared to each other and short-term risk measures can be derived from them. Moreover, a well-designed scale can serve as a warning signal that the market is entering a turbulent state.

15.7.1 Defining the Scale of Market Shocks

The broad range of trading horizons of market agents from market makers to speculators, investors, treasurers, and central bankers affect the way they react to news and other financial market events. Even though the market participants all track the same market, they do so inside a broad range of time horizons. The different frequencies at which they intervene in the markets explain why market participants perceive market volatility differently. These different market perceptions are accounted for by the characteristic time scale τ of the volatility $\sigma[\tau]$. Real shocks happen when all market participants become involved, namely when the volatilities at various time scales τ all become large.

The volatility is essentially given by Eq. (2.2) with $\sigma[\tau] = \sigma[\delta t_\sigma = \tau, \delta t_r = \tau/16]$. The scale of market shocks $S(t)$ at time t is defined by the integration over the characteristic time τ of the volatility of returns measured at time horizon τ:

$$S(t) = \gamma \int d\ln\tau \, \mu(\ln\tau)\sigma[\tau](t). \tag{15.15}$$

The integration measure $\mu(\ln\tau)$ fixes the weight associated with the contribution of each time horizon τ. This measure has a certain arbitrariness but should be bell-shaped and with a unit total mass. A smooth function centered at τ_{center} and with an exponential decay on both sides is a convenient choice:

$$\mu(\ln\tau) = ce^{-x}\left(1 + x + x^2/2\right) \qquad \text{with} \quad x = \left|\alpha\ln(\tau/\tau_{\text{center}})\right|. \tag{15.16}$$

The constant c is adjusted so that μ is a unit measure, i.e., $\int d\ln\tau\,\mu(\ln\tau) = 1$. The parameter α controls the decay of the measure, and $\alpha = 2$ gives good results. As any time series indicator, S has an inherent response time, which in this case is induced by the integration measure. The parameter τ_{center} controls the response time of the SMS indicator. In order to have a fast enough response, the integration measure is centered at one day, corresponding to a sum of squared returns measured on price movements over 1.5 hours.

These short time intervals result in the indicator having a quick intra-day response, and an accurate timing of the events. Practically, the integral is computed with τ ranging from 1 hour to 42 days, on a logarithmic grid. As the measure $\mu(\ln\tau)$ is small at the border of integration, the integral in Eq. (15.15) is independent of the chosen integration limits. The constant γ has the value $\gamma = 29.1$ (as explained in [164]) and with the particular forms of Eqs. (15.15) and (15.16) is chosen to relate the scale with the likelihood of the movements. The distribution of volatility is analyzed in [164], showing that the SMS for large events is inversely proportional to the probability of them to occur.

The pointwise volatility definition (2.2) is computationally heavy, in particular with high-frequency data. Moreover, the volatility value depends on the prices at the particular sampling points, whereas a mean price around the sampling point is more appropriate. In order to remedy both issues, the volatility is measured instead with an equivalent formula based on iterated exponential moving averages (EMA):

$$\sigma^2[\tau; x](t) = \mathrm{MA}\big[\tau, 8; \big(D[\tau/16; x]\big)^2\big](t). \tag{15.17}$$

The operator $D[\tau]$ is a differential operator that computes an annualized return with a smooth kernel based on EMA:

$$D[\tau; x] \simeq \sqrt{1y/\tau}\big(x(t) - x(t-\tau)\big).$$

The MA[τ] operator computes a moving average with a rectangular-like kernel, but the kernel decays smoothly toward zero around $t - \tau$, and the index 8 in MA[τ, 8] denotes that this operator is built from 8 iterated EMAs. This operator has already been introduced to build the Market-ARCH process in (2.14). The exact definitions of the operators are discussed in details in [170], where a set of convenient high-frequency operators on time series are presented. From the computational point of view, the operators used in Eq. (15.17) can easily be computed for tick-by-tick data with a set of recursion formulae. Finally, the time series are deseasonalized using the dynamical business time developed in [32]. This part is important as the main ingredient is the volatility, a quantity subject to large daily and weekly seasonalities.

A formal analogy with the Richter scale is developed in [164], using the similarity between the mechanical energy and the price volatility. In particular, this analogy makes explicit the relationship between the probability of earthquakes as measured by the Richter scale and the probability of large movements in a financial market. In order to construct a scale inversely proportional to the likelihood of events for a particular asset, the statistical properties of each asset need to be taken into account. This leads to the construction of another "adaptive" scale of market shocks (see [164]).

15.7.2 Empirical Properties of the Scale of Market Shocks

In order to illustrate the behavior of the SMS indicator and its relation to the volatility at different time horizons, these quantities are plotted in Fig. 15.4 for the exchange rate USD/JPY during the year 1997. This year was particularly eventful for the Japanese economy. The background color corresponds to the volatilities at different time range and illustrates that there can be very different volatility values depending on τ. On the scale of market shocks curve, the peaks stand out clearly from the background fluctuations. They occur when the background color scheme turns red, i.e., when the market participants with different time scales perceive high levels of volatility. It should also be emphasized that the volatilities used to compute $S(t)$ are mostly not visible on the price curve. The volatilities with the largest weight correspond to intra-day fluctuations, and this graph is built with six points per day. Therefore, the scale of market shocks really brings out new information, invisible in a simple price curve.

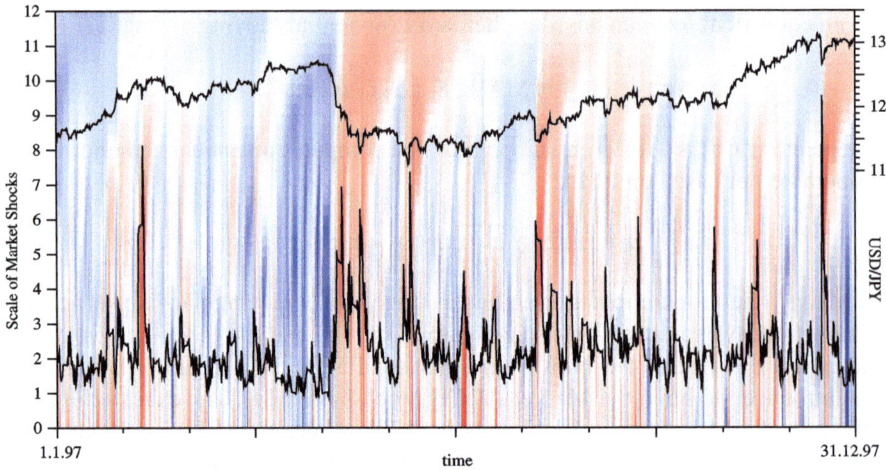

Fig. 15.4 The Scale of Market Shocks for USD/JPY (*left scale*) and the corresponding price (*right scale*) for the year 1997. The *horizontal axis* is divided into 12 equal intervals, approximating one month. The *background color* corresponds to the volatilities at the different time range τ on a logarithmic scale, with $\tau = 1$ h on the lower border and $\tau = 42$ days on the upper border. The value $\delta t_\sigma = 1$ day is in the *middle of the graph* and corresponds to the maximum of the integration measure. The *coloring* is done according to the value of $\ln(\sigma[\tau]/\overline{\sigma})$ with $\overline{\sigma} = 7.6\ \%$. The *blue* (*red*) corresponds to *lower* (*higher*) than average volatility

Yet, it remains to relate large values for the SMS index with major events. There is a substantial body of literature on the relations between news and market movements. Most of these papers use low-frequency data (daily, sometimes open, high, low, and close), look at price differences (return), and focus on economic news. The economic news like unemployment rates or GDP are natural candidates because already given as numbers, namely they are naturally quantified. For these events, the remaining major difficulty is to obtain an a priori quantitative estimate of the expected and the unexpected parts in announcements. For general news like political events, natural disasters, or major bankruptcies, they are not naturally quantified, and indeed difficult to translate into a single magnitude number. Generally, the outcome of these event studies is that the relation is found to be very small or nonexistent (see, e.g., [5] and references therein).

The present approach is different since focusing on high-frequency volatility at different time horizons. Thus, clear signals can be obtained. In [164], an event study is done using the FX rates USD/JPY and USD/DEM from January 1997 to September 1998. These particular months were full of events in East Asia and, by contrast, relatively calm in Europe. Interestingly, all large values for the index can be related to particular news and major events that hit the market. In particular, rumors of the Japanese central bank interventions (publicized in Reuters headlines) had a significant impact on the scale for the Yen. A comparison of the SMS index with either an implied volatility or a volatility computed using the RiskMetrics simple EMA applied to daily data shows that only the SMS is able to pick-up clearly major events.

Both other measures are not as efficient to distinguish the background fluctuations of the market from the response to the external events, pointing out to the difficulty of separating abnormal turbulences from normal behavior in financial time series. It is only by making full use of the high-frequency data that one is able to track these events and to really quantify them. Returning to the analogy with the Richter scale, there is a sharp difference between tectonic movements and finance. The former shows long loading periods that are clearly separated from dramatic relaxation events. Subsequently, "one" earthquake can be clearly defined, and measures like the total energy released during the earthquake are well defined. This is clearly different from financial time series which are always moving randomly. These makes much more difficult the task to isolate the response to an external event.

The scale of market shocks can be constructed in principle for any market: the index is computed from the price time series. An extension to the stock market was done by [93]. They used the scale to show that the events of 11 September 2001 proved to be a major shock for the French market, in particular compared to the Russian and Asian crises. Because the volatility is impacted strongly by the daily and weekly cycles of human activity, an important difficulty in such computations is to discard properly the seasonalities. The computations in [164] use an intrinsic business time; another possibility would be to do the computations directly in tick time. Another interesting approach is presented in the "scale of market quake" by [20], where the authors build the index from tick-by-tick price directional changes at different magnitudes. The computation of the index involves the magnitude of Fourier components, in effect introducing a volatility like measure.

Chapter 16
Option Pricing

16.1 Processes and the Price of Options

Generally, an option is a contract with a value depending on the price of one or several other financial contract(s) or index at some specified future time. The base contract is called the *underlying*, and the specified time the *maturity*. The interesting and difficult problem is to price such contracts before the maturity. Clearly, the price of the option is depending on the possible future behavior of the underlying. This behavior is described by a random process, which is a model for the future possible values taken by the underlying.

The reference option pricing model takes a constant volatility Gaussian random walk for the underlying price process, leading to the celebrated Black–Scholes–Merton model [22, 106]. The importance of this approach originates in several properties of the equations or their solutions.

- This is the simplest model, as based on the simplest process. Because the Gaussian distribution is stable under aggregation, the solution for a logarithmic random walk with constant volatility can be computed explicitly and involves only the (Gaussian) distribution of the price changes up to the maturity. Because the process can be integrated, the final formula is simple and can be evaluated with an elementary program or spreadsheet. This simplicity contributed to the popularity of the approach.

  The stability under aggregation is also at the core of a continuum limit for this process. The continuum limit is convenient as the process time increment δt disappears, making the computation simpler by using the Itô calculus. This powerful tool allows for many convenient analytical computations for more complex options, or for some more complex process for the underlying. Yet, all are based on normal innovations and/or on processes which have proper continuum limits.
- The solution captures just the essential part of the problem and is therefore very robust. Essentially, the option price is depending on the diffusive character of the underlying. period. Because the B–S solution gets the basics right, many approaches to obtain more accurate prices start from this solution.

G. Zumbach, *Discrete Time Series, Processes, and Applications in Finance*,
Springer Finance, DOI 10.1007/978-3-642-31742-2_16,

- The terms in the formula are either known from the contract (strike, maturity) or from the market (underlying price, interest rate), except for the volatility σ of the underlying. This is indeed the difficult parameter in the computation. An approach fateful to the constant volatility process is to compute the volatility on a long historical sample. Yet, a constant volatility approach is difficult to maintain today. A simple engineer solution is to replace the σ in the formula by a volatility forecast (more explicitly, to replace $\sigma^2 \Delta T$ by a forecast for the integrated variance between the present time and the maturity of the option). Such a solution can be partly justified by a computation but cannot be derived from first principle using a better process for the underlying.
- On the market, option prices differ from the theoretical B–S prices. There are many reasons for these differences, originating mainly in the shortcoming of the model (fat-tailed distribution, heteroscedasticity, leverage, etc.), but also from market considerations (liquidity, inventory, cost of hedging, difference between long and short hedgings, risk of "real life" hedging, etc.). As the volatility is the only "free" parameter, it can be adjusted to market option prices. This approach leads to the implied volatility surface, namely the volatility in the B–S equation is adjusted so that the market price is obtained. As the option price is depending on the strike and maturity, an implied volatility surface is obtained. The implied volatility computation can be summarized in this critical manner: the wrong number (the implied volatility) is used in the wrong formula (the B–S equation) so as to obtain the correct number (the market price).

  The implied volatility is in broad use in option trading desk and is actually the intuitive quantity that is traded. The B–S formula is then used as a dictionary to map back and forth between implied volatilities and prices. The reasons to give more importance to the implied volatility are many, but essentially it gives a good summary of the properties of the underlying up to maturity, while removing the specificity of actual options (put, call, exercise type, etc.). Therefore, the same surface can be used to price various options on the same underlying.
- Gaussian diffusions have many special properties. Among them, the option pricing problem can be mapped to a partial differential equation (PDE) whose leading character is similar to a heat equation. The differential equation allows for numerical solutions and to natural modifications in order to obtain more realistic option prices. Yet, the relation between diffusion and PDE can be derived only for Gaussian processes.
- Beyond the implied volatility adjustment factor, it is difficult to modify the B–S approach. In particular, modifications of the process in order to be more realistic break the analytical tractability. This is the case for fat-tailed innovations and for ARCH heteroscedasticity, although some stochastic volatility can be included in a PDE approach. Moreover, a perturbative approximation using an expansion in a small parameter around the B–S solution seems difficult, as there is no small parameter in the problem.
- The normal distribution of the innovations can be extended to include jumps. This extension allows one to have a normal diffusion part captured by the Gaussian and infrequent large events captured by the jumps. This approach requires some

advanced mathematics and is the subject of the excellent book [44]. Yet, the distribution for the innovations cannot include a fat-tailed distribution like Student, and heteroscedasticity cannot be included. Another shortcoming of this model is that the density of jumps is large when estimated on empirical time series. This can be understood intuitively by the jumps making up for the deficiency of the normal distribution with respect to large events.

- With option pricing comes the question of the measure, namely the physical measure $\mathbb{P}$ versus the risk-neutral measure $\mathbb{Q}$. The price process for the underlying is set in the unique physical measure $\mathbb{P}$, and the process parameters are estimated in this measure. The introduction of a risk free bond and of the noarbitrage argument leads to a risk-neutral measure $\mathbb{Q}$, and the option price corresponds to the expected discounted payoff at maturity in $\mathbb{Q}$. For a constant volatility Gaussian process, the natural change of measure is given by the exponential of the random variable, in effect introducing only a shift for the location of the Gaussian innovations expressed in the unique $\mathbb{Q}$ measure. Because the innovation distribution is the same up to a simple shift for its center, it is simpler to perform the computations in the $\mathbb{Q}$ measure. In this way, the $\mathbb{Q}$ measure acquires a central status, and the physical measure can be put aside.

  Clearly, the properties of the normal distribution is central in this computation. The distribution is used also in the selection of the analytical form of the change of measure, so as to keep analytical tractability. As a side effect, this makes $\mathbb{Q}$ unique. Changing the innovation distribution say for a fat-tailed distribution breaks completely the scheme.
- The wish list of desirable properties in 1970 and 2010 is clearly different with respect to analytical tractability. It was important to have a simple formula in 1970 so that explicit computations of prices can be done. With the cheap computational power available today, more brute force computations can be done, and in particular full Monte Carlo option pricing can be achieved. Consequently, the requirements on an option pricing scheme move from analytical simplicity to numerical accuracy against market prices.

The list above should make clear that the B–S option pricing is fairly singular: the choice for the process makes the problem solvable, but changing most ingredients breaks the scheme. For this reason, many propositions to obtain better option prices are plausible modifications, but they cannot be justified rigorously. Clearly, some ingredients have to be abandoned in order to make progress toward a fundamental option pricing.

The solution found so far is to state the problem with discrete time increment and to abandon the continuum limit and the Itô calculus. In this direction, a first principle computation of option prices based only on a realistic process for the underlying is slowly emerging. This is a much tougher framework, with slow progresses mainly due to [55, 80], and recently due to [40] (CEFJ hereafter). The work of CEFJ is an important contribution, separating cleanly the issues related to the equivalent martingale measure from the utility function. The framework allows for very general heteroscedastic dynamics, and the distribution for the innovations can in principle be fairly arbitrary.

The CEFJ computations use a logarithmic random walk. The difficulties of this setup have been discussed already in Chap. 6, and they appear for the same reasons in the option pricing computation. In particular, the proofs involve the characteristic function of the innovation distribution, namely

$$E[\exp(\sigma\epsilon)]$$

where the expectation is over the distribution $p(\epsilon)$ for the innovations. This integral diverges for any fat-tailed distribution. The solution has already been presented and is to use geometric processes which can elegantly accommodate fat-tailed distributions.

This chapter presents the main ideas required to price option in a realistic framework, namely to use a process for the underlying which can include at least fat-tailed innovations and heteroscedasticity and to derive the option price based on an arbitrage argument. The basic step is to construct a set of equivalent measures. The introduction of a bond allows us to select one equivalent measure in which the process is a martingale. With this equivalent martingale measure, the standard argument about no arbitrage carries over. Then, an expansion in small δt allows us to perform analytically the integral over the innovation distributions. This chapter is based mostly on [40, 116], and [166], but the notation for the time dependency is compatible with the other parts of this book.

16.2 The Stock Process

Processes with relative returns have already been presented. The core reason to use this formalism in option pricing is to have well-defined integrals over the innovations, including with fat-tailed distributions. This also simplifies some issues related to the construction of equivalent martingale measures and to the choice of an utility function, but the cost is some technical conditions related to positivity.

Under the physical measure $\mathbb{P}$, the geometric price process $S(t)$ for the stock is given by

$$\begin{aligned} r_{\text{rel}}(t+\delta t) &= \frac{S(t+\delta t)-S(t)}{S(t)} = \mu(t) + z(t+\delta t) - 1 \qquad (16.1)\\ z(t+\delta t)|\mathcal{F}(t) &\sim D\big(1, \sigma(t)^2\big), \qquad z(t+\delta t) > 0 \end{aligned}$$

where $S(t)$ is the stock price at time t, and $\sigma^2(t)$ is the conditional variance of the relative return in the period $(t, t+\delta t)$. The expected gross rate of return is $\mu(t)$, and $E[r_{\text{rel}}(t+\delta t)] = \mu(t)$. The variance σ^2 and the rate of return μ are taken at the scale δt, namely both are proportional to δt using the usual diffusion scaling. The weak technical conditions are that $\mu(t)$ and $\sigma(t)^2$ are $\mathcal{F}(t)$-measurable. This allows for very general specifications, for example, the variance can be taken as $\sigma^2(t) = \sigma^2_{\text{eff}}(t)$ for all the ARCH processes and stochastic volatility processes presented in this book. In this chapter, the random innovation ϵ has been replaced by $z = 1 + \sigma\epsilon$, with the corresponding changes in the mean and variance of the distribution for z. The rationale for this change is that z is a positive random variable, and keeping positivity is important in the proofs presented below. For the small δt expansion hereafter, we will return to the innovation ϵ for the random variable.

The process starts at the current time t_0, and the option matures at T. The time t and t' denote times between t_0 and T. The time interval up to maturity is given by $\Delta T = T - t_0$. For a process with time increment δt, the number of steps to reach maturity is given by $n = \Delta T/\delta t$.

16.3 The Equivalent Martingale Measure

In order to price a derivative, an equivalent martingale measure (EMM) should be constructed. This is done in two steps. First, a sufficiently large family of equivalent measures is constructed. This family is depending on a function $f(z)$ and on a time-dependent parameter $\nu(t)$. Second, a bond is introduced, and the measure is selected so that the price process is a martingale. This EMM condition gives an equation for $\nu(t)$, depending on the chosen function f.

Let us introduce the risk aversion function f over the positive real line

$$f : \mathbb{R}^+ \to \mathbb{R}^+ \tag{16.2}$$

and the "normalization" function

$$\psi(t, \nu) = E_t\big[f\big(\nu z(t+\delta t)\big)\big] \qquad \text{for} \quad \nu > 0. \tag{16.3}$$

The positivity conditions on z, ν, and f ensure that $\psi(t) > 0$. The function f encapsulates the risk aversion of the investor by weighting differently the events $z(t + \delta t)$. In practice, this should be a decreasing function, and it can also be viewed as the derivative of a utility function. Technically, the mapping f should be such that $\psi(t) < \infty$. Notice that the function f can be made time dependent, but should be known at t. In this case, the normalization function becomes $\psi(t, \nu) = E_t[f_t(\nu z(t + \delta t))]$.

For a given risk aversion function f, a change of measure can be constructed, and this quantity is called a Radon–Nikodym derivative. Essentially, the random variable z has a distribution specified in $\mathbb{P}$, and its distribution in another measure $\mathbb{Q}$ is obtained by multiplying by the change of measure from $\mathbb{P}$ to $\mathbb{Q}$. For a predetermined sequence $\{\nu(t)\}$ with $\nu(t) > 0$, the Radon–Nikodym derivative is

$$\frac{dQ}{dP}\bigg|_{F_T} = \prod_{t=t_0}^{T} \frac{f(\nu(t)z(t+\delta t))}{\psi(t, \nu(t))}. \tag{16.4}$$

This construction is such that the distribution for z in the measure $\mathbb{Q}$ is a proper probability distribution, namely is positive and with a unit mass (for each time t).

The conditioning by the filtration on the left-hand side makes explicit that the realization for $z(t)$ should be known up to T, and in particular the (time-dependent) variance and drift. Intuitively, this introduces a dependency on the path followed by the stock $S(t)$ up to T. For the proof of the statement (16.4), the positivity conditions are such that $\frac{dQ}{dP}|F_t > 0$. We also need to show that

$$E^{\mathbb{P}}_{t_0}\left[\frac{dQ}{dP}\middle|F_T\right] = 1. \tag{16.5}$$

Using the law of iterated expectations, the desired results is obtained:

$$\begin{aligned}
&E^{\mathbb{P}}_{t_0}\left[\frac{dQ}{dP}\middle|F_T\right] \\
&= E^{\mathbb{P}}_{t_0}\left[E^{\mathbb{P}}_{t_0+\delta t}\cdots E^{\mathbb{P}}_{T}\left[\frac{dQ}{dP}\middle|F_T\right]\right] \\
&= E^{\mathbb{P}}_{t_0}\left[E^{\mathbb{P}}_{t_0+\delta t}\cdots E^{\mathbb{P}}_{T-\delta t}\left[\prod_{t=t_0}^{T-\delta t}\frac{f(\nu(t)z(t+\delta t))}{\psi(t,\nu(t))}E^{\mathbb{P}}_{T}\left[\frac{f(\nu(T)z(T+\delta t))}{\psi(T,\nu(T))}\right]\right]\right] \\
&= E^{\mathbb{P}}_{t_0}\left[\frac{f(\nu(t_0)z(t_0+\delta t))}{\psi(t_0,\nu(t_0))}\right] = 1.
\end{aligned}$$

In the B–S setup, the changes of measure are given by the exponential function $f(u) = \exp(-u)$. This particular choice ensures that the probability distribution in the measure $\mathbb{Q}$ is also a normal distribution, but with a different location. With this choice, the continuum limit can be taken for the process in $\mathbb{P}$ and in $\mathbb{Q}$, with the convenient Itô calculus that follows for both measures. In our setup, the process time increment δt is fixed, and therefore more freedom in the choice for f is allowed. With relative returns, the natural setup is of products of positive random variables, leading to the product in (16.4). With processes defined using logarithmic returns, the function f is naturally an exponential, and the formula (16.4) takes the form of the exponential of a sum.

This Radon–Nikodym derivative can now be used to specify an equivalent martingale measure (EMM). A risk-free bond process $B(t)$ is introduced, with $B(t+\delta t) = B(t)(1 + r_{\text{rf}}(t))$ and where r_{rf} is the risk-free rate of return. Notice that this definition is consistent with the process definition as both use a multiplicative setup. The technical condition is that $r_{\text{rf}}(t)$ if $\mathcal{F}(t)$-measurable. The price process is a martingale in the measure $\mathbb{Q}$ defined by the Radon–Nikodym derivative (16.4) when $E^{\mathbb{Q}}[S(t+\delta t)/B(t+\delta t)|\mathcal{F}(t)] = S(t)/B(t)$, or equivalently

$$E^{\mathbb{Q}}\left[\frac{S(t+\delta t)}{S(t)}\bigg/\frac{B(t+\delta t)}{B(t)}\middle|\mathcal{F}(t)\right] = 1.$$

This condition leads to the equation for $\nu(t)$

$$\frac{E_t[z(t+\delta t)f(\nu(t)z(t+\delta t))]}{E_t[f(\nu(t)z(t+\delta t))]} = 1 - \mu(t) + r_{\text{rf}}(t) = 1 - \phi(t)\sigma(t)^2 \tag{16.6}$$

where the *market price of risk*

$$\phi(t) = \frac{\mu(t) - r(t)}{\sigma(t)^2} \tag{16.7}$$

has been introduced in the last equality. This last definition can be rewritten in the form $\mu(t) = r(t) + \phi(t)\sigma(t)^2$, namely the return of the bond $\mu(t)$ is the risk-free rate r_{rf} added with a risk premium proportional to the variance. Let us emphasize that $\nu(t)$ is depending on the path followed by the process, and Eq. (16.6) should be solved along each trajectory. Equation (16.6) follows from the computation

$$\begin{aligned}
&E^{\mathbb{Q}}\left[\frac{S(t+\delta t)}{S(t)} \Big/ \frac{B(t+\delta t)}{B(t)} \middle| \mathcal{F}(t)\right] \\
&= E^{\mathbb{P}}\left[\left(\frac{\frac{dQ}{dP}|F_{t+\delta t}}{\frac{dQ}{dP}|F_t}\right) \frac{S(t+\delta t)}{S(t)} \Big/ \frac{B(t+\delta t)}{B(t)} \middle| \mathcal{F}(t)\right] \\
&= E^{\mathbb{P}}\left[\frac{f(\nu(t)z(t+\delta t))}{\psi(t,\nu(t))} \frac{\mu(t) + z(t+\delta t)}{1 + r_{\text{rf}}(t)} \middle| \mathcal{F}(t)\right] \\
&= \frac{1}{1 + r_{\text{rf}}(t)}\left(\mu(t) + \frac{E^{\mathbb{P}}[z(t+\delta t) f(\nu(t)z(t+\delta t))]}{\psi(t,\nu(t))}\right) \\
&= 1.
\end{aligned}$$

This computation shows that when (16.6) is satisfied, the probability measure $\mathbb{Q}$ makes the stock discounted by a risk-free bond a martingale.

As $\mu(t) > r(t)$ (or equivalently the market price of risk $\phi(t)$ is positive), the right-side in (16.6) is smaller than 1. For a solution of the (16.6) to exist, the left-hand side should also be smaller than 1, imposing some restrictions on f. Essentially, f should decay fast enough so that the numerator is smaller than the denominator. An exponential decreasing risk aversion function $f(x) = \exp(-x)$ seems to be a convenient choice, but other dependencies are possible. This freedom in the choice for f shows that different EMM can be constructed, equivalent to a choice of an utility function (see [40] for a discussion on this point). The function $f(x) = 1/(1 + \exp(x))$ will be used bellow.

16.4 The Valuation of European Style Contingent Claims

Christoffersen et al. [40] show that the no-arbitrage price of a European contingent claim $C(t)$ is given by the discounted expectation under the EMM measure $\mathbb{Q}$ of the terminal value

$$C(t) = E^{\mathbb{Q}}\left[C_T\big(S(T)\big)\frac{B(t)}{B(T)} \middle| \mathcal{F}(t)\right] = E^{\mathbb{P}}\left[\frac{D\mathbb{Q}}{D\mathbb{P}} C_T\big(S(T)\big)\frac{B(t)}{B(T)} \middle| \mathcal{F}(t)\right] \tag{16.8}$$

and the Radon–Nikodym derivative between t and T is given by

$$\frac{D\mathbb{Q}}{D\mathbb{P}} = \frac{\frac{d\mathbb{Q}}{d\mathbb{P}}|_{\mathcal{F}(T)}}{\frac{d\mathbb{Q}}{d\mathbb{P}}|_{\mathcal{F}(t)}} = \prod_{t'=t}^{T-\delta t} \frac{f(\nu(t')z(t'+\delta t))}{E_{t'}[f(\nu(t')z(t'+\delta t))]}. \tag{16.9}$$

Because the process is known in the $\mathbb{P}$ measure, the expectation in $\mathbb{P}$ must be evaluated. The subtle point embedded in the term $D\mathbb{Q}/D\mathbb{P}$ is a dependency on the price process through the variance $\sigma^2(t)$ and on the rate of return $\mu(t)$. Therefore, the expectation involves the realizations for the random paths between t and T.

This formula is important as it allows one to price a contingent claim for very general processes for the underlying. In the B–S scheme, the option price can be obtained with many formulas that are equivalent thanks to the peculiar properties of the Gaussian random walk (for example, as a solution of a PDE). In the present general setting, the expected discounted payoff is the proper generalization, but the expectation and change of measure are path dependent. As such, the option pricing formula is exact and complete (and the derivation does not involve any small δt expansion). But it is still not a practical scheme, as it involves integrals that should be computed and implicit equations that should be solved. In order to turn these mathematical formulas into a convenient algorithm, a small δt expansion should be performed so as to compute the expectations analytically, as explained in Sect. 16.6. But before, the option pricing formula needs to be justified, using a no-arbitrage argument.

The computations below are done formally in the $\mathbb{Q}$ measure when evaluating expectations, with the understanding that the integrals are indeed evaluated in the $\mathbb{P}$ measure and include the Radon–Nikodym change of measure. In the previous section, the EMM was constructed so that the stock discounted by the bond is a martingale. The forthcoming computations are made simpler by using the bond as the numeraire. The discounted stock is defined as $S^B(t) = S(t)/B(t)$, and the same definition is used for other quantities. As a bond has a positive value $B(t) > 0$ for all t, this definition is legitimate. With this definition, the discounted stock S^B is a $\mathbb{Q}$ martingale, and the discounted option price becomes

$$C^B(t) = E^{\mathbb{Q}}\big[C_T^B(S(T))|\mathcal{F}(t)\big]. \tag{16.10}$$

The law of iterated expectations implies for $s > t$ that

$$\begin{aligned} C^B(t) &= E^{\mathbb{Q}}\big[C_T^B(S(T))|\mathcal{F}(t)\big] \\ &= E^{\mathbb{Q}}\Big[E^{\mathbb{Q}}\big[C_T^B(S(T))\,|\,\mathcal{F}(s)\big]\,\Big|\,\mathcal{F}(t)\Big] \\ &= E^{\mathbb{Q}}\big[C^B(s)\,|\,\mathcal{F}(t)\big]. \end{aligned}$$

This computation shows that C^B is also a $\mathbb{Q}$ martingale.

The next step requires to introduce the trading strategies, in our case between the underlying S, the option C, and the bond B. The positions in these assets are denoted respectively by ρ, γ, and β, and the value of the portfolio is

$$V(t) = \rho(t)S(t) + \gamma(t)C(t) + \beta(t)B(t). \tag{16.11}$$

The positions are assumed to be known at t, namely to be included in the filtration $\mathcal{F}(t)$. The one-step gain between t and $t+\delta t$ resulting from the change in values of the assets is given by

$$\begin{aligned} dG(t+\delta t) &= \rho(t)\big(S(t+\delta t) - S(t)\big) + \gamma(t)\big(C(t+\delta t) - C(t)\big) \\ &\quad + \beta(t)\big(B(t+\delta t) - B(t)\big) \\ &= \rho(t)\cdot dS(t+\delta t) + \gamma(t)\cdot dC(t+\delta t) + \beta(t)\cdot dB(t+\delta t) \end{aligned} \tag{16.12}$$

and the total gain up to t is

$$G(t) = V(0) + \sum_{t'=\delta t}^{t} dG(t') \tag{16.13}$$

with $G(0) = V(0)$.

A trading strategy is *self-financing* when no cash is added or taken in the portfolio. This condition implies that the possible trades at t between the three assets are such that

$$\begin{aligned} V(t) &= \rho(t-\delta t)S(t) + \gamma(t-\delta t)C(t) + \beta(t-\delta t)B(t) \\ &= \rho(t)S(t) + \gamma(t)C(t) + \beta(t)B(t). \end{aligned} \tag{16.14}$$

For a self-financing strategy, the changes in value of the portfolio are given by the one-step gains, namely

$$dV(t+\delta t) = V(t+\delta t) - V(t) = dG(t+\delta t). \tag{16.15}$$

By summing the above relation, the value of the portfolio is given by the total gain

$$V(t) = G(t). \tag{16.16}$$

The value and gain processes can be expressed in the numeraire B by defining $V^B = V/B$ and $G^B = G/B$, as done above for the stock and option. For a self-financing strategy, the one-step discounted gain can be computed from the above definitions:

$$\begin{aligned} dG^B(t+\delta t) &= G^B(t+\delta t) - G^B(t) = V(t+\delta t)/B(t+\delta t) - V(t)/B(t) \\ &= +\rho(t)\frac{S(t+\delta t)}{B(t+\delta t)} + \gamma(t)\frac{C(t+\delta t)}{B(t+\delta t)} + \beta(t)\frac{B(t+\delta t)}{B(t+\delta t)} \\ &\quad - \rho(t)\frac{S(t)}{B(t)} - \gamma(t)\frac{C(t)}{B(t)} - \beta(t)\frac{B(t)}{B(t)} \\ &= \rho(t)\cdot dS^B(t+\delta t) + \gamma(t)\cdot dC^B(t+\delta t). \end{aligned} \tag{16.17}$$

Essentially, the discounted gain is the sum of the gain in the discounted stock and discounted option, while there is no gain in the bond when using the bond as numeraire. As shown above, S^B and C^B are martingales, and therefore $E_t[dS^B(t + \delta t)] = 0$ and $E_t[dC^B(t + \delta t)] = 0$. Because $\rho(t)$ and $\gamma(t)$ are included in the filtration $\mathcal{F}(t)$, we have $E_t[dG^B(t + \delta t)] = 0$, and therefore G^B is also a martingale. Hence, the portfolio value V^B is also a martingale, namely $V^B(t) = E_t^{\mathbb{Q}}[V^B(T)]$.

A self-financing trading strategy is an arbitrage if $V(0) < 0$ and $V(T) \geq 0$ a.s. or if $V(0) \geq 0$ and $V(T) < 0$ a.s. For a martingale, the condition $V(T) \geq 0$ implies $V^B(0) = E_0[V^B] \geq 0$, which is in contradiction with the arbitrage condition $V^B(0) < 0$ (equivalent to $V(0) < 0$ since $B > 0$). The second no-arbitrage condition leads to a contradiction following the same argument. This computation shows that no arbitrage can be constructed when the discounted portfolio V^B is a martingale. The whole line of arguments can be summarized as follows: the definition of the option price as an expectation implies that the option price process is a martingale and that all portfolios constructed from martingales are also martingales. Finally, no arbitrage trading strategy can be constructed from martingales.

This result is very general as it holds regardless of the self-financing trading strategy which might be implemented in ρ, γ, and β. It is essentially due to (a) the definition of the change of measure so that S^B is a martingale, and (b) the definition of the option price by a discounted expectation which implies that C^B is a martingale. In particular, this derivation does not use the Itô formula or a small δt expansion. Notice also that the proof relies on the existence of an EMM, but not on the explicit form for the change of measure. Therefore, this result is valid for processes defined either with logarithmic returns or with relative returns. The above derivation is also very simple and does not require the subtleties introduced in the continuum (difference between martingale and semi-martingale, conditions for the left and right continuity or limit, integrability, Itô formula, etc.; see, e.g., [44] for a complete presentation).

16.5 Minimal Variance Hedging

The no-arbitrage argument presented in the previous section does not use any particular form for the self-financing trading strategy. For example, it is valid for a portfolio with one naked short option corresponding to $\rho = 0$ and $\gamma = -1$. The expected value of this portfolio corresponds to its initial value, but its *variance* is very large. For an option issuer (i.e., $\gamma = -1$), the less risky self-financing trading strategy corresponds to minimizing the variance of the final portfolio value. Therefore, the goal in this section is to find $\rho(t)$ and $\beta(t)$ such that $E_t^{\mathbb{Q}}[(V^B(T) - V^B(t))^2]$ is minimal.

The computations are fairly standard and only summarized. The final portfolio value can be expressed as the sum over the one-step gains. After expanding the square in the variance, the martingale property makes all the cross terms with $t \neq t'$ to vanish. Therefore, the variance of the final portfolio is given by a sum of the one-step variances for the gain, and the minimum is reached when each term is minimal (no transaction cost is assumed here). The variance for the discounted one-step gain at time t is

$$\begin{aligned}E_t^{\mathbb{Q}}\left[\left(dG^B(t+\delta t)\right)^2\right] &= \rho^2(t)E_t^{\mathbb{Q}}\left[\left(dS^B(t+\delta t)\right)^2\right]\\ &\quad + 2\rho(t)\gamma(t)E_t^{\mathbb{Q}}\left[dS^B(t+\delta t)\cdot dC^B(t+\delta t)\right]\\ &\quad + \gamma^2(t)E_t^{\mathbb{Q}}\left[\left(dC^B(t+\delta t)\right)^2\right].\end{aligned}$$

The minimization with respect to ρ leads to the solution

$$\rho(t) = -\gamma(t)\frac{E_t^{\mathbb{Q}}[dS^B(t+\delta t)\cdot dC^B(t+\delta t)]}{E_t^{\mathbb{Q}}[dS^B(t+\delta t)\cdot dS^B(t+\delta t)]}. \tag{16.18}$$

The changes in the option and the underlying are clearly related. A Taylor expansion at lowest order is used to extract the leading dependency

$$C^B(t+\delta t) = C^B(t) + \frac{\partial C}{\partial S}(t)\,dS^B(t+\delta t) + \cdots. \tag{16.19}$$

Inserting this expansion into the solution for ρ leads to the usual δ-hedged portfolio with

$$\rho(t) = -\gamma(t)\frac{\partial C}{\partial S}(t). \tag{16.20}$$

Therefore, the δ-hedge replication strategy minimizes the risk for the option issuer. But it replicates the option payoff only in average, as for any other self-financing strategy. In the present scheme, the option cannot be replicated exactly, contrarily to the B–S scheme where the δ-hedge trading strategy produces always the option payoff. In the realistic model used in this chapter, the option is not a redundant asset. Issuing options carry a risk, and the market is not complete in this sense. The risk for the option issuer is computed explicitly in Sect. 16.9 using Monte Carlo simulations, and a corresponding risk premium added to the no-arbitrage option price.

16.6 Small δt Expansion

In the previous derivations, the dependency in δt is hidden in $\sigma \simeq \sqrt{\delta t}$ and $\mu \simeq \delta t$. Systematic Taylor expansions can be done in $\sqrt{\delta t}$, neglecting terms of order $\delta t^{3/2}$ and higher. The δt expansion is more natural in the ϵ random variable, using $z = 1 + \sigma\epsilon$. In order to alleviate the notation, the rate of return and risk-free rate are assumed to be time independent, but it is easy to reintroduce time dependency in these terms.

For a general distribution for z, an expansion of the integrals around the mean can be performed. When the covariance σ^2 is small enough, the terms of order higher than σ^2 in the expansion can be neglected. The properties $E[z] = 1$ and

$E[(z-1)^2] = \sigma^2$ are equivalent to $E[\epsilon] = 0$ and $E[\epsilon^2] = 1$. Dropping the t dependency in the intermediate computations, we have

$$\begin{aligned} f(\nu z) &= f\big(\nu(1+\sigma\epsilon)\big) \\ &= f + \sigma\epsilon\nu f' + \frac{1}{2}\sigma^2\epsilon^2\nu^2 f'' + O\big(\sigma^3\big) \\ &\simeq f\left(1 + \sigma\epsilon\nu\frac{f'}{f} + \frac{1}{2}\epsilon^2\sigma^2\nu^2\frac{f''}{f}\right) \\ &\simeq f\left(1 - \epsilon\sigma\varphi_1 + \frac{1}{2}\epsilon^2\sigma^2\varphi_2\right) \end{aligned} \tag{16.21}$$

where the following definitions are used:

$$\varphi_1(\nu) = -\nu\frac{f'}{f} \tag{16.22a}$$

$$\varphi_2(\nu) = \nu^2\frac{f''}{f} \tag{16.22b}$$

and where f, f', and f'' are evaluated at ν. The function $\varphi_1(\nu)$ is positive, provided that $f(\nu)$ is a decreasing function. Taking the expectation, we have

$$E\big[f(\nu z)\big] \simeq f\left(1 + \frac{1}{2}\sigma^2\varphi_2\right) \tag{16.23}$$

and

$$\frac{f(\nu z)}{E[f(\nu z)]} \simeq 1 - \epsilon\sigma\varphi_1 + \frac{1}{2}\sigma^2(\epsilon^2 - 1)\varphi_2. \tag{16.24}$$

A similar computation gives

$$E\big[zf(\nu z)\big] = E\big[(1+\sigma\epsilon)f\big(\nu(1+\sigma\epsilon)\big)\big] \tag{16.25}$$

$$\begin{aligned} &= E\left[(1+\sigma\epsilon)\left(f + \sigma\epsilon\nu f' + \frac{1}{2}\sigma^2\epsilon^2\nu^2 f'' + O\big(\sigma^3\big)\right)\right] \\ &= E\left[f + \sigma\epsilon\big(\nu f' + f\big) + \sigma^2\epsilon^2\left(\frac{1}{2}\nu^2 f''(\nu) + \nu f'\right) + O\big(\sigma^3\big)\right] \\ &= f + \sigma^2\left(\frac{1}{2}\nu^2 f''(\nu) + \nu f'\right) + O\big(\sigma^3\big) \\ &\simeq f\left(1 + \sigma^2\left(\frac{1}{2}\varphi_2 - \varphi_1\right)\right). \end{aligned} \tag{16.26}$$

At the same order, (16.6) becomes

$$\frac{E_t[z(t+\delta t)f(\nu(t)z(t+\delta t))]}{E_t[f(\nu(t)z(t+\delta t))]} = 1 - \sigma^2(t)\varphi_1(t) = 1 - (\mu - r_{\mathrm{rf}}) \tag{16.27}$$

and the equation to be solved for $\nu(t)$ is

$$\varphi_1\big(\nu(t)\big) = \frac{\mu - r_{\mathrm{rf}}}{\sigma^2(t)} = \phi(t). \tag{16.28}$$

The left-hand side is a function of $\nu(t)$ only, while the right-hand side is a number depending on the process up to t through the variance $\sigma^2(t)$. The right-hand side is positive, and the left-hand side is positive, provided that f is decreasing. For $f(x) = \exp(-x)$, the term $-f'/f$ is equal to 1, leading to an explicit solution $\nu(t) = \phi(t)$.

With a fat-tailed distribution, some cares need to be taken in the Taylor expansion. A naive application of the Taylor expansion introduces terms in ϵ^p with increasing p. For a large enough p, the integral $E[\epsilon^p]$ is diverging due to the slow decay in the tails of the probability distribution. The cautionary step is to expand $f(z)$ using a second-order Taylor polynomial with an explicit remainder given in the Lagrange form. The remainder is of order $(\sigma\epsilon)^3$, and the corresponding moment is assumed to be finite. After integration, the remainder is of order σ^3, which can be neglected in a small δt expansion (see [166] for more details).

The price of the underlying at expiry needs to be evaluated because the option payoff depends on it. The number of process steps between t_0 and T is n, with $\Delta T = n\delta t$. Using the definition of the stock price process (16.1), the price at the option expiry is

$$\begin{aligned}
S(T) &= S(t_0) \prod_{t=t_0}^{T-\delta t} \big(1 + \mu + \sigma(t)\epsilon(t+\delta t)\big) \\
&= F(T) \prod_{t=t_0}^{T-\delta t} \left(\frac{1 + \mu + \sigma(t)\epsilon(t+\delta t)}{1 + r_{\mathrm{rf}}}\right)
\end{aligned} \tag{16.29}$$

with $F(T) = S(t_0)(1 + r_{\mathrm{rf}})^n$ the forward price at time T. This formula can be used directly in a numerical evaluation.

Using a Taylor expansion at order $O(\delta t)$, the price can be approximated by

$$\begin{aligned}
S(T) &= F(T) \prod_{t=t_0}^{T-\delta t} \big(1 + \mu - r_{\mathrm{rf}} + \sigma(t)\epsilon(t+\delta t) + O\big(\delta t^2\big)\big) \\
&\simeq F(T)\Bigg(1 + n(\mu - r_{\mathrm{rf}}) + \sum_t \sigma(t)\epsilon(t+\delta t) \\
&\quad + \sum_{t<t'} \sigma(t)\epsilon(t+\delta t)\sigma\big(t'\big)\epsilon\big(t'+\delta t\big)\Bigg)
\end{aligned}$$

$$\simeq F(T)\left(1+R+\frac{1}{2}\left(\sum_t \sigma(t)\epsilon(t+\delta t)\right)^2 + n(\mu - r_{\text{rf}}) - \frac{1}{2}\sum_t \sigma(t)^2\epsilon^2(t+\delta t)\right)$$

$$\simeq F(T)\left(1+R+\frac{1}{2}R^2 + n(\mu - r_{\text{rf}}) - \frac{1}{2}\sigma^2_{\text{integrated}} - \frac{1}{2}y\right) \tag{16.30}$$

with

$$R = \sum_{t=t_0}^{T-\delta t} \sigma(t)\epsilon(t+\delta t) \tag{16.31a}$$

$$y = \sum_{t=t_0}^{T-\delta t} \sigma^2(t)\left\{\epsilon^2(t+\delta t) - 1\right\} \tag{16.31b}$$

$$\sigma^2_{\text{integrated}} = \sum_{t=t_0}^{T-\delta t} \sigma^2(t). \tag{16.31c}$$

The stock path innovation R is of order $\sqrt{n\delta t} = \sqrt{\Delta T}$, while the term y and the integrated variance $\sigma^2_{\text{integrated}}$ are of order $\Delta T = n\delta t$. This computation shows that the leading term is the total return $R \sim \sqrt{\Delta T}$, but there are corrections of order ΔT compared to the B–S scheme.

Using the same expansion, the Radon–Nikodym derivative (16.9) becomes

$$\frac{D\mathbb{Q}}{D\mathbb{P}} = \prod_{t=t_0}^{T-\delta t} \frac{f(v(t)z(t+\delta t))}{E[f(v(t)z(t+\delta t))]} \tag{16.32}$$

$$\simeq \prod_{t=t_0}^{T-\delta t} \left[1 - \varphi_1(t)\sigma(t)\epsilon(t+\delta t) + \frac{1}{2}\varphi_2(t)\sigma^2(t)\left\{\epsilon^2(t+\delta t) - 1\right\}\right].$$

This formula can be directly used in a numerical evaluation. Notice that at this order

$$E^{\mathbb{Q}}[1] = E^{\mathbb{P}}\left[\frac{D\mathbb{Q}}{D\mathbb{P}}\right] = 1. \tag{16.33}$$

As for the price above, this formula can be expanded up to order δt to give

$$\frac{D\mathbb{Q}}{D\mathbb{P}} \simeq 1 - \sum_{t=t_0}^{T-\delta t} \varphi_1(t)\sigma(t)\epsilon(t+\delta t) + \sum_{t<t'} \varphi_1(t)\sigma(t)\epsilon(t+\delta t)\varphi_1(t')\sigma(t')\epsilon(t'+\delta t)$$

$$+ \frac{1}{2}\sum_t \varphi_2(t)\sigma^2(t)\left\{\epsilon^2(t+\delta t) - 1\right\}$$

$$
\begin{aligned}
&\simeq 1 - \sum_t \varphi_1(t)\sigma(t)\epsilon(t+\delta t) + \frac{1}{2}\Big(\sum_t \varphi_1(t)\sigma(t)\epsilon(t+\delta t)\Big)^2 \\
&\quad - \frac{1}{2}\sum_t \varphi_2(t)\sigma^2(t) - \frac{1}{2}\sum_t \big(\varphi_1^2(t) - \varphi_2(t)\big)\sigma^2(t)\epsilon^2(t+\delta t) \\
&\simeq 1 - \widetilde{R} + \frac{1}{2}\widetilde{R}^2 - \frac{1}{2}\widetilde{\sigma}^2_{\text{integrated}} - \frac{1}{2}\widetilde{y}
\end{aligned}
\tag{16.34}
$$

with

$$
\widetilde{R} = \sum_{t=t_0}^{T-\delta t} \varphi_1(t)\sigma(t)\epsilon(t+\delta t), \tag{16.35a}
$$

$$
\widetilde{y} = \sum_{t=t_0}^{T-\delta t} \big(\varphi_1^2(t) - \varphi_2(t)\big)\sigma^2(t)\epsilon^2(t+\delta t) \tag{16.35b}
$$

$$
\widetilde{\sigma}^2_{\text{integrated}} = \sum_{t=t_0}^{T-\delta t} \varphi_2(t)\sigma^2(t). \tag{16.35c}
$$

This computation shows that the change of measure is given at first order by $D\mathbb{Q}/D\mathbb{P} \simeq \exp(-\widetilde{R})$, similarly to the B–S case. But as for the price, corrections of order ΔT enter the formula for the change of measure, and $\widetilde{R}$ is depending on the process path up to T through $\sigma(t)$ and $\varphi_1(t)$.

When the underlying is described by a process with constant volatility, as in the B–S scheme, the equations are much simpler. At order $\sqrt{\delta t}$, the price change is $R = \sigma_\infty \sum_t \epsilon(t)$, and the change of measure is $D\mathbb{Q}/D\mathbb{P} \simeq \exp(-\widetilde{R}) = \exp(-\varphi_1 R)$. Therefore, there is a simple relation between the total price change between t_0 and T and the change of measure. Moreover, because the process is not depending on the previous values of ϵ, the path dependency in the change of measure is given only by R, up to a multiplicative coefficient. Using an exponential risk aversion and a normal distribution for the innovation simplifies further the equations.

As shown in this section, the expectations over the distributions for the innovations ϵ can be computed in a small δt expansion. This leads to a practical framework to price European options on a fundamental basis, which can include heteroscedasticity, fat-tailed distributions, and possibly other stylized facts. The user can chose its favorite process for the underlying, validated against empirical data. Monte Carlo simulations are used to draw paths for the underlying, and the option pricing formula (16.8) is used as an expectation over the paths with the Radon–Nikodym derivative for each path given by (16.32). Although numerically intensive, such a procedure is not difficult with today's computational power.

16.7 Moneyness and Implied Volatility

It is easier to understand the properties of the option prices with the implied volatility surface, using the Black–Scholes formula as a dictionary between prices and implied volatilities (IV). Essentially, the prices go as the vols, namely a higher price translates into a higher implied volatility. But the implied volatility removes the dominant features of the option price surface and concentrates on the difference against the B–S model. For a Gaussian random walk with constant volatility, the implied volatility is flat and at the level of the specified volatility. Then, any deviation from a constant indicates a departure from the reference model.

In order to obtain a good representation of the problem, the coordinates to use on the implied volatility surface are important. In the time-to-maturity direction, the natural coordinate is $\log(\Delta T)$. In the moneyness direction, the natural measure is the distance between the price and the strike, with a yard stick given by the diffusion of the underlying process. The Black–Scholes analytical formula already contains such a measure (with the return defined as logarithmic price difference):

$$m = \frac{\ln(F/K)}{\sqrt{\Delta T}\tilde{\sigma}} \tag{16.36}$$

where F is the forward price at the option maturity, and K is the strike. In the B–S framework, the volatility is constant, and the term $\sqrt{\Delta T}\sigma$ measures the integrated volatility until the option expiry. With heteroscedasticity, the proper measure for the diffusion until the option maturity is a volatility forecast up to $t + \Delta T$, assuming the information at the current time t. In formula (16.36), the term $\sqrt{\Delta T}\tilde{\sigma}$ is the best estimate at t of the standard deviation of the underlying process until the option maturity. For ARCH processes, the volatility forecast can be derived analytically, as done in Sect. 7.7. Strictly speaking, the formulas are derived for microscopic processes, but it is a reasonable approximation to use the same weights for aggregated processes, even when the leverage effect has been added in the volatility measures σ_k. With this forecast, the optimal dependency on the time-to-maturity is included, with increasing weights given to long-term components for increasing forecast horizons. In this way, the available information about the expected volatility relevant for the option is incorporated in the measure of moneyness. As an order of magnitude, a moneyness of $m = 1$ means a one σ distance between price and strike (regardless of ΔT), which is clearly in the money for a call option.

Simpler measures of moneyness are defined with F/K or $\ln(F/K)$. These definitions are easier to work with as they do not depend on the underlying, and the computations between moneyness, strike and underlying are elementary. Yet, the yardstick provided by the underlying diffusion is missing, leaving in the moneyness a dependency on the underlying history up to t. For example, a moneyness $\ln(F/K) = 1$ for a call option can be close to the money or deep in the money, depending on the time to maturity and volatility. This dependency translates into a complex dynamics in the volatility surface, where, for example, the smile and smirk are proportional to the volatility. Instead of a complex IV surface dynamics, it is conceptually better to incorporate the predictable dynamics into the definition of moneyness as in (16.36), and subsequently to have simpler IV dynamics with mostly static smiles and smirks.

16.8 Application for ARCH Processes

The option scheme presented in this chapter was investigated extensively in [166], using ARCH processes of increasing complexity and for empirical data on the SP500 European options. An important advantage of the present option pricing is to be independent of the analytical form for the processes. Therefore, the various ingredients can be inserted one-by-one in the equations, so as to show their impact on the IV surface. The main features of the option prices using various processes can be analyzed on Fig. 16.1. For all processes, the mean annualized volatility is $\sigma_\infty = 20~\%$, and the annualized implied volatility is plotted as a function of the moneyness for increasing time to maturity.

For constant volatility and normal innovation, the IV is at the level of σ_∞, except for deep-in-the-money put option. This deviation is due to the relative returns definition in the process, instead of the log-normal definition in the Black–Scholes model. For short time, the difference between both definitions is very small, but the differences are increasing with the simulation length. The difference occurs in the far left tail, corresponding to a slightly larger probability in the very small price region. The larger probability translates in a higher price for long-term and deep-in-the-money put option and hence to a higher implied volatility.

Using constant volatility and Student innovation places a smile for short time to maturities (top right plot). This smile vanishes with increasing time to maturities, due to the aggregation and convergence toward a normal distribution as induced by the central limit theorem (approximately, because of the relative returns used in the process).

The smile corresponding to the GARCH(1, 1) process is plotted in the middle line. With normal innovation, the smile is increasing with the time to maturity, due to the dynamics of the heteroscedasticity. Time is required for the GARCH(1, 1) dynamics on the volatility to induce changes and to produce a significant deviation from the normal distribution used for the innovation. The GARCH(1, 1) process with Student innovation displays a smile for all time to maturities. This is due to the conjunction of the Student innovation for short time to maturities and of the volatility dynamics for long time to maturities.

The long-memory ARCH processes is displayed on the bottom line. With Student innovations, the smile is slightly stronger compared to the GARCH(1, 1) dynamics. The difference is larger for long time to maturity, due to the longer characteristic times present in the process. Leverage has been added for the bottom right plot, using Eq. (14.1). The leverage intensity is $\lambda_{\text{leverage}} = 0.95$, corresponding to a very large correlation between the historical returns and realized volatilities. The goal is to replicate the European options on the SP500, and the indexes show a larger leverage effect compared to individual stocks. The figure shows the characteristic

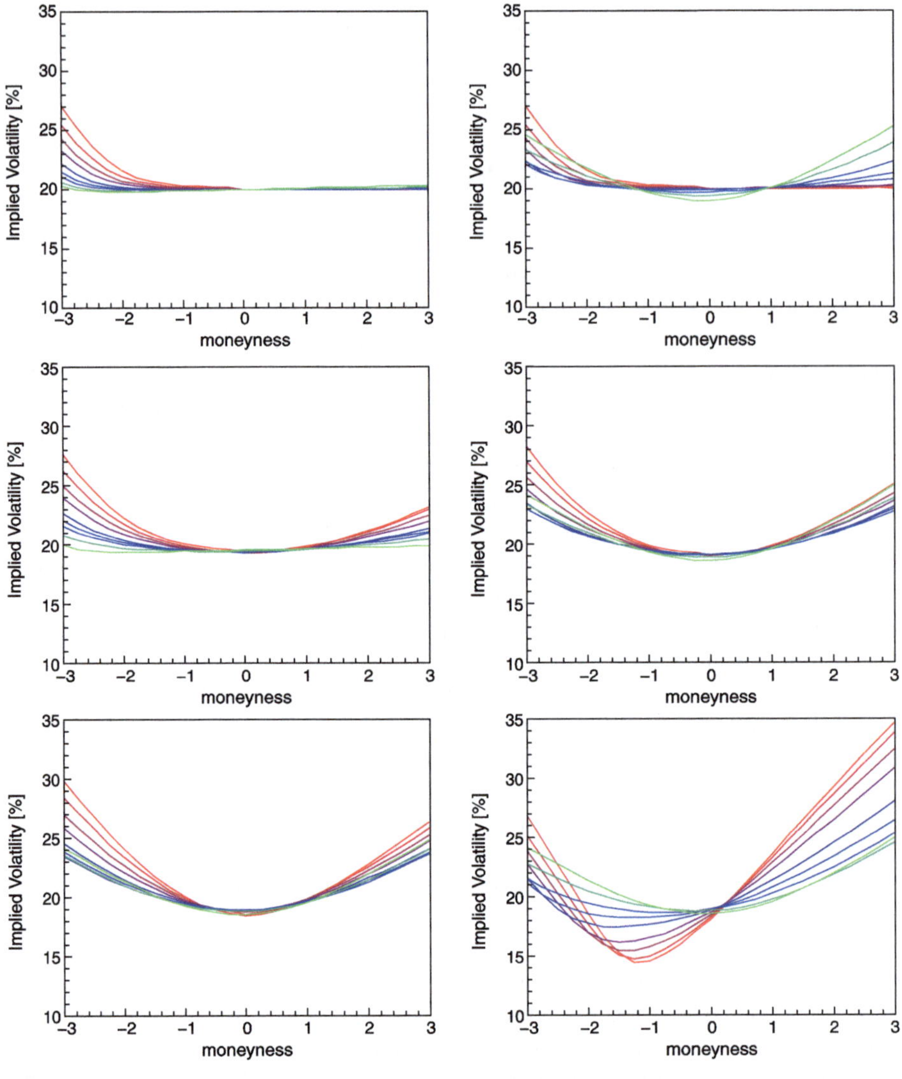

Fig. 16.1 The implied volatility surfaces for various processes, for put options. The time-to-maturities are 1, 2, 5, 10, 21, 65, 130, 260, and 520 days. The *colors* are according to the maturity, with a linear interpolation in the logarithm of the time-to-maturity through the *colors green* (1 day) to *blue* (2 weeks) to *red* (2 years). The simulation is done with 10^6 paths. The processes are constant volatility with normal and Student innovations (*top left* and *right*), GARCH(1, 1) with normal innovations and Student innovation (*middle left* and *right*), LM-Agg-ARCH with Student innovations (*bottom left*), same + leverage (*bottom right*)

smirks, with an increasing effect for increasing time to maturities. For this process, the innovations have a symmetric distribution, leading to a symmetric smile at 1 day. The asymmetry is induced by the dynamics, leading to an increasing smirk.

Clearly, the dominant features of the implied volatility smile are due the dominant features included in the processes. The observed smiles on the markets are close to the bottom right figures, in agreement with our findings about the properties of the underlying time series. Therefore, a fairly sophisticated process is required in order to price accurately options.

A similar study can be done for different risk aversion functions $f(z)$. The salient outcome is that changing the risk aversion has essentially no influence on the smile, as can be understood by the following argument. In the small δt expansion, Eq. (16.28) fixes explicitly the value for $\varphi_1 = \phi(t)$ in terms of the drift and volatility. This equation should be solved for ν, which will in turn give the value of φ_2. In this chain of computations, only the value of φ_2 is depending on the choice for f, as φ_1 is depending only on the process up to t through the market price of risk ϕ. In the equation for the change of measure, the term of order $\sqrt{\delta t}$ is depending only on $\varphi_1 = \phi$. As this gives the leading dependency, the implied volatility smile at this order is not depending on f. At the next order in the δt expansion, the term φ_2 introduces a weak dependency on the risk aversion. Therefore, although in principle the option prices are depending on the risk aversion, this is essentially not the case in practice.

An extensive comparison with European put and call options on the SP500 is done in [166]. Overall, the empirical implied volatility surfaces are reproduced very well by the theoretical computations, in particular the smirks, smiles, and term structures. The noticeable exception is the overall level, where the IV from the market is systematically higher compared to the pure arbitrage theoretical computations. The difference is higher before 2001 and seems to level down around that time. Likely, the option issuers could price at their advantages the options, up to the advent of large hedge funds that were able to arbitrage them.

16.9 Cross-Product Approximation of the IV Surface

An important feature of the numerical computation is the mostly static shape of the smile, when expressed with the moneyness defined in (16.36). Indeed, the smile is depending quantitatively on the process and its parameters, which are fixed for the simulations. The process parameters are set so as to reproduce a panel of statistics for the underlying time series, without using the options. The Monte Carlo simulations are done by initializing the process in the mean volatility state, thus removing the dynamics in the backbone direction. Therefore, a process with its parameters is associated to a given volatility surface called $\mathrm{IV}_{\mathrm{ref}}(m, \Delta T)$, as shown, for example, in Fig. 16.1. This static smile is roughly in agreement with the empirical implied volatility surface on the SP500, where the backbone changes with time, whereas the smirk and smile are mostly static.

With historical data up to a given day, the process initialization should be done with the historical volatility σ_k measured at the relevant time horizons. Essentially, this step uses the historical filtration up to t in order to initialize the process. The variance between t and the option maturity $t + \Delta T$ is then given by the variance forecast as derived in Sect. 7.7:

$$\widetilde{\sigma^2}(t, t + \Delta T) = \sum_k w(k, \Delta T)\sigma_k^2(t) \tag{16.37}$$

where the weights $w(k, \Delta T)$ are derived analytically. For the options, this variance forecast is setting the level and ΔT shape of the implied volatility surface.

The difference in the dynamic behaviors between the backbone and the smile is the basis for the cross-product approximation of the implied volatility surface [165]

$$\mathrm{IV}(m, \Delta T, t) = \mathrm{backbone}(\Delta T, t) \times \frac{\mathrm{IV}_{\mathrm{ref}}(m, \Delta T)}{\sigma_\infty}. \tag{16.38}$$

The backbone is computed using the volatility forecast given in (16.37). The shape $\mathrm{IV}_{\mathrm{ref}}$ is computed once using Monte Carlo simulations with the initial conditions $\sigma_k = \sigma_\infty$ and sampled on a fine enough grid in the moneyness and maturity directions. The mean volatility σ_∞ used to compute the reference surface is not important, and a typical mean value is good enough.

With the reference IV surface available, the procedure for the computation of an option price is the following. The volatility forecast up to the option maturity ΔT can be computed using (16.37), giving the backbone of the IV surface. Using the option's strike and volatility forecast, the option's moneyness m can be computed. The reference implied volatility $\mathrm{IV}_{\mathrm{ref}}(m, \Delta T)$ is evaluated at the option's coordinates (with a linear interpolation for example). The final implied volatility $\mathrm{IV}(m, \Delta T, t)$ is obtained using (16.38), and this value is inserted into the Black–Scholes formula in order to compute the option price. Only explicit expressions are involved in the computations, and no Monte Carlo simulations should be performed.

Furthermore, the Greeks can be estimated using the formula for composite functions. For example, the delta required to compute the hedging portfolio is given by

$$\Delta = \frac{\partial O}{\partial S} = \Delta_{\mathrm{BS}} + \frac{\partial O}{\partial \mathrm{IV}} \cdot \frac{\partial \mathrm{IV}}{\partial m} \cdot \frac{\partial m}{\partial S} \tag{16.39}$$

where Δ_{BS} is the Black–Scholes delta, and O the option price (put or call). The other sensitivities can be computed similarly. Again, all the formulas are explicit and can be evaluated very quickly. Notice that the equivalent evaluation using Monte Carlo simulations would require significant computational resources. For example, the computation of Δ along the price paths in Monte Carlo simulations is an MC^2 problem (because the evaluation of Δ requires Monte Carlo simulations).

With this efficient estimation for delta, the replication strategy can be simulated, with the daily rebalancing of the position in the underlying. At the beginning of the strategy, the premium from selling the option is used to buy an amount of stocks given by delta, and the rest is invested in the cash account. Everyday during the life of the option, the hedge is adjusted so that delta stocks are owned, while the proceeds are put or taken on/from the cash account accruing over time. In this computation,

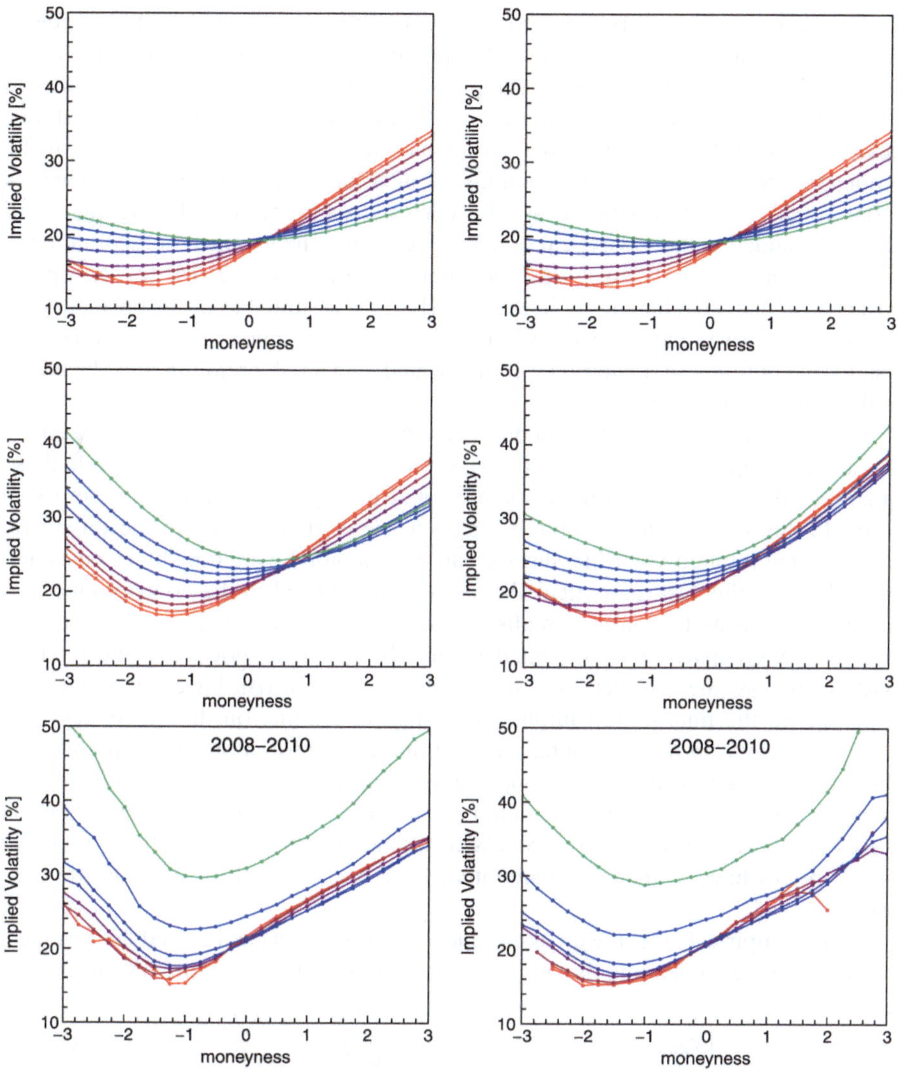

Fig. 16.2 The reference and empirical quotient IV surfaces. The put (call) are on the *left* (*right*) *column*, the time-to-maturities are 2, 5, 10, 21, 63, 130, 260, and 520 days. The *colors* are set according to the time-to-maturity, from *dark green* (2 days) to *blue* (21 days) to *red* (520 days). The *top line* shows the theoretical IV for a long-memory ARCH model, with leverage and fat tail innovations, as already given at the *bottom right* in Fig. 16.1 (with a different vertical scale). The *middle line* presents the same model, but with a transaction cost and a risk premium for the option replication. The *bottom line* shows the empirical coset IV surfaces computed over the years 2008 to 2010 for the options on the SP500

realistic transaction costs can be easily included. At the option's maturity, all the available stocks are sold in order to unwind the hedge, and the option's payoff is delivered and settled in cash. The remainder on the cash account is the P/L of the replication strategy, cost included.

This computation can be done along any underlying price path, whether generated by a Monte Carlo simulation with a process or for the actual values observed for the SP500. This allows us to compute either statistical quantities in a Monte Carlo setting or the corresponding realized values when using the historical values of the SP500. Three quantities are of particular interest. First, the P/L of the replication. Second, the cumulative transaction volume for the underlying, which is proportional to the transaction cost. This corresponds to the cost of hedging, which should be added to the theoretical option price. Third, the variance of the total portfolio (option + underlying + cash account) corresponds to a risk for the option writer. Because the rebalancing in the underlying is assumed to occur once per day, the risk is positive (and not zero as in the exact B–S continuous-time method). This risk for the option writer should be compensated, and a risk premium must be added to the theoretical option price.

For the option issuers, the theoretical arbitrage price should be increased by the average replication cost and the risk premium for the replication. All these quantities depend on the time-to-maturity and moneyness, and are therefore surfaces. In this way, a realistic option price surface is obtained. From these realistic prices a new implied volatility surface can be computed, which includes costs and risks. In turn, this last theoretical IV surface can be used in the cross-product approximation so as to compute realistic option prices, as well as replication cost and risk. A comparison is presented in Fig. 16.2, which shows that the realistic ARCH process, the transaction cost, and the replication risk have important contributions to the final option implied volatility, depending on the moneyness and time to maturity. In this computation, the transaction cost is of 5 basis points, and the risk is proportional to the standard deviation of the final payoff with a coefficient $\beta = 0.25$ (see [165] for the details). The large difference between the option prices computed with the bare process or including cost and risk shows that real option prices have to include quantitatively the cost and risk of hedging the contracts.

On the empirical side, the cross-product decomposition can be used "in reverse", so as to remove the leading dynamics and to reconstruct a "quotient" IV surface

$$\mathrm{IV}_{\mathrm{quotient}}(m, \Delta T, t) = \sigma_\infty \times \frac{\mathrm{IV}_{\mathrm{market}}(m, \Delta T, t)}{\mathrm{backbone}(\Delta T, t)}. \tag{16.40}$$

The quotient surface has a very weak time dependency, leading to a better convergence for the empirical expectations. A graph of the mean coset surface computed over 3 years is given on the bottom line in Fig. 16.2. Clearly, a very good agreement is obtained between the expectation of the empirical quotient IV surfaces and the theoretical reference IV surfaces with costs and risk (except at very short maturities). More details about the computations can be found in [165], including a decomposition of the option prices into the respective contributions.

At this point, realistic option prices and implied volatility surfaces can be derived, including the replication costs and risks, thanks to the cross-product decomposition. The scheme includes most of the required features for the underlying (heteroscedasticity with long memory, fat-tailed innovations, leverage, risk premium in the drift) and for the option pricing (risk aversion function, risk-free measure, no-arbitrage replication, replication cost and risk). Although the evaluations have to be numeric, the inclusion of realistic features is a substantial step beyond the idealization of the Black–Scholes formula. The limitation of the cross-product approximation is that it can only be applied to European options, and further extensions are required for American options.

Chapter 17
The Empirical Properties of Large Covariance Matrices

17.1 Introduction

So far, the empirical properties of single time series have been established. The most salient feature is the dynamics of the volatility, with a similar decay for the lagged correlations of all empirical time series. This remarkable stylized fact will be very useful when moving from a univariate world to a multivariate universe. In the univariate case, this is used to define simple ARCH processes that capture the leading empirical properties (volatility clustering and fat-tailed distributions). Through these processes, a natural definition of "the" volatility emerges, namely the best one-step volatility forecast, corresponding to the σ_{eff} in the processes. Furthermore, quadratic process structures lead naturally to volatility forecasts which can be used in several contexts.

We want now to apply a similar program in the multivariate case. The natural path is to extend the univariate variance to the multivariate covariance, replacing the univariate quadratic term r^2 by the multivariate term $r_\alpha r_\beta$. The univariate weights in the variance can be used directly in the multivariate covariance. Following these extensions, the road looks straight and smooth. Yet, as in any good story plot, new difficulties are awaiting us after the first curve.

In a given investment universe, the covariance matrix encapsulates a wealth of information. Part is related to the volatility of each component, and part to the mutual dependencies of the components as measured by the correlations. Aiming at practical applications, the number of time series N should be large, up to one hundred or more. The corresponding covariance is of size $N \times N$, resulting in a huge amount of information. Moreover, the dynamics of the covariance are important, as the volatilities and correlations evolve with time. Therefore, we are faced with $O(N^2)$ time series. This is clearly difficult to grasp, and the raw covariance matrix should be transformed into more palatable quantities.

Because the covariance matrix is symmetric, a natural tool is an eigen-decomposition. With the time dependency, the eigenvectors and eigenvalues are themselves time series which contain the same information as the raw covariance. For large N, the eigenvalues have some generic properties that allow us to better

G. Zumbach, *Discrete Time Series, Processes, and Applications in Finance*, Springer Finance, DOI 10.1007/978-3-642-31742-2_17,

understand, and possibly to summarize, the information contained in the covariance (or correlation) matrices.

The new difficulties raised by the multivariate extension can now be briefly summarized. The first one is related to the small eigenvalues of the covariance matrix. These small eigenvalues are creating problems when inverting the covariance, like in a portfolio allocation or in a process estimation. The second difficulty is related to the lack of stability of the eigenvectors. This instability prevents the separation of the global asset space into a leading stable subspace and a noise part. The third difficulty is that more quadratic extensions of the variance can be defined, introducing new free parameters into the covariance definition. This chapter investigates the first two points, while the last point is discussed in the next chapter.

The covariance matrix appears in many computations in finance. Common examples are in Monte Carlo schemes for instrument pricing or for risk evaluation, in optimal portfolio allocation or in process inference using a log-likelihood maximization. For the last two examples, the *inverse* of the covariance is needed (the inverse square root for process inference). As the number of time series N becomes large, the covariance contains null eigenvalues. In this case, the inverse is not defined, and a proper regularization scheme should be used. This occurs always when N is larger than the memory length $i_{\max}$ used to compute the covariance. For many practical applications, the memory length is of the order of one to two years ($i_{\max} = 260$ to $i_{\max} = 520$), whereas the number of time series can be of the order of thousands. Yet, similar practical problems occur even when the covariance should be (mathematically) regular. This happens because the spectrum decays exponentially fast toward zero, leading to very large values in the inverse. Essentially, for N large, the presence of many small (or null) eigenvalues makes the inverse covariance badly defined, and a regularization scheme should be used even when the covariance is in principle regular.

It is natural to define the regularization scheme using the eigen-decomposition. An important issue is then the number of eigenvalues to include in the inverse covariance. In order to select a cutoff, the "meaningful" eigenvalues should be separated from the "noise-induced" eigenvalues. This is essentially the line followed by Principal Components Analysis (PCA) and by factor models. Another idea is to use random matrix theory to find the edge of the noise-induced spectrum. As we will see, no simple threshold can be found: the covariance structure is more of a gradual crossover from meaningful to noise-induced eigenvalues.

When estimating covariance matrices, the question of the best covariance estimator arises, as well as of the best forecast. Several paths can be followed to define the covariance matrix. First, the covariance can be computed as the natural extension of the univariate case, essentially replacing r^2 in the univariate formulas by the product $r_\alpha \cdot r_\beta$ in the multivariate case. This is the approach taken in this chapter, using slow decaying weights that produce the best univariate forecast. This route imposes the least structure on the covariance.

Second, the volatility and correlation could be separated, and each part evaluated with a specific and optimal formula. The underlying idea is that the volatilities have fairly fast dynamics, while the correlations are more stable. Correspondingly, the optimal estimators for both quantities should be different. This idea has been pursued in the GARCH context by [23] and subsequent papers which use a constant correlation matrix or a correlation with slow dynamics. The validity of this approach can be formulated as the identification of two different characteristic time scales in the covariance: a short one for the fast evolution of the volatilities (from a few days to a few months) and a long one corresponding to the correlation (at least one year). This chapter investigates the characteristic time of the fluctuations using lagged correlations for the covariance and correlation matrices, and for the projectors on the leading subspaces.

Third, more structure can be imposed on the covariance, for example, by a factor model or by a Bayesian method with a given prior for the covariance matrix [86]. A comparison of various structures for the covariance matrix is presented in [34] for a universe of UK equities, with the aim at volatility forecasting. This route implies that some extra information exists about the covariance matrix, and this information is dominant and stable through time. The present chapter investigates the stability of the structures present in the covariance matrix by analyzing the dynamics of the eigenspaces of increasing ranks.

This chapter is mostly based on the paper [160]. It focuses on the empirical properties of the covariance and correlation for large N, using the spectrum, the spectral density, and the subspaces span by the eigenvectors. The present study is done with a dataset of size 330, with daily time increment. This allows us to extract generic features of the covariance. The following chapter uses these properties to study various processes definitions and the regularization of the inverse covariance. Let us emphasize that these sizes are small by today's standard but sufficient to explore the generic properties of large matrices, and increasing the sizes will only make the problems more acute. For the present study, the computations are fairly demanding in terms of computational time, and it is not possible to increase substantially the size of the matrices.

17.2 Theoretical Framework

For a time series $x(t)$ with a daily time increment δt, the daily (logarithmic) return is defined as

$$r(t) = x(t) - x(t - \delta t). \qquad (17.1)$$

The mapped price x is $x(t) = \ln(p(t))$ for stock, stock indexes, FX, and commodities. For interest rates, x corresponds to the rate at a fixed time to maturity.[1] In this

[1] More precisely, for the interest rate R, the mapped price is $x = \log(1 + R/R_0)$ with $R_0 = 4\ \%$. This mapping decouples the volatility from R; see [154]. This mapping introduces a small correction on the returns for interest rates.

chapter, we consider only the class of covariance matrices Σ_{eff} that are the cross-products of the past return vectors

$$\Sigma_{\text{eff}}(t) = \sum_{i=0}^{i_{\max}} \lambda(i)\mathbf{r}(t - i\delta t)\mathbf{r}'(t - i\delta t) \tag{17.2}$$

with $\mathbf{r}$ a column vector and $\mathbf{r}'$ its transpose. The weight for the past returns $\lambda(i) \geq 0$ obeys the sum rule $\sum_i \lambda(i) = 1$. Three common choices for the weights are equal weights, exponential weights, and long-memory weights. Because the long-memory shape better describes the data, we selected the corresponding weights with the parameter $\tau_0 = 1560$ days. Yet the detailed shape of the kernel has a minor impact on the salient results.

For symmetric matrices, the eigen-decomposition is

$$\Sigma_{\text{eff}} = \sum_{\alpha=1}^{N} e_\alpha \mathbf{v}_\alpha \mathbf{v}'_\alpha \tag{17.3}$$

where the eigenvalues $e_\alpha = e_\alpha(t)$ and eigenvectors $\mathbf{v}_\alpha = \mathbf{v}_\alpha(t)$ are time dependent, and the eigenvectors $\mathbf{v}_\alpha$ are orthogonal. For convenience, the eigenvalues are ordered by decreasing values, so that $e_\beta \leq e_\alpha$ for $\beta > \alpha$.

The spectrum of the covariance matrix is investigated in Sect. 17.7. The spectral density in the interval $[\lambda - \delta\lambda, \lambda + \delta\lambda]$ is computed by

$$\rho(\lambda) = \frac{1}{2\delta\lambda} \frac{1}{NT} \sum_{t=1}^{T} \sum_{\alpha} \chi\big(\lambda - \delta\lambda < e_\alpha(t) < \lambda + \delta\lambda\big) \tag{17.4}$$

where $\chi(x)$ is the characteristic function of x (1 if x is true, 0 if x is false). The spectral density $\rho(\lambda)$ measures the time average density of the eigenvalues around λ. The normalization is defined such that $\int \rho(\lambda)\, d\lambda = 1$. In the theoretical computation related to random matrices, the time average is replaced by an ensemble average.

The spectrum and eigenvectors of the covariance matrix need to be studied in order to understand the properties of such large matrices. The canonical picture in finance is that a few leading eigenvalues dominate, corresponding to broadly defined directions, like "the market", "value versus growth", or the overall "interest rate level". In mathematical terms, the canonical picture is that the leading eigenvalues e_α correspond to stable directions for the corresponding vectors v_α. In order to measure this stability, the mean eigenvectors are not interesting as they converge to zero. The cause is a sign indeterminacy for the eigenvector: $\mathbf{v}_\alpha$ and $-\mathbf{v}_\alpha$ have the same eigenvalue. A second problem is the crossing of the eigenvalues, leading to an apparent abrupt change of the eigenvector with a given rank. The object that is better behaved for our purpose is the projector on the leading subspace. The projectors $\mathsf{P}_k(t)$ on the leading subspace of rank k is

$$\mathsf{P}_k = \sum_{\alpha=1}^{k} \mathbf{v}_\alpha \mathbf{v}'_\alpha. \tag{17.5}$$

For a given rank k, the mean projector is defined by the time average

$$\langle \mathsf{P}_k \rangle = \frac{1}{T} \sum_{t=1}^{T} \mathsf{P}_k(t). \tag{17.6}$$

The trace is preserved by the average trace$\langle \mathsf{P}_k \rangle = k$, but $\langle \mathsf{P}_k \rangle$ is not a projector as its eigenvalues are between 0 and 1 (and not either 0 or 1). If the projector is essentially static, $\langle \mathsf{P} \rangle \simeq \mathsf{P}(t)$, the projector has k eigenvalues 1 and $N-k$ eigenvalues 0. In the other direction, if the projector dynamics explores fully the available space, the mean projector is proportional to the identity: $\langle \mathsf{P}_k \rangle \simeq k/N\mathbb{I}$ with all eigenvalues given by k/N.

Factor models and principal Components Analysis (PCA) are widespread in the investment business (see [63] for the original reference, [142, 145] for recent books). One approach is to use the spectral decomposition of the covariance matrix to reduce the dimensionality of the original universe to a small number of uncorrelated factors. A large number of empirical studies (mostly on equities) on this dimension reduction technique have been done and tend to show that only a small number of eigenvalues are significant. The picture conveyed by PCA is of a stable leading subspace. For the mean projector, this picture translates into a set of eigenvalues close to one corresponding to the relevant and stable eigenvectors, with a clear gap separating eigenvalues close to zero corresponding to the idiosyncratic noise. As the trace is constant, the choice of the rank for the projector alters the eigenvalues. The relevance of this picture is investigated in Sect. 17.8.

17.3 The Datasets for the Empirical Studies

The role of null eigenvalues in the covariance matrix is important to understand, and they occur when the size of the covariance matrix N is larger than the historical depth $i_{\max}$ in (17.2). The reference data set used in this chapter contains 340 time series that covers majors asset classes and world geographical areas. A history depth of one year (260 business days) has been used regardless of the memory kernel, leading to 80 null eigenvalues (and many small eigenvalues). This set is called ICM (International Capital Market) and is divided as follows: 19 commodities (metal and gas futures), 78 foreign exchange rates, 52 equity indices, 127 interest rates, with maturities at one day, one month, one year, and ten years, and 54 individual stocks from USA, France, and Switzerland.

In [160], two other datasets are analyzed with sizes 55 and 54, covering respectively part of the G10 and USA universe. All results are consistent between the three datasets, despite that both smaller sets have no zero eigenvalues. All time series contain daily prices from 1 January 1999 to 1 January 2008 (nine years), corresponding to a length of 2515 days. The in-sample investigations are done from 1 January 2000 to 1 January 2008, and one year ($i_{\max} = 260$ business day) is used to evaluate the effective covariance Σ_{eff}. All the FX rates are against the USD in order to avoid the degeneracy induced by cross rates. The FX values are a snapshot at a given GMT time, while the values for the other time series correspond to the closing of the market. Therefore, a part of the correlations can be induced by the asynchronous nature of the data and the circulation of the information across the globe. Unfortunately, it is difficult to obtain data with such a large world coverage, with a sufficient time span, and taken at the same world time. For the USA dataset, all the time series are obtained at the closing of the US market and are therefore synchronous. The results for the USA set are in line with both world sets, confirming that the asynchrony is negligible for such an analysis.[2]

17.4 The Dynamics of the Covariance Spectrum

The dynamics of the covariance spectrum are plotted in Figs. 17.1. The top figure shows that the core of the spectrum is essentially independent of the time, except for a slow global dynamics measured, for example, by the mid eigenvalue $e_{N/2}(t)$. The global slow dynamics has a time scale of a few years, while the peaks of volatility on the first eigenvalues have essentially no influence on the deep eigenvalues. Only two events—namely the 11 September 2001 attack and the start of the subprime crisis (August 2007)—have had an influence on the core eigenvalues. Interestingly, the largest eigenvalue of the ICM dataset shows no peculiar behavior on 11 September 2001, and is in fact related to the Argentina currency crisis in the same year. The top ten eigenvalues have a richer structure (bottom figure). The first five eigenvalues have a large and independent dynamics, while the next five eigenvalues move much more in synch. Several eigenvalue crossings can also be observed, hinting at different subspaces with changing ranks. The bottom line is that only a handful of the largest eigenvalues have meaningful dynamics, and the rest can be described by a static distribution around a slow moving mean.

By analyzing the eigenvectors, the major peaks on Fig. 17.1 can be identified. They correspond to crises related to a single time series, and the stock indexes often do not belong to the first three eigenvectors. Overall, this analysis shows that the leading eigenvectors are often changing direction and that it is not possible to select a priori stable leading directions. The analysis of the leading eigenvectors and eigenvalues shows that they are dominated by particular events and crises. On the other hand, the deeper eigenvalues only contain information about the overall volatility of the market corresponding to the selected investment space. For example, the middle eigenvalue can be a good simple measure of the overall base market volatility. In turn, this eigenvalue can serve as a baseline for the volatility, and the differences between the top eigenvalues and the baseline illuminate peculiar crises related to a few time series.

[2] If lagged correlations would have been a issue, they can be treated in the covariance computation by replacing $\mathbf{r}(t)\mathbf{r}'(t)$ by $\mathbf{r}(t)\mathbf{r}'(t) + \mathbf{r}(t)\mathbf{r}'(t+\delta t) + \mathbf{r}(t)\mathbf{r}'(t-\delta t)$. See [127].

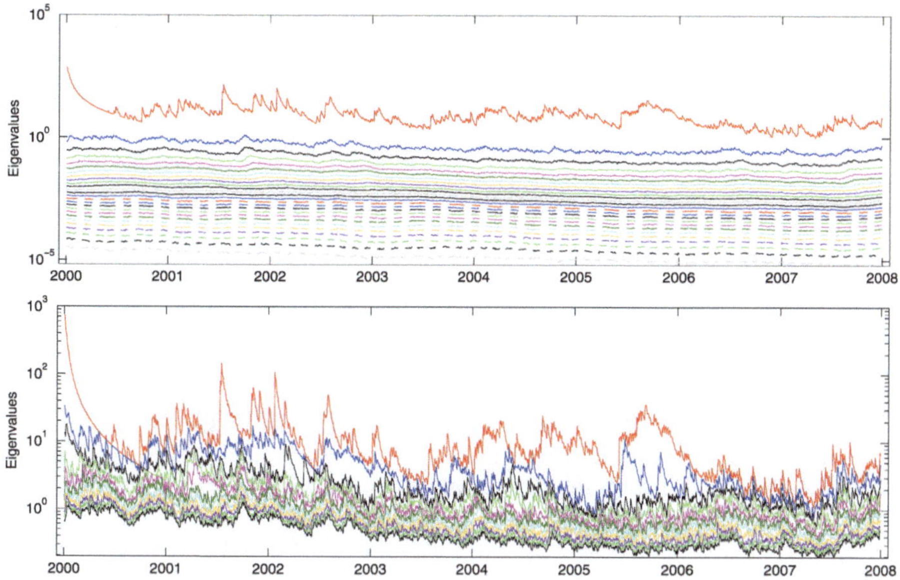

Fig. 17.1 Eigenvalues of the daily covariance matrix. The *top figure* displays each 10th eigenvalue (i.e., $e_1, e_{11}, e_{21}, e_{31}, \ldots$) for the ICM dataset. The *bottom figure* displays the largest eleven eigenvalues (i.e., $e_1, e_2, \ldots, e_{11}$)

17.5 The Dynamics of the Correlation Spectrum

In a multivariate context, the correlation matrix ρ is defined elementwise by

$$\rho_{\alpha,\beta} = \frac{\Sigma_{\alpha,\beta}}{\sqrt{\Sigma_{\alpha,\alpha}\Sigma_{\beta,\beta}}} \tag{17.7}$$

or using a matrix expression by

$$\rho = \Sigma_d^{-1/2} \cdot \Sigma \cdot \Sigma_d^{-1/2} \tag{17.8}$$

where Σ_d is the diagonal of the covariance matrix. The eigenvalues of the correlation matrix have to obey more constraints: in particular, the spectrum is bounded by $0 \le e_\alpha \le N$ and $\sum_\alpha e_\alpha = N$. Moreover, the upper bound $e_1 = N$ for the first eigenvalue is reached for a perfectly correlated system with $\rho_{\alpha,\beta} = 1$.

The same dynamical analysis can be carried out for the spectrum of the correlation matrix. Overall, the correlation spectrum is very static, in agreement with the common lore that the volatilities capture the largest part of the dynamics. This is also consistent with the existence of a unique limit distribution of eigenvalues in large random correlation matrices. Only the first three to five eigenvalues have an interesting dynamics, as reported in Fig. 17.2.

By comparing Figs. 17.1 and 17.2, the similarity of the dynamics for the largest eigenvalues is clear. The volatilities have larger moves, but the time scales involved

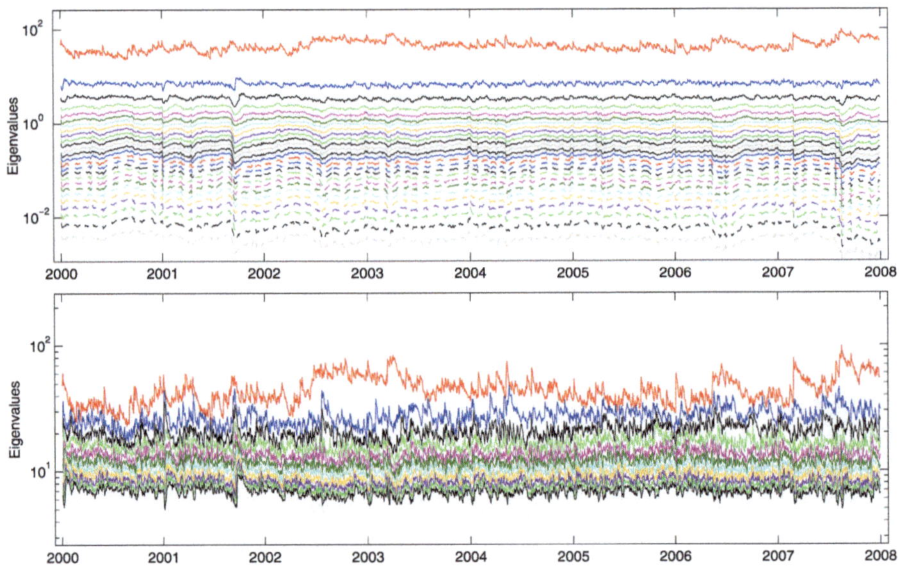

Fig. 17.2 Eigenvalues of the daily correlation matrix. The *top figure* displays each 10th eigenvalue (i.e., $e_1, e_{11}, e_{21}, e_{31}, \ldots$) for the ICM dataset. The *bottom figure* displays the largest eleven eigenvalues (i.e., $e_1, e_2, \ldots, e_{11}$). The upper vertical limit is N_{pos}, the largest possible value for the first eigenvalue, corresponding to perfectly correlated time series

in both graphs are similar, roughly of the order of one month. This finding supports a joint evaluation of the covariance including the correlation and volatility, and against a model with a simple static behavior for the whole correlation matrix.

17.6 Spectral Density of the Correlation Matrix

The base description of financial time series is a (Gaussian) random walk, where the returns $r_\alpha(t)$ are i.i.d. (Gaussian) random variables. When using constant weights $\lambda(i) = 1/T$ and $T = i_{\max}$, the covariance elements are a simple sum of products of random variables with

$$\Sigma_{\alpha,\beta} = \frac{1}{T} \sum_{1 \leq t \leq T} r_\alpha(t) r_\beta(t). \tag{17.9}$$

When the distribution of returns have a unit variance, the covariance matrix is equal (in average) to the correlation. The set of matrices built from independent returns with unit variance is called the Wishart ensemble, and the spectrum of these matrices has been studied extensively, going back to the work [143]. In this ensemble, Marchenko and Pastur (1967) [100] derived the spectral density of correlation matrices in the limit of large matrix size N and large historical sample size T for a fixed ratio $q = N/T$. The condition is that r_α should have a unit variance, but the distribution of r is otherwise unspecified (it is a central limit like result). The M–P spectral density is given by

$$\rho(\lambda) = \frac{\sqrt{4\lambda q - (\lambda + q - 1)^2}}{2\pi\lambda q} = \frac{\sqrt{(\lambda^+ - \lambda)(\lambda - \lambda^-)}}{2\pi\lambda q} \qquad (17.10)$$
$$\lambda \in [\lambda^-, \lambda^+] \qquad \lambda^\pm = (1 \pm \sqrt{q})^2.$$

When $q > 1$, the spectrum also contains a set of $N - T$ zero eigenvalues. This spectrum has bounded support and is generated purely by the random nature of the uncorrelated returns. The idea here is to apply the theoretical upper bound of the M–P spectral density to empirical correlation matrices, in order to separate the significant spectrum (which is driven by true time series dependencies) from the noise-induced spectrum (which is described by the M–P density).

References [85, 121, 122] have introduced the idea to use random matrix theory and the M–P spectral density in order to separate the meaningful eigenvalues from the noise-induced eigenvalues. On the analytical side, the theoretical computation of Marchenko and Pastur has been extended in a few directions. Reference [118] have computed the spectral density for an exponential weighting scheme, while [19] have introduced a Wishart–Student ensemble, both leading to implicit equations for the spectral density. Reference [128] review these computations and investigate the implications for portfolio construction. Unfortunately, the M–P spectral density and its extensions are not precisely applicable to our case. These analytical computations are done in the limit $N, T \to \infty$, whereas the empirical investigation is done for a fixed kernel, corresponding to a constant T. The computations in finance are better described by the limit $N \to \infty$, but with a fixed (long-memory) correlation kernel (the equivalent of T). Unfortunately, there are no analytical results in this limit. A second point is that the spectral density has been derived only for constant weights and exponential weights, whereas the long-memory weights are more interesting because related to processes and forecasts.

The empirical mean spectral density can be computed from the time evolution of the spectrum (in effect, replacing an ensemble average in the theoretical computation by the time average of the empirical spectrum). The spectral densities of the correlation matrices are shown in Figs. 17.3 and 17.4, respectively with constant and long-memory weights. For each weighting scheme, the similarity of the spectral densities between the three datasets is striking, particularly since the ratio $q = N/T$ varies by a factor of six across the data sets. This similarity points to a large N limit at fixed T (or at fixed weights $\lambda(i)$). With constant weights used for Fig. 17.3, the M–P spectral densities do not describe well the lower part of the empirical densities, particularly for $q \simeq 0.2$ (values closer to one give better overall fits). The shoulder in the spectral density $\rho(\lambda)$ around $\lambda \sim 3$ seems to separate the noise-induced density from the meaningful eigenvalues, in reasonable agreement with an M–P upper limit for $q \simeq 1$. With long-memory weights used for Fig. 17.4, this shoulder is smeared out, and there is no feature left to separate the spectrum.

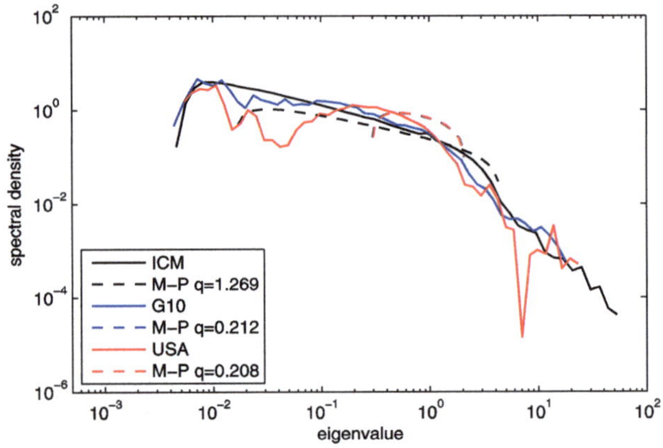

Fig. 17.3 Mean spectral density of the correlation matrix ρ evaluated with constant weights. The *dashed curves* are the Marchenko–Pastur spectral densities (they overlap for the G10 and USA data sets). For the USA data set, the dip around $\lambda \sim 7$ is created by the large gap between the leading market eigenvalue $e_1(t) \simeq 8$ to 25 and the next eigenvalue $e_2(t) \simeq 2$ to 7

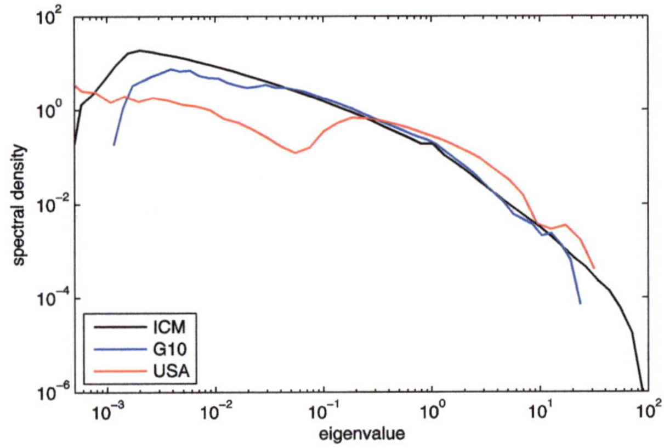

Fig. 17.4 Mean spectral density of the correlation matrix ρ evaluated with long-memory weights. The range for the axes are identical to Fig. 17.3

17.7 Spectrum and Spectral Density of the Covariance

For financial applications, we are interested mostly in covariance matrices. The key difference with the correlation matrix is the volatility, which has an important dynamic and serial correlations. For the covariance (with an appropriate assumption about the distribution for the volatilities), there seems to be no known analytical results about the spectral density. However, we can expect to have the most important eigenmodes related to the volatility dynamics, first of the market, then of the most important market factors. We can also expect to have a similar dense part for the lower section of the spectrum, corresponding essentially to random noise.

The spectrum of the covariance is clearly depending on the volatility distribution of each time series included in the dataset. Similarly to the return distributions that are very similar for all financial time series, the volatility distribution has a fairly universal shape, close to a log-normal distribution. For one time series, the volatility distribution is essentially given with a single parameter that fixes the mean volatility. For most financial time series (IR, FX, stocks, commodities, etc.), the

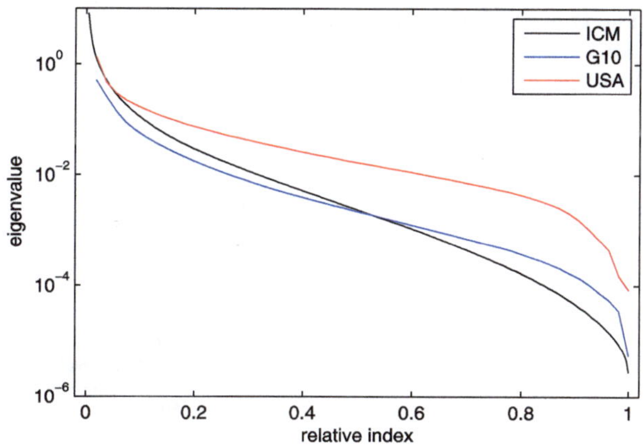

Fig. 17.5 Logarithmic mean spectrum of the covariance matrix Σ_{eff} plotted as a function of the relative index $\alpha' = \alpha/N_{\mathrm{pos}}$. The logarithmic mean is computed as $\exp(\frac{1}{T}\sum_{t=1}^{T}\log(e_\alpha(t)))$

mean annualized volatilities can vary between 2 to 100 %, with the typical mean volatility in the 5 to 20 % range. Provided that the choice of time series entering the covariance is "random" enough, the overall volatility and eigenvalue distributions should be fairly universal for large matrices, with a single parameter measuring the mean volatility for the whole set (a second possible parameter is the variance of the mean volatilities). This argument is essentially validated by the empirical analysis in this section and justifies an approach by random matrices also for the covariance matrices.

The mean spectrums of the covariance for the three datasets are plotted on Fig. 17.5. Despite the fairly different sizes and universes, their overall shapes are very similar. The core of the spectrum is well described by a linear relation between the logarithm of the eigenvalues and their relative index. This linear dependency shows that the eigenvalues decay exponentially fast toward zero:

$$e_\alpha \simeq \exp\left(-\frac{a\alpha}{N}\right) \qquad \alpha = 1, \ldots, N. \tag{17.11}$$

This fast decay is leading to certain troubles when computing the inverse of the covariance.

Up to a normalization by N, the index α is essentially the cumulative density of state, with

$$\alpha = \alpha(e) = N \int_e^{e_{\max}} de' \rho(e') \tag{17.12}$$

and where $e_{\max}$ is the largest eigenvalue. The density of states is obtained by derivation of the above equation:

$$\rho(e) = -\frac{1}{N}\frac{\partial\alpha}{\partial e} = -\frac{1}{e}\frac{1}{N\partial \ln e/\partial\alpha}. \tag{17.13}$$

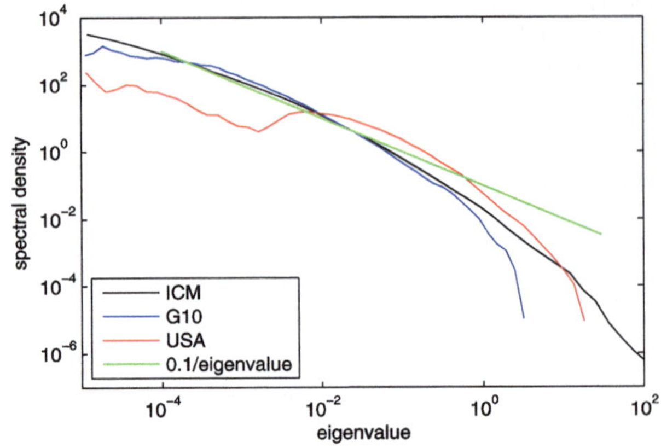

Fig. 17.6 Mean spectral density of the covariance matrix Σ_{eff}. The *green line* corresponds to $\rho(\varepsilon) = 0.1/\varepsilon$

The Ansatz for the spectrum leads to a scale free density of states

$$\rho(e) = \frac{1}{ae} + \text{curvature correction.} \tag{17.14}$$

Similarly to the spectral density for the correlation, the spectral density for the covariance is given in Fig. 17.6. The three densities are very similar (with a displacement toward higher eigenvalues for the USA dataset because it contains mostly stocks that are more volatile). The green line corresponds to $\rho(\varepsilon) = 0.1/\varepsilon$ as in (17.14), equivalent to a linear shape for the logarithmic spectrum. This very simple form describes well the empirical density of states over two to three orders of magnitude, while the upper part of the densities clearly decays faster. Notice that there is again no special feature that would clearly separate the leading significant eigenvalues from a bulk density described by random matrices.

17.8 Mean Projector

The spectrum of the covariance gives only a very partial view of the dynamical behavior of a matrix time series. In particular, the directions and subspaces associated with the eigenvalues are important. In order to gain an insight into the behavior of the leading subspaces, the mean projector is defined in (17.6). Figure 17.7 shows the spectrum of the mean projector for increasing projector rank. As the rank of the projector increases, the mean projector changes from the null matrix to the identity matrix. For the intermediate ranks, the eigenvalues decrease gradually, without gaps in the spectrum. Only cusps can be seen in the spectrum behavior with increasing rank, around ranks two and seven, with the smaller rank eigenvalues being close to one and the larger rank eigenvalues decreasing rapidly. Globally, the image is more of a cascade of subspaces with regularly decreasing importance, while the leading eigenvalue (corresponding to the market) is dominant for small ranks. Similar results are obtained for the three datasets.

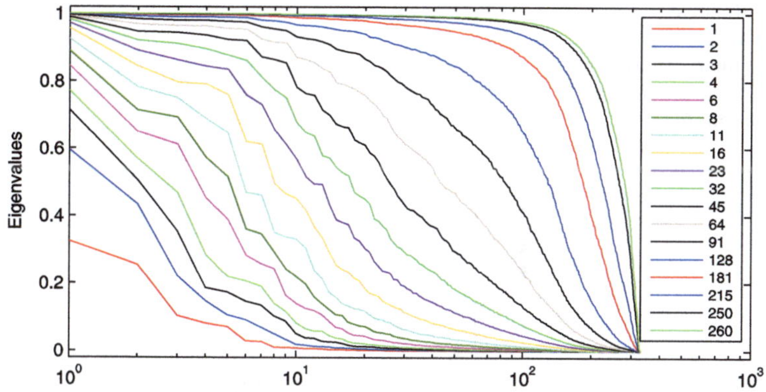

Fig. 17.7 Spectrum of the mean projector as a function of the projector rank (for the ICM dataset)

The properties of the mean projector spectrum show that it is not possible to identify clearly a dominant stable leading subspace. Instead, the picture is of subspaces with regularly decreasing importance, without a threshold that would separate a significant fixed subspace from a random subspace. Even though the covariance spectrum and correlation spectrum are very static (except for a handful of top eigenvalues), the eigenvectors of the covariance have important dynamics deep in the spectrum. This goes somewhat against the PCA picture [142], which explains the dominant stock dynamics by a few stable eigenmodes. Similarly, the picture conveyed in factor models, as initiated by [63], is an oversimplification of the complex dynamic occurring between the time series.

17.9 Projector Dynamics

In order to have a better understanding of the dynamics of the subspaces associated with the projectors $\mathsf{P}_k(t)$, a fluctuation index and a scalar lagged correlation are defined. The fluctuation index γ essentially measures the difference between $\mathrm{trace}\langle \mathsf{P}^2\rangle$ and $\mathrm{trace}(\langle \mathsf{P}\rangle^2)$, with a convenient normalization

$$\gamma = \frac{\mathrm{trace}\langle \mathsf{P}^2\rangle - \mathrm{trace}(\langle \mathsf{P}\rangle^2)}{\mathrm{trace}(\langle \mathsf{P}\rangle)} = 1 - \frac{1}{k}\,\mathrm{trace}\big(\langle \mathsf{P}\rangle^2\big). \tag{17.15}$$

If the projector is mostly constant $\langle \mathsf{P}\rangle \simeq \mathsf{P}(t)$, then $\langle \mathsf{P}\rangle^2 \simeq \langle \mathsf{P}\rangle$, and the fluctuation index is close to zero. In the other direction, if the projector dynamics explores fully the available space $\langle \mathsf{P}_k\rangle \simeq k/N\mathbb{I}$ and the fluctuation index reaches its maximal value $\gamma_{\max} = 1 - k/N$. The fluctuation index can be computed for increasing ranks k and for projectors associated with the covariance or correlation matrices.

For large ranks k, the matrices are dominated by the random noise, and large fluctuations are expected. For small ranks (say for eigenvalues larger than a random matrix threshold), the eigenvectors are expected to correspond to global factors corresponding to the overall market and its main subcomponents. Figure 17.8 displays

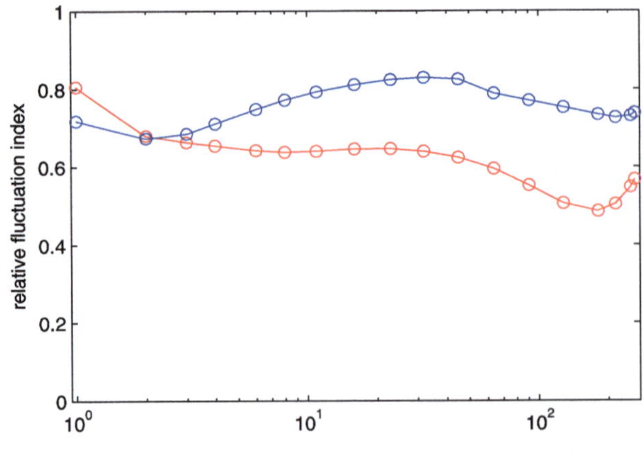

Fig. 17.8 Relative fluctuation index $\gamma/\gamma_{\max}$ as a function of the projector rank. Projectors of covariance (*red*) and correlation (*blue*)

the relative fluctuation index $\gamma/\gamma_{\max}$ as a function of the projector rank. The fluctuation indexes are at least of the order of 50 % of the maximal value, and typically larger. Moreover, the behavior is similar for the covariance and correlation. This clearly shows that the eigenvectors always have an important dynamic behavior, although the eigenvalues can be fairly static. Interestingly, the relative fluctuation indexes show a minimum for small sizes. Another feature is the smaller fluctuation index for the projectors derived from correlation matrix than for the covariance matrix, for rank one and two. These lower values show that the two first eigenvectors of the correlation are more stable than for the covariance. These are the only indication of some increased stability for selected small ranks.

The lagged correlations for different matrices (covariance, correlation, and the related projectors with increasing ranks) is used in order to assess more quantitatively the dynamics. For a time series of matrices $\mathsf{X}(t)$, the scalar lagged correlation is defined by

$$\rho(\tau) = \frac{\mathrm{trace}\langle(\mathsf{X}(t) - \langle\mathsf{X}\rangle)(\mathsf{X}(t+\tau) - \langle\mathsf{X}\rangle)\rangle}{\mathrm{trace}\langle(\mathsf{X} - \langle\mathsf{X}\rangle)^2\rangle}. \tag{17.16}$$

Essentially, $\rho(\tau)$ quantifies the overall linear dependency between the matrices $\mathsf{X}(t)$ and $\mathsf{X}(t+\tau)$. In order to avoid correlations created by the long-memory weights, a compact kernel for $\lambda(i)$ is used for this particular computation when computing Σ_{eff}: the weights $\lambda(i)$ are constant with $i_{\max} = 21$. The value $i_{\max} = 21$ allows us to measure lagged correlations at shorter lags and with a larger effective sample size, but it is rather low to obtain reliable estimates of the correlation matrix. The results have been checked with $i_{\max} = 42$, without significant differences.

Figure 17.9 displays the lagged correlations. For all curves, the regular drop between lags 1 to 21 is simply due to the overlap of the volatility estimated with 21 historical points. The lag correlation for the covariance (left graph, blue curve) shows the long memory of the volatility, as visualized by the slow decay of the lagged correlation. The corresponding projectors (red curves) have also a long memory, but with a regularly decreasing intensity for increasing ranks. This shows that the first eigenvectors have also a long-memory dependency, but the lagged dependency is

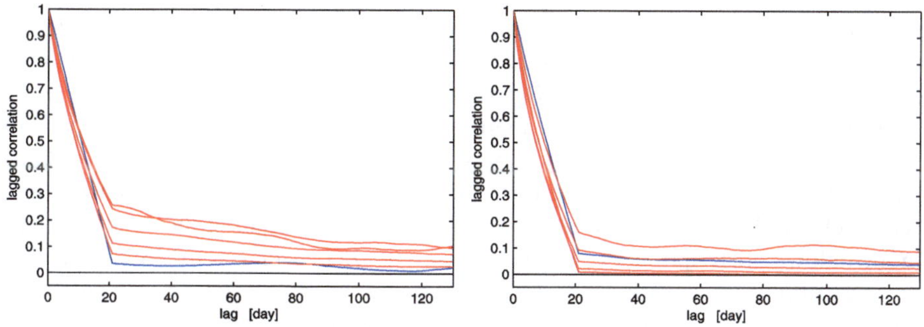

Fig. 17.9 Lagged correlations for the covariance (*left*) and correlation (*right*). Results for full matrix (*blue*) and the projectors of increasing size (*red*, *curves* lying lower for increasing ranks)

getting weaker for increasing rank. The right graph shows the same quantities, but for the correlation. The curves have the same qualitative behaviors, but with a lower overall level for the correlations. The lower overall level can be expected as part of the lagged dependency in the covariance is due to the (single component) volatilities. This clearly shows that the covariance and correlation matrices have the same qualitative behavior: both have important long-memory dynamics, mainly for the eigenvectors.

17.10 Putting All the Pieces Together

Bringing together the pieces above, we see a better picture of the dynamic behavior of the covariance and correlation matrices emerging. The eigenvalues of the covariance are mostly uninformative, except the first three to ten eigenvalues. Essentially, only a handful of eigenvalues show meaningful dynamics, while the bulk of the spectrum is very static and well described by a simple distribution. The transition from significant to noisy eigenvalues is gradual, and the spectrum and spectral density show no particular features that would separate them. On the other hand, the corresponding eigenvectors do not span clear invariant subspaces corresponding to the main market modes. Overall, the resulting image is of dynamic subspaces with regularly decreasing importance across the whole spectrum.

The dynamics of the correlation eigenvalues are even more uniform, indicating that the volatility dynamics are dominant over the correlation. Yet the top three to five eigenvalues have a clear and distinct time evolution, and an approximation by a constant spectrum seems inappropriate. The corresponding projectors have a very similar behavior compared to the projectors derived from the covariance, with an important dynamics. Beside, the time scales for the correlation fluctuations are very similar from the time scales in the covariance fluctuations, as both exhibit the signature of long memory as measured by the slow decay of the lagged correlations. These similar properties show that a simple covariance estimator measures correctly both volatilities and correlations, and separate estimators are unnecessary. For the mathematical description of multivariate time series, this points to a multivariate GARCH structure and not to the separation of volatilities and correlations as initiated by [23] with the CCC-GARCH process.

The definition of the covariance matrix that is presented in depth corresponds to weights that decay logarithmically slowly, in order to build a good one-step volatility forecast based on the long memory for the volatility observed in financial time series. Other shapes can be used for the weights $\lambda(i)$, including the "equal weight in a window" prescription, or exponential weights $\lambda(i) = (1 - \mu)\mu^i$. Subsequently, the spectrum of the covariance is dependent on the weights $\lambda(i)$. Yet, the empirical properties are not specific to the long-memory kernel, and a uniform or an exponential kernel shows essentially the same features as presented in this chapter. In particular, regardless of the weights, the spectrum decreases exponentially fast toward zero (the pace for the decay is depending on the kernel), and an important dynamics for the projectors is observed. This means that these key properties are generic, and only their quantitative aspects are influenced by the weights $\lambda(i)$.

Chapter 18
Multivariate ARCH Processes

18.1 Introduction

The univariate time series analysis shows the importance of the volatility dynamics and of fat-tailed distributions. A large number of processes have been proposed to model the univariate empirical data, but only the ARCH family is able to reproduce some statistical properties (long memory, time reversal non invariance, etc.). This careful analysis combining multiscale statistics applied on processes and data has narrowed down the candidate processes to half a dozen, presenting various degrees of complexity, analytical tractability, and accuracy with respect to the data. The previous chapter moved to the empirical properties of large multivariate system, studying the main properties of the covariance matrix. Besides heteroscedasticity, the salient points are the occurrence of small or zero eigenvalues, the lack of stability of the leading eigenspaces, and the similar dynamics of the covariance and correlation matrices. This analysis points to the direction we have to follow to write large multivariate processes. Yet, new difficulties arise, and a careful stepwise approach should to be followed.

Essentially, the useful univariate linear ARCH processes have a variance σ^2 proportional to the squared returns r^2. The natural multivariate extension is to replace the variance by the covariance matrix Σ and to use an exterior product of the vector of returns $\mathbf{r}\mathbf{r}'$. Yet, several quadratic forms can be written along this line, with different properties for the eigenvalues of the covariance. The analysis of bilinear covariance is the focus of the next section, introducing only a few new parameters in the covariance estimation. The introduction of the affine term, setting properly the long-term properties of the processes, is the subject of Sect. 18.4. The affine term changes significantly the problem, as a number of parameters proportional to N^2 has to be introduced. Let us emphasize that large multivariate systems are still an area of active research, where significant progresses can be expected in the forthcoming years.

In a process setup, the covariance is used to transform the (independent) innovations into (correlated) returns. To validate a process using historical data, the realized returns are transformed into innovations (or residuals) using the inverse covariance.

G. Zumbach, *Discrete Time Series, Processes, and Applications in Finance*,
Springer Finance, DOI 10.1007/978-3-642-31742-2_18,

If the process describes correctly the empirical time series, the innovations should be uncorrelated white noises by hypothesis. For the univariate long-memory ARCH process, the statistical properties of the residuals are presented in the Chap. 13. In particular, a Student distribution with 5 degrees of freedom gives a good description of the empirical residual distributions for most time series, and the serial correlations of the innovations are fairly low. These properties validate the choice of the univariate process. This chapter extends the study of the innovations for the large multivariate case. In particular, specific measures of quality are introduced that quantify how good the inverse (square-root) covariance is at producing white noise innovations.

The key difference in the multivariate setting is that the inverse (square root) of the covariance matrix should be computed, and not simply the inverse of the scalar volatility. In Chap. 17, the time-dependent covariance and correlation matrices are analyzed in detail. This analysis allows us to better understand the specificity of the multivariate problem, in particular the role of the largest eigenvalues and eigenvectors, as well as the small and possibly null eigenvalues. This last point is particularly critical for the computed innovations as the inverse square root of the covariance is needed. For our purpose, the key property of the covariance is that the eigenvalues of the covariance matrix decrease exponentially fast toward zero, even when the covariance is mathematically not singular. The accumulation of very small or zero eigenvalues is creating problems when computing the inverse volatility, even when the covariance should be mathematically well behaved. A very similar problem appears in portfolio optimization in a mean-variance scheme, as the inverse of the covariance determines the optimal allocations. In both cases, a proper regularization should be introduced.

18.2 General Setup for Multivariate Processes

The different time series are denoted by the Greek indices α or β that run from 1 to N. In a multivariate context, the usual convention is to use the covariance matrix $\Sigma_{\alpha,\beta}$ (instead of the univariate volatility σ). In a process, the vector of returns is

$$r_\alpha(t+\delta t) = \sum_\beta \Sigma^{1/2}_{\mathrm{eff},\alpha,\beta}(t)\epsilon_\beta(t+\delta t)$$

or in vector form

$$\mathbf{r}(t+\delta t) = \Sigma^{1/2}_{\mathrm{eff}}(t)\cdot\boldsymbol{\varepsilon}(t+\delta t). \tag{18.1}$$

The multivariate innovations ϵ are i.i.d. with the following properties:

$$\begin{aligned} E[\epsilon_\alpha(t)] &= 0 \\ E\left[\epsilon^2_\alpha(t)\right] &= 1. \end{aligned} \tag{18.2}$$

Sometimes, the condition of independence is replaced by the weaker condition $E[\epsilon_\alpha(t)\epsilon_\beta(t')] = \delta_{\alpha,\beta}\delta_{t,t'}$.

18.3 Multivariate Linear ARCH Processes

This long section focuses on the broad statistical properties of the innovations for linear processes. The motivation to use first a linear process is to have much less parameters to deal with. With the knowledge already gathered with univariate ARCH processes, only a few more parameters are introduced in the multivariate case. Crucially, the number of parameters is not proportional to the portfolio size N, or to a power of N. The case of affine processes will be studied in Sect. 18.4, where $O(N^2)$ parameters have to be introduced. The innovation properties are postulated to be the independence and unit variance, following (18.2). Starting from empirical return time series, can such properties for the empirical innovations be achieved for a given process structure? If this is the case, the information related to the multivariate correlations and to the heteroscedasticity has been correctly captured. Subsequently, inferences for the process parameters can be done, typically using a log-likelihood loss function for a given analytical distribution such as a Gaussian or a Student. As we will see, this is not the case as the postulated properties for the innovations cannot be achieved. This section is based mostly on [161], where more details about the empirical investigations can be found.

18.3.1 The Linear Covariance Matrix

In general, we consider the class of processes where the (time-dependent) covariance matrix $\Sigma_{\mathrm{eff}} = \Sigma_{\mathrm{eff}}(t)$ is given by a bilinear function of the past returns (i.e., the affine term is absent from the covariance). The simplest extension of the univariate process is given by definition (17.2) used in the previous chapter:

$$\Sigma_{\mathrm{eff}} = \sum_{i=0}^{i_{\max}} \lambda(i)\mathbf{r}(t - i\delta t)\mathbf{r}'(t - i\delta t)$$

with $\mathbf{r}$ a column vector and $\mathbf{r}'$ its transpose. Only the microscopic definition of the volatility is used in this chapter, and the return vector $\mathbf{r} = \mathbf{r}[1\ \mathrm{d}]$ is computed over an implicit time difference of 1 day. The weight for the past returns $\lambda(i)$ obeys the sum rule $\sum_i \lambda(i) = 1$.

A similar form is obtained by using a combination of historical covariance Σ_k measure at the respective time scale τ_k. The matrices of historical covariances are

$$\Sigma_k(t) = \mu_k \Sigma_k(t - \delta t) + (1 - \mu_k)\mathbf{r}(t)\mathbf{r}'(t) \tag{18.3}$$

with $\mu_k = \exp(-\delta t/\tau_k)$ defined as in the univariate case, and with k running over the number of time components included in the process. For the linear processes, the effective covariance is a convex combination of the historical covariance

$$\Sigma_{\mathrm{eff}}(t) = \sum_{k=1}^{k_{\max}} w_k \Sigma_k(t) \tag{18.4}$$

with the constraint that $\sum w_k = 1$. By reordering the summation in this last expression, the form (17.2) is obtained. When using historical covariance, the extension to aggregated volatility measures is straightforward, but the effective covariance does not reduce to the simple form (17.2).

This extension is minimal in terms of the number of parameters, as no new parameters are introduced. For example, an EMA multivariate process needs only one parameter, while a long-memory process requires only the parameter τ_0 (and two cut-offs). Moreover, the parameters can be estimated from a univariate study and directly exported in a multivariate world. Such an approach has been used extensively in market risk measurement (see Chap. 15).

In the class of linear equations between covariance and return squares, other extensions of the univariate process can be written. An interesting first extension is similar to the shrinkage of the covariance matrix [86, 87], but applied on the correlation. A simple shrinkage of the correlation matrix with the shrinkage parameter γ is

$$\rho(\gamma) = (1-\gamma)\rho + \gamma \mathbb{I}_N \tag{18.5}$$

where ρ is the correlation matrix corresponding to Σ_{eff} with elements $\rho_{\alpha,\beta} = \Sigma_{\alpha,\beta}/\sqrt{\Sigma_{\alpha,\alpha}\Sigma_{\beta,\beta}}$. The rationale for using a shrinkage is to allow more fluctuations for the return across assets than what the historical correlation structure would impose. The natural prior for the correlation is the identity corresponding to the condition imposed on the residuals $E[\varepsilon_\alpha \varepsilon_\beta] = \delta_{\alpha,\beta}$. The corresponding equation for the shrinkage of the covariance matrix is

$$\Sigma_{\text{eff}}(\gamma) = (1-\gamma)\Sigma_{\text{eff}} + \gamma \Sigma_{\text{eff}}|_{\text{diag}} \tag{18.6}$$

where $\Sigma_{\text{eff}}|_{\text{diag}}$ is the diagonal part of Σ_{eff}. Essentially, this equation shrinks only the off-diagonal part by $1-\gamma$, whereas the diagonal part is given by the volatility of the respective time series. Intuitively, increasing γ reduces the off-diagonal coupling, resulting in the eigenvalues becoming closer to the diagonal terms of the covariance.

A second interesting extension consists in shrinking the spectrum of the covariance toward a multiple of the identity matrix

$$\begin{aligned}\Sigma_{\text{eff}}(\gamma,\xi) &= (1-\xi)\Sigma_{\text{eff}}(\gamma) + \xi\langle\sigma^2\rangle\mathbb{I}_N \\ &= (1-\xi)\big((1-\gamma)\Sigma_{\text{eff}} + \gamma\Sigma_{\text{eff}}|_{\text{diag}}\big) + \xi\langle\sigma^2\rangle\mathbb{I}_N\end{aligned} \tag{18.7}$$

and with the mean variance across all assets defined by

$$\langle\sigma^2\rangle = \frac{1}{N}\operatorname{trace}(\Sigma_{\text{eff}}). \tag{18.8}$$

The covariance $\Sigma_{\text{eff}}(\gamma,\xi)$ has been defined so as to preserve the mean variance across assets

$$\operatorname{trace}\big(\Sigma_{\text{eff}}(\gamma,\xi)\big) = \langle\sigma^2\rangle \qquad \forall\gamma,\xi. \tag{18.9}$$

It is easy to check that, if e_α is an eigenvalue of $\Sigma_{\text{eff}}(\gamma)$, the corresponding eigenvalue of $\Sigma_{\text{eff}}(\gamma, \xi)$ is $(1-\xi)e_\alpha + \xi\langle\sigma^2\rangle$. In particular, the addition of the identity matrix changes the spectrum of the covariance by setting the minimal eigenvalue at $\xi\langle\sigma^2\rangle$. Therefore, the regularization parameter ξ allows us to compute a well-defined inverse for Σ_{eff}. This modification is very similar to the regularization of the (inverse) covariance given below by a full rank regularization (18.13). The intuition for this term is to have the same number of (significant) sources of randomness as of time series. In contrast, zero eigenvalues project out the corresponding source of randomness, and very small eigenvalues act similarly in practice. The addition of the identity matrix introduces a minimal level of fluctuations corresponding to each source of randomness that would be missing otherwise. As the parameters for the long-memory kernel are fixed, the process defined with the covariance (18.7) has only the parameters γ and ξ that need to be studied empirically, as done in Sect. 18.3.6. Moreover, for $N = 1$, the dependency on γ and ξ disappears, and the variance reduces to the univariate case.

The extension $\Sigma_{\text{eff}}(\gamma, \xi)$ of the covariance can be evaluated with simple matrix computations, and in particular no eigenvalue decomposition should be done. This is a significant advantage compared to the regularization defined using the spectrum, like the "projected" and "full rank" regularization (18.13) defined in the next section.

Ledoit and Wolf [86, 87] introduced a different shrinkage corresponding to a constant correlation $\rho_{\alpha,\beta} = \bar{\rho}$ for any pair of time series. In their article, the authors compute an optimal value for their shrinkage parameter by minimizing a root mean square error with an assumed theoretical covariance, leading to a biased but more efficient estimator. This computation assumes a stationary system, whereas financial time series display heteroscedasticity. A similar approach could be used with the covariance (18.7), leading to estimators for ξ and γ through a trade-off between bias and efficiency for the covariance. Instead, we are using an approach through ARCH processes. With a process, the optimal values of the parameters are such that the innovations are the closest to white noises. This method is better, as it allows one (1) to capture the heteroscedasticity, (2) to study the specificity of large multivariate systems, and (3) to obtain consistent covariance forecasts for all forecast horizons.

18.3.2 The Inverse Covariance Matrix

Equation (18.1) is formulated as a process, with the returns depending on the random innovations. Using historical data, we want to validate this model by inverting (18.1), namely

$$\boldsymbol{\varepsilon}(t+\delta t) = \Sigma_{\text{eff}}^{-1/2}(t)\mathbf{r}(t+\delta t). \tag{18.10}$$

The spectral decomposition $\Sigma_{\text{eff}} = \sum_{\alpha=1}^{N} e_\alpha \mathbf{v}_\alpha \mathbf{v}'_\alpha$ has been introduced in the previous chapter in Eq. (17.3). Provided that the eigenvalues are strictly positive, the inverse square root covariance, or inverse volatility, is

$$\Sigma_{\text{eff}}^{-1/2} = \sum_{\alpha=1}^{N} \frac{1}{\sqrt{e_\alpha}} \mathbf{v}_\alpha \mathbf{v}'_\alpha. \tag{18.11}$$

For $\Sigma_{\text{eff}}(\gamma = 0, \xi = 0)$ and for systems of practical interest, the covariance matrix is singular, and some eigenvalues are nil. This always occurs when the number N of time series is larger than the memory length $i_{\max}$ used to compute the covariance. In such a case, the number of strictly positive eigenvalues is given by $N_{\text{pos}} = \min(N, i_{\max})$. For many practical applications, the memory length is of the order of one to two years ($i_{\max} = 260$ to $i_{\max} = 520$), whereas the number of time series can be of the order of thousand(s). However, even for systems that are non-singular according to the mathematical criterion, the eigenvalues of the covariance matrix decay exponentially toward zero, leading to very large values in the inverse volatility. Clearly, this will impact the computed residuals. Therefore, except for systems of very small dimensions, the computation of the inverse volatility at $\gamma = \xi = 0$ needs to be regularized, even when $N < i_{\max}$.

With a singular covariance, several schemes can be used to define an inverse volatility with an appropriate cut-off. A first possibility is to use only the leading eigenvalues, namely to invert the covariance in the leading k subspace

$$\Sigma_{\text{proj},k}^{-1/2} = \sum_{\alpha=1}^{k} \frac{1}{\sqrt{e_\alpha}} \mathbf{v}_\alpha \mathbf{v}'_\alpha \tag{18.12}$$

and the "cut-off parameter" k is chosen so that $e_k > 0$. We call this scheme "projected". A second possibility is to complement the previous operator so that it has full rank

$$\Sigma_{\text{full rank},k}^{-1/2} = \sum_{\alpha=1}^{k} \frac{1}{\sqrt{e_\alpha}} \mathbf{v}_\alpha \mathbf{v}'_\alpha + \frac{1}{\sqrt{e_{k+1}}} \sum_{\alpha=k+1}^{N} \mathbf{v}_\alpha \mathbf{v}'_\alpha \tag{18.13}$$

and we call this scheme "full rank". A multiplicative constant can be inserted in front of both definitions so as to preserve the trace. In practice, the optimal rank k should be chosen large enough (see below), making essentially irrelevant this normalization constant.

The singularity related to the inverse of a covariance matrix appears in many places in finance. A common example is the computation of the most efficient portfolio in a mean-variance scheme and the related definition of an efficient frontier. Other examples are the computations of the factor loadings in a factor model, the inference in a multivariate process as explained above, or the computation of a conditional Gaussian distribution in a multivariate setting. Depending on the application, the choice of the cut-off can differs, with the "projected" definition being possibly the most widely used. As the empirical analysis below shows, the "full rank" scheme is better. Yet, the regularization through a modification of the covariance by using $\Sigma_{\text{eff}}(\gamma, \xi)$ turns out to be the most efficient method.

18.3.3 Quantifying the Properties of the Returns and the Innovations

The sample correlations for the returns measure the dependencies present in the dataset while the correlations for the residuals measure how effective is $\Sigma_{\text{eff}}(\gamma, \xi)$ at whitening the residuals. The main correlation is $\rho(\mathbf{r}, \mathbf{r})$, but more correlations can be studied between $\mathbf{r}$, $\mathbf{r}^2$, $L[\mathbf{r}]$, and $L[\mathbf{r}^2]$. The correlations matrices for the residuals $\boldsymbol{\varepsilon}$ are computed similarly but also depend on the regularization used to compute $\Sigma_{\text{eff}}^{-1/2}$. For the returns, all these correlations are very interesting to study as they summarize the information about the market structures and their dynamics. For the innovations, if they are effectively i.i.d., the correlation matrices should be either proportional to the identity matrix (for $\rho(\boldsymbol{\varepsilon}, \boldsymbol{\varepsilon})$ and $\rho(\boldsymbol{\varepsilon}^2, \boldsymbol{\varepsilon}^2)$) or zero (for all the other cross-correlations). Simple overall measures of these relationships are given by

$$q^2 = \frac{1}{N(N-1)} \sum_{\alpha \neq \beta} \rho_{\alpha,\beta}^2 \tag{18.14}$$

for $\rho(\mathbf{r}, \mathbf{r})$ and $\rho(\mathbf{r}^2, \mathbf{r}^2)$ and by

$$q^2 = \frac{1}{N^2} \sum_{\alpha,\beta} \rho_{\alpha,\beta}^2 \tag{18.15}$$

for the other correlation matrices. Essentially, the whitening quality $q(\mathbf{r}, \mathbf{r}) = q(\rho(\mathbf{r}, \mathbf{r}))$ measures the linear independence of all the pairs of residuals. The unit variance of the residuals $E[\varepsilon_\alpha^2] = 1$ should still be tested, and a simple measure is given by

$$q^2(\boldsymbol{\varepsilon}^2) = \frac{1}{N} \sum_\alpha \left(\langle \varepsilon_\alpha^2 \rangle - 1 \right)^2. \tag{18.16}$$

Empirically, the variances $E[\varepsilon_\alpha^2]$ have a similar behavior for all α, and an informative quantity is the mean residual variance

$$\frac{1}{N} \sum_\alpha \langle \varepsilon_\alpha^2 \rangle. \tag{18.17}$$

18.3.4 The Sample Correlations for the Returns

From the seven possible correlations between return, squared return, and their lag one time series, the largest correlation is $\rho(\mathbf{r}, \mathbf{r})$, while the correlation between volatility $\rho(\mathbf{r}^2, \mathbf{r}^2)$ is the second largest. The overall return correlation is presented on Fig. 18.1 with the time series ordered as described on page 261. Essentially, this figure measures the dependencies in today's financial world, and several features are worth noticing on a qualitative basis.

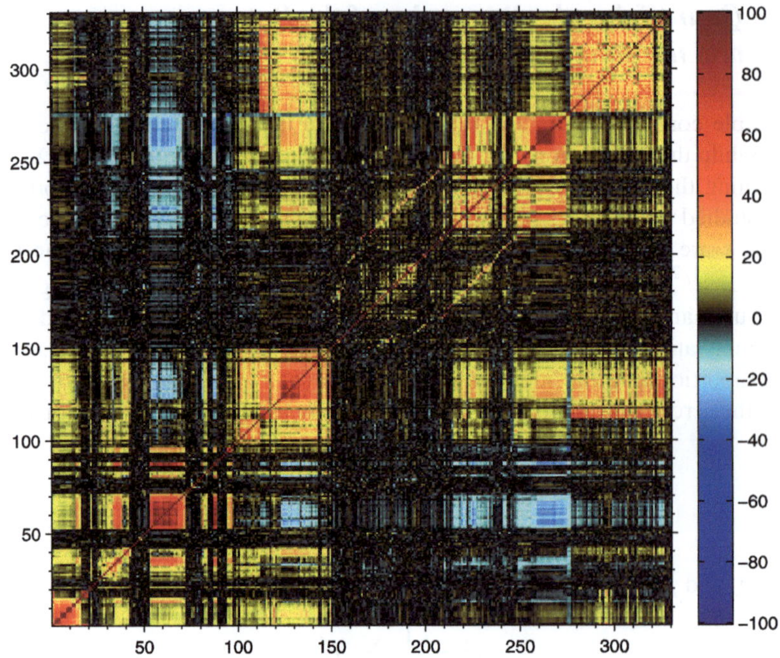

Fig. 18.1 The correlation $\rho(\mathbf{r}, \mathbf{r})$ for the returns (for the ICM dataset)

1. By and large, there is one single cluster of connected markets. The cluster contains all developed economies from Asia, North America, Europe, and South Africa. The disconnected countries are in Asia, central America, most of South America, Central Africa, and the former Soviet Union.
2. In the FX sector, there is a strongly correlated cluster around the Euro (red patch between indexes 53 to 70), related to another cluster containing Canada, Australia, Singapore, and Japan (indexes 33 to 38). The rest of the Asiatic currencies are less correlated.
3. In the stock index sector, the correlations are overall fairly large, with three clusters corresponding respectively to Pacific/Asia (index 98 to 111), North America (112 to 116), and Euroland (123 to 142).
4. The interest rates show a strong difference between the short maturities at 1 day (indexes 150 to 165) and 1 month (indexes 166 to 207) and the long maturities at 1 year (indexes 208 to 245) and 10 years (indexes 246 to 276). Probably, this is created by the segmentation of the market into the short-term "money market" and the long-term "bond market". The short maturities are essentially disconnected from all other time series, and have only small correlations with 1 year interest rates, except for a larger correlation with the same yield curve at 1 year, visible as an off diagonal line.
5. In the commodities sector, the metals (spot and future, indexes 1 to 15) behave very similarly to the FX, while the gas futures (indexes 16 to 19) have no correlation except with some American stocks in the energy and petroleum business.

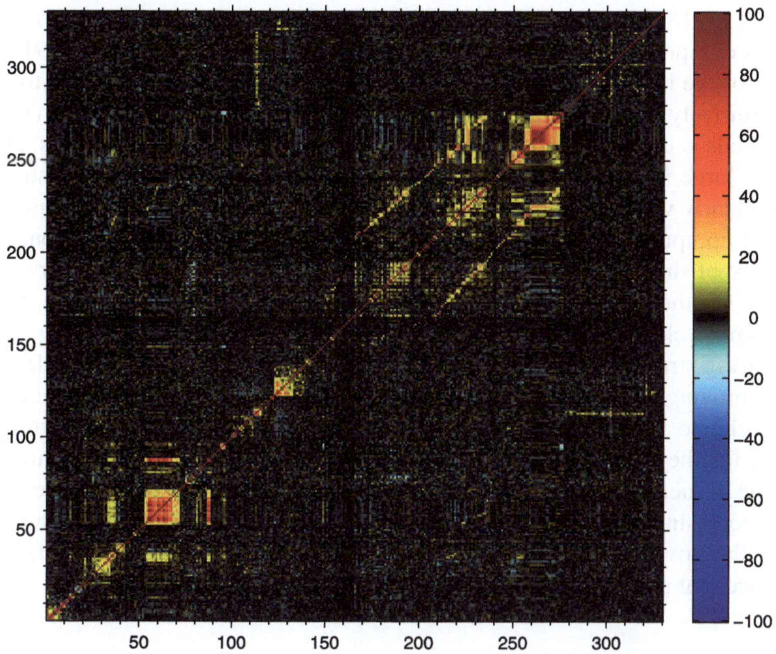

Fig. 18.2 The correlation $\rho(\boldsymbol{\varepsilon}, \boldsymbol{\varepsilon})$ for the residuals at $\gamma = \xi = 0$ and cut-off parameter $k = 91$

6. The correlation between European currencies and European stock indexes is negative. This indicates that when the dollar gets stronger (i.e., the FX rate USD/EUR decreases), the stock indexes increase. This is probably due both to the prices of European stocks appearing cheaper to US investors, as well as American stocks appearing more expensive to European investors.
7. The IR with long maturities show clear positive correlations with stock indexes. These correlations indicate moves in the same directions for both asset classes and go somewhat against the view of alternating cycle between bonds and equities. The time scales are however very different, as the correlations between daily price changes are used here, whereas the business cycles appear at a time scale of years and for the prices.
8. The IR with long maturities show clear negative correlations with FX. This is probably due to larger (smaller) US long-term IR making the dollar stronger (weaker), coupled with the strong positive correlations for long-term IR.

18.3.5 The Sample Correlations for the Residuals

The empirical sample covariances and correlations for the residuals are computed similarly as for the returns, on the same sample. The residuals $\varepsilon(t)$ are computed according to (18.10). The covariance is given by $\Sigma_{\text{eff}} = \Sigma_{\text{eff}}(\gamma = 0, \xi = 0)$, and the inverse is computed using the "projected" regularization (18.12) with $k = 91$. These choices for the inverse covariance make the interpretation easier. In order to understand intuitively the statistical properties of the residuals, the correlation $\rho(\boldsymbol{\varepsilon}, \boldsymbol{\varepsilon})$ for the residuals is displayed in Fig. 18.2, similarly to Fig. 18.1 for the returns. Despite the fairly large number of eigenvalues included in the inverse volatility, structures are still clearly visible, remnant of the correlation for the returns. This figure makes clear that computing the inverse volatility using only the leading eigenspace for k small (say of the order of 3 to 15) is leaving most of the return structures in the residuals. On the other hand, the noise in the background is also visible in the form of random speckles. The same figure plotted for increasing cut-off k shows slowly disappearing structures while the background noise is increasing. The trade-off between removing the correlation structures and keeping the noise small leads to an optimal choice for the cut-off k or for the regularization ξ. The other important criterion for the statistical properties of the residuals is that they have a unit variance. The influence of the regularization parameter on the residual variance is large, with a decreasing variance for an increasing regularization parameter ξ. In order to validate the postulated multivariate process, the goal is to find parameters for the covariance that produces independent innovations with unit variances.

18.3.6 Whitening of the Residuals

The overall quality measures q for the "whitening" of the residuals are given by (18.14), (18.15) and (18.16). They are plotted on Fig. 18.3. The horizontal axes give the regularization parameter ξ, and the black curves correspond to no shrinkage $\gamma = 0$. The curves with the color changing from black to blue correspond to the increasing shrinkage parameter $\gamma = 0.0, 0.05, 0.1, 0.2, 0.4$. The black horizontal line gives the value of the corresponding quantity for the returns, while the pair of horizontal red lines corresponds to the 5 % and 95 % quantiles of the distribution of the whitening quality for uncorrelated Student innovations (see the text below). Essentially, the empirical whitening qualities should lie between both extremes. On the horizontal axis for the regularization parameter ξ, the point at 10^{-5} corresponds to $\xi = 0$ (that would be otherwise at $-\infty$ on a logarithmic axis). The inverse volatility is computed with a floor on the spectrum of 10^{-12} in order to avoid numerical overflows due to the zero eigenvalues.

The spectrum of the covariance is given on the top left panel in Fig. 18.3. For this panel, the black curves correspond to $\gamma = 0$, for different values of ξ. The effect of the regularization can be clearly seen, with a spectrum that goes from singular for $\xi = 0$ (with zero eigenvalues for rank larger than 260) to a constant spectrum given by the mean volatility ($\langle\sigma^2\rangle \simeq 0.1$) for $\xi = 1$. The spectra for increasing shrinkage γ are drawn with colors that go from black for $\gamma = 0$ to blue for $\gamma = 0.4$. The shrinkage effect on the small eigenvalues is very clear, with a less singular spectrum.

The mean sizes of the residuals are shown on the top right panel. There is no particular feature around a unit variance. Instead, the dominant feature is the very

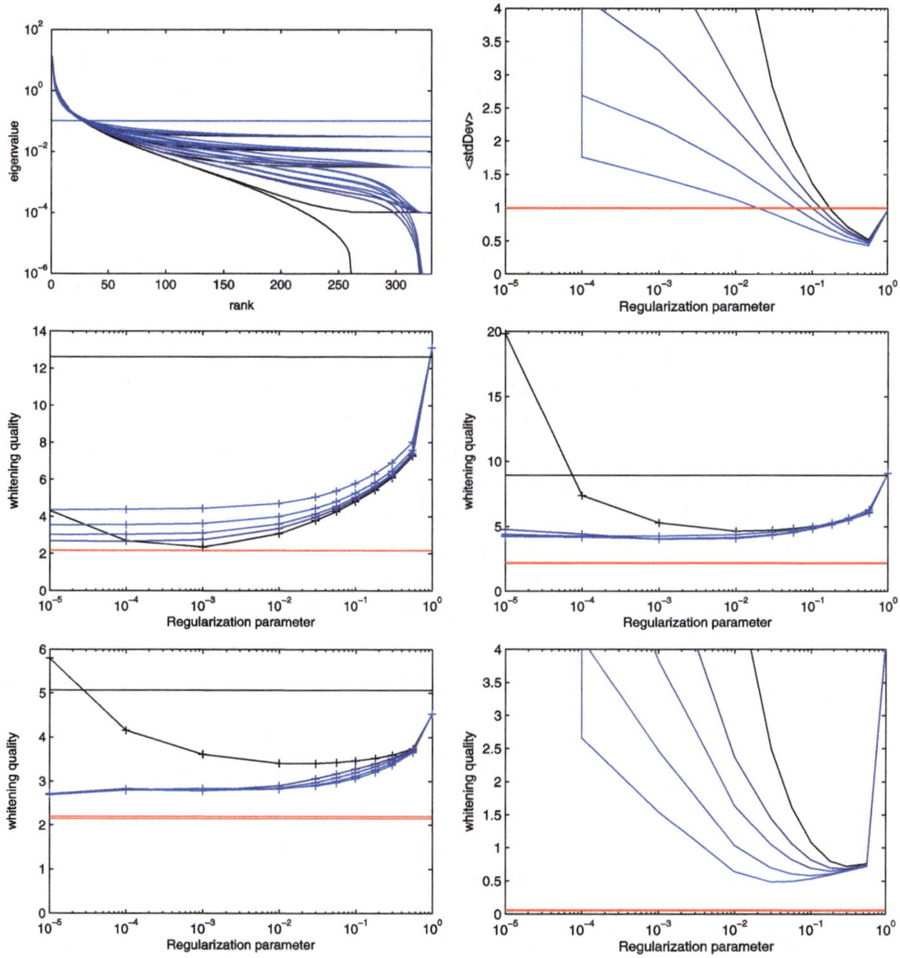

Fig. 18.3 The most important whitening measures as functions of the regularization parameter ξ and shrinkage parameter $\gamma = 0.0$ (*black*), 0.05, 0.1, 0.2, 0.4 (*blue*). *Upper left panel*: the spectrum as a function of the eigenvalue rank. *Upper right panel*: the mean magnitude of the residual $\langle \boldsymbol{\varepsilon}^2 \rangle$ defined by (18.17). *Center left*, *center right*, and *bottom left panels* display respectively the whitening quality $q(\boldsymbol{\varepsilon}, \boldsymbol{\varepsilon})$, $q(\boldsymbol{\varepsilon}^2, \boldsymbol{\varepsilon}^2)$, and $q(L[\boldsymbol{\varepsilon}^2], \boldsymbol{\varepsilon}^2)$. *Bottom right*: the whitening quality $q(\boldsymbol{\varepsilon}^2)$ for the unit variance of the residuals

strong increase of the residual size for decreasing values of the regularization ξ. Increasing the shrinkage parameter γ alleviates the problem, but there is no plateau around $\langle \boldsymbol{\varepsilon}^2 \rangle = 1$.

The whitening qualities of the residuals are displayed on the four other panels. The largest correlations are between the contemporaneous quantities $\rho(\mathbf{r}, \mathbf{r})$ and $\rho(\mathbf{r}^2, \mathbf{r}^2)$; the corresponding whitening qualities are plotted in the center panels. The next largest correlation is $\rho(L[\mathbf{r}^2], \mathbf{r}^2)$, corresponding to the heteroscedasticity, and is displayed in the bottom right panel. For these three measures of quality, the best results are achieved for parameters in the range $\gamma \in [0.05, 01]$ and $\xi \in [10^{-3}, 10^{-2}]$. The four measures of quality related to the other correlations have a similar behavior, but with a smaller magnitude. The bottom right panel shows the whitening quality for the magnitude of the residuals according to (18.16). For this measure, the optimal values for the parameters are larger, with $\gamma \simeq 0.2$ and $\xi \simeq 10^{-1}$. With this dataset, it seems difficult to have optimal parameter values according to all the whitening qualities.

Monte Carlo simulations are used in order to obtain confidence bounds around the null hypothesis of zero correlation for the residuals. Independent residuals are drawn from a Student distribution with five degrees of freedom, and the measure of quality computed for the same number of time series and sample length. Let us emphasize that it is only when computing the confidence bounds that a distributional assumption is made. This procedure is repeated 1000 times, and the 5 % and 95 % quantiles are extracted from the distribution for the measure of quality. Both values are plotted as horizontal red lines. For all the measures of quality, the empirical values are clearly above the 95 % confidence limits. This points to the misspecification of the covariance $\Sigma_{\mathrm{eff}}(\gamma, \xi)$, regardless of the parameter values. Despite this negative result, the most substantial part of the dependencies are removed by the covariance. For example, for $q(\boldsymbol{\varepsilon}, \boldsymbol{\varepsilon})$ (center left panel), the measure of quality for the returns is slightly below 13 %, for the residuals, they are in the 3 % range, while perfectly uncorrelated residuals have a value around 2 %. Clearly, the less satisfactory quantitative results are for the magnitude of the residuals, with empirical values all above 0.7, while the perfectly uncorrelated 95 % quantile is at 0.07.

Figure 18.3 corresponds to the ICM dataset, but the same quantities for the G10 and USA datasets show similar properties (the details can be found in [161]). Despite the fact that these smaller data sets relate to nonsingular covariance matrices and are quite different in their composition, very similar patterns emerge compared to the presented ICM dataset. This similarity shows that the model with $\gamma = \xi = 0$ is misspecified regardless of the dataset (and its size), while adding shrinkage and regularization improves the situation but is still not perfect. Interestingly, the optimal values for the parameters are similar for the three datasets, pointing to a general feature of the underlying data generating process. The residuals are the closest to the postulated properties with values around $\gamma \simeq 0.05$ and $\xi \simeq 10^{-3}$. An intuitive picture for this result can be built as follows. With this regularization, the bottom of the covariance spectrum is at $10^{-3}\langle\sigma^2\rangle$, leading to an inverse volatility in the range $32/\sqrt{\langle\sigma^2\rangle}$. The fluctuations of the returns in the corresponding direction get multiplied by this large factor, leading to large residuals. With an increasing matrix size, the number of very small covariance eigenvalues increases with N, in turn increasing the residual sizes.

In principle, the large residual problem can be solved by using large enough shrinkage and regularization parameters, but then the covariance is modified too much and leaves dependencies between residuals. Therefore, one cannot find optimal values for the parameters that would both lead to small residual dependencies and to unit variances. The later issue is essentially due to the fast decrease toward zero of the spectrum and to the possible zero eigenvalues. Indeed, this is the signature of the limited information that can be extracted from such a multivariate system. Corresponding to a given kernel $\lambda(i)$, there is an effective number of time series beyond which the multivariate information gets very tenuous, as signaled by the very small or null eigenvalues. The obvious solution would consist in using a memory kernel that allows for more information to be harvested, but there are fundamental limitations in this direction. First, the long-memory kernel is optimal at capturing the dynamics of the univariate volatility clustering, and using another kernel leads to worst (univariate) decorrelation of the residuals. Second, a rectangular kernel such that $i_{\max} > N$ is not always possible due to the limited availability of historical data. Moreover, such a solution would clearly miss the short-term dynamics of the volatility. Therefore, there is a fundamental limitation to the empirical multivariate information that can be extracted from a set of time series. This limitation leads to small eigenvalues, and to the large variance for the residuals. Intuitively, the large residuals "make up" for the small eigenvalues, so that fluctuations along all possible directions do occur. The regularization by $\xi > 0$ is such that all sources of randomness ε contribute to the fluctuations of the returns. Notice that this is fundamentally different from a misspecification, as the process could be specified correctly, but our inference power is limited for N large.

18.3.7 Comparing Different Covariance Kernels

The performances of different kernels at producing i.i.d. residuals is investigated in Table 18.1. All kernels are evaluated using 260 days of history, with shapes that are equal weights, exponential with decay 0.94, long memory, and long memory with shrinkage and regularization. The performances of the first three kernels are somewhat similar, while the added parameters related to shrinkage and regularization make the fourth kernel often the best. Yet, there is no absolute ranking. The most salient feature is the difficulty in fulfilling the criterion $\langle \varepsilon^2 \rangle = 1$ with increasing size N, with the shrinkage and regularization in the fourth kernel helping significantly. In this respect, the exponential kernel shows the worst performance, which can be understood as follows. Due to the fast decay of the kernel, the amount of multivariate information harvested by the exponential kernel is the smallest, leading to the fastest decays toward zero in the spectrum. In turn, these small eigenvalues lead to the largest inverse volatilities and therefore to the largest residual sizes. In contrast, the long-memory kernel leads to residuals six times smaller, with a somewhat similar overall shape.

It would then be tempting to evaluate the correlations with a much longer kernel, for example, with equal weights (corresponding to the usual formula for the correlation). Yet, the empirical study in the previous chapter shows that the correlation has clear dynamics, with a long memory. Using a long kernel with equal weights will wash out this structure, and therefore it will miss the short-term correlation moves (this will also produce bad volatility estimators). This points to a fundamental limitation of the information that can be extracted from a multivariate system, when

Table 18.1 The whitening qualities for the ICM data set, for different kernel shapes $\lambda(i)$

Whitening quality	Return	Equal weights	Exponential	Long memory	LM + regularization	White noise
$\rho(\boldsymbol{\varepsilon}, \boldsymbol{\varepsilon})$	12.6	3.3	3.1	2.7	3.4	2.2
$\rho(\boldsymbol{\varepsilon}^2, \boldsymbol{\varepsilon}^2)$	8.9	10.6	5.9	7.4	4.1	2.2
$\rho(\boldsymbol{\varepsilon}, \boldsymbol{\varepsilon}^2)$	3.5	4.0	3.0	3.4	3.0	2.3
$\rho(L[\boldsymbol{\varepsilon}], \boldsymbol{\varepsilon})$	4.8	2.5	2.4	2.4	2.5	2.2
$\rho(L[\boldsymbol{\varepsilon}^2], \boldsymbol{\varepsilon})$	2.8	2.8	2.5	2.6	2.4	2.2
$\rho(L[\boldsymbol{\varepsilon}], \boldsymbol{\varepsilon}^2)$	3.1	2.8	2.5	2.6	2.4	2.2
$\rho(L[\boldsymbol{\varepsilon}^2], \boldsymbol{\varepsilon}^2)$	5.1	5.6	3.9	4.2	2.9	2.2
$\langle \boldsymbol{\varepsilon}^2 \rangle = 1$		105	691	129	2.4	0.060

The column "return" gives the empirical values for the whitening qualities for the returns, while the column "white noise" gives the average values for an uncorrelated Student white noise. The column "LM + regularization" corresponds to a long-memory covariance with parameters $\gamma = 0.05$ and $\xi = 0.01$. The equal weights, exponential, and long-memory kernels are regularized using $\xi = 0.0001$

the dynamics of the correlation have also to be captured correctly. This limitation explains the strong similarities of the key properties for the three data sets, even though their sizes vary by a factor six, and two of them are nondegenerate. Essentially, the exponential decay of the eigenvalues (without regularization) is shared by all datasets. This decay makes the history cut-off $i_{\max}$ mostly irrelevant.

18.3.8 "Projected" and "Full Rank" Regularizations

In order to compute the inverse volatility, a widely used method is to use the cross-product covariance $\Sigma_{\text{eff}}(\gamma = 0, \xi = 0)$ and to regularize the inverse volatility by using the "projected" inverse (18.12) or the "full rank" inverse (18.13). It is interesting to contrast these two approaches with the regularization of the covariance with γ and ξ. Figure 18.4 shows the whitening quality $q(\boldsymbol{\varepsilon}, \boldsymbol{\varepsilon})$. The data corresponding to $\Sigma_{\text{eff}}(\gamma, \xi)$ are mapped into an equivalent projector rank by using the mean spectrum of the covariance at $\gamma = 0$, $\xi = 0$. These curves show clearly that a "projected" regularization (black line) is not good for small ranks. This finding goes clearly against a common practice consisting in using only the leading eigenvalues/eigenvectors for computing the inverse of the covariance. The "full rank" regularization is clearly better than the "projected" scheme for all choices of rank. For the regularization on the covariance (red curves), the similarities of the "full rank" and "regularized" qualities are due to the similar modifications of the spectrum made by both schemes. Overall, the regularized method using $\Sigma_{\text{eff}}(\gamma, \xi)$ is slightly superior.

Figure 18.5 compares the measure of quality for the norm of the residuals. For this quality measure, the regularization of the spectrum performs better for the "pro-

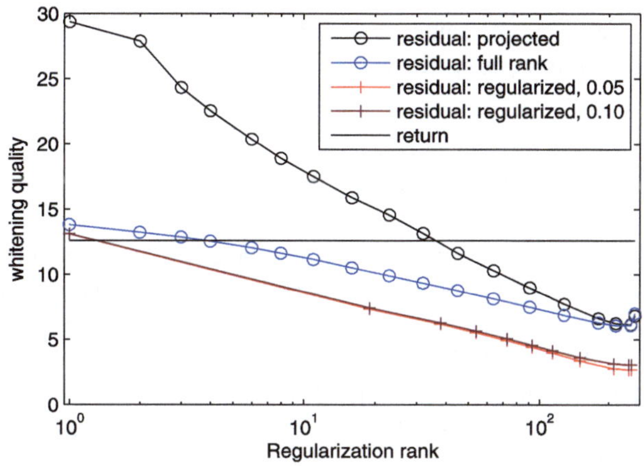

Fig. 18.4 The whitening quality $q(\boldsymbol{\varepsilon}, \boldsymbol{\varepsilon})$ as a function of the regularization rank. The regularized quality is computed with $\gamma = 0.05$ (*red*) and $\gamma = 0.1$ (*brown*), and the regularization parameter ξ is mapped to a "plausible" equivalent rank k by using the mean spectrum

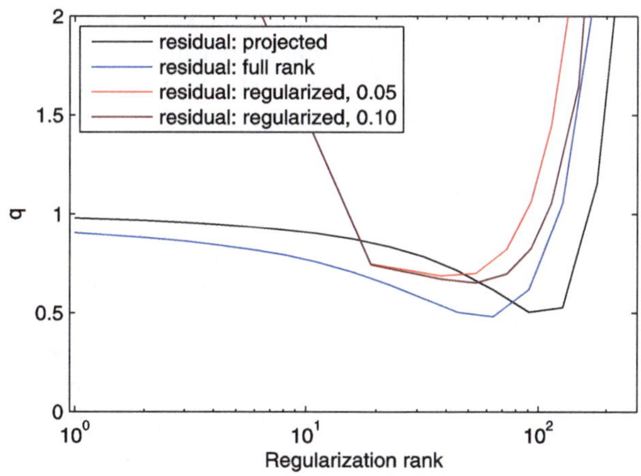

Fig. 18.5 The whitening quality $q(\boldsymbol{\varepsilon}^2)$ for the norm of the residuals according to (18.16) as a function of the regularization rank. The regularized qualities are computed with $\gamma = 0.05$ (*red*) and $\gamma = 0.1$ (*brown*), and the regularization parameter ξ is mapped to a "plausible" equivalent rank k by using the mean spectrum

jected" and the "full rank" schemes. However, the best measure occurs for a quite narrow range (in the logarithmic scale) of projector ranks, and it is not clear how to select a priori the best rank. As a first rule of thumb, a regularization parameter k between 30 % to 60 % of the rank of the spectrum seems to give good results. The common argument for choosing the number of eigenvalues is that the few largest ones are enough to capture the covariance, i.e., $k \simeq 1$ to 10, but this argument is clearly at odd with the properties of the covariance.

In practice, the evaluation of the inverse covariance occurs mostly for portfolio optimization. In this context, the mechanism of the problem is similar to the process context, namely small eigenvalues correspond to risk-free combination of assets. As formulated, the optimal investment strategy is to maximize the allocation along those directions, depending on the expected returns and on the investment constraints. Yet, the small eigenvalues are created by the limited information gathered by the simple "cross-product" covariance computation (and possibly by the rank deficiency when $N > i_{\max}$), and fluctuations will occur along those directions.

Therefore, the same regularization applied to portfolio optimization should lead to better asset allocation.

18.3.9 Summarizing the Situations

In a multivariate process setting, the covariance defined by a simple cross product as in (17.2) is misspecified. A shrinking term and a regularization term as in (18.7) lead to residuals with empirical properties that are closer to the postulated ones. With a regularization, the spectrum of the covariance is always nonsingular, and the computed residuals have statistical properties that are closer from being uncorrelated with unit variances. This is clearly a step in the right direction for large multivariate systems.

Yet, it seems not possible to obtain residuals that are both independent and with unit variances. For a given covariance kernel $\lambda(i)$, the number of very small or null eigenvalues is increasing with increasing size N, leading to very large values in the inverse volatility. The realized returns can have fluctuations along these small eigenvalue directions, in turn inducing large realized innovations. The shrinkage and regularization terms act so as to introduce more fluctuations along all possible directions and therefore lead to better properties for the innovations. Yet, these terms ought to be large enough to result in innovations of unit sizes, but such values wash out the correlation structure captured by the historical covariance. For the process, the directions corresponding to the small eigenvalues are suppressed, effectively reducing the number of sources of randomness. When using historical time series, return fluctuations do occur along the suppressed directions, and very large innovations are needed to compensate for the small eigenvalues. The fluctuations in the directions corresponding to small eigenvalues are directly related to the absence of stable subspace spanned by the eigenvectors, as demonstrated empirically in the previous chapter. Let us emphasize that this mechanism occurs regardless of the memory kernel $\lambda(i)$ used in estimating the covariance, as the eigenvalues always decrease exponentially fast toward zero. Together, the fast decay of the eigenvalues and the absence of stable subspaces imply a fundamental limitation in our capacity to describe large multivariate systems.

As shown in this chapter, it is not possible to strike a balance between the various terms so that the innovations have all the desired properties. For our parsimonious linear process, the parameter space in only two-dimensional, and the statistical properties of the innovations can be fully explored. Two other commonly used regularizations of the covariance matrix display similar empirical properties. Even if optimal parameter values cannot be found, these new terms in the covariance clearly lead to a better description of the empirical time series compared to the simplest multivariate cross-product extension. An open question is whether other terms in the spirit of the present shrinkage and regularization terms can be added in the covariance so that the innovations have the desired properties.

The connection between quadratic ARCH processes and volatility forecasts has already been used several times in the previous chapters. As for the univariate

ARCH processes, we have chosen a covariance $\Sigma_{\text{eff}}(\gamma, \xi)$ which is bilinear in the returns. This implies that volatility forecasts can be evaluated, namely all the quadratures can be computed analytically. For a general quadratic structure, the covariance forecast at t for the time $t + n\delta t$ takes the form

$$E\big[\Sigma_{\alpha,\beta}(t + n\delta t)\big|\Omega(t)\big] = \sum_{\alpha',\beta'} \sum_{i=0}^{i_{\max}} \widetilde{\lambda}_{\alpha,\beta;\alpha',\beta'}(n, i) r_{\alpha'}(t - i\delta t) r_{\beta'}(t - i\delta t). \quad (18.18)$$

The equation for the weights $\widetilde{\lambda}_{\alpha,\beta;\alpha',\beta'}(n, i)$ is a recursion equation in n which can be derived from the process equations and which can be implemented numerically. Therefore, the multivariate volatility forecast can be evaluated, regardless of the forecast horizon $t + n\delta t$, for all values of γ and ξ. The implication for multivariate volatility forecasts is that these additional terms in the process covariance make the long-term forecast of the covariance closer to a diagonal matrix, capturing the decay of the historical covariance due to the fluctuations in all directions. Although optimal parameter values cannot be found, both shrinkage and regularization lead to a better description of the multivariate time series and therefore should lead to better volatility forecasts. Such modifications have clear implications for portfolio optimizations and for risk evaluations, as for both the key ingredient is a multivariate volatility forecast up to the desired horizon. Further studies along these lines, extending the works of [86] in a heteroscedastic setting, should lead to better volatility forecasts and to better portfolio allocations.

18.4 Multivariate Affine ARCH Processes

For the affine processes, the effective covariance is a convex combination of the historical covariance and the mean covariance, with several possibilities depending on the structure puts on w_∞. We assume that the time components included in the affine processes are similar to the univariate processes or to the multivariate linear processes. In this way, only the "coupling constant" w_∞ of the mean covariance need to be studied. The simplest form is given by

$$\begin{aligned} \Sigma_{\text{eff}}(t) &= \sum_{k=1}^{n} w_k \Sigma_k(t) + w_\infty \Sigma_\infty \\ &= \sum_{l \geq 0} w(l) \mathbf{r}(t - l\delta t) \mathbf{r}(t - l\delta t) + w_\infty \Sigma_\infty \end{aligned} \quad (18.19)$$

with the constraint that $\sum w_k + w_\infty = 1$. The process equations and innovation properties are as for the linear processes (Eqs. (18.1) and (18.2)). Using the process equations and constraints, it is easy to show that the unconditional average is given by

$$E\big[r(t) r'(t)\big] = \Sigma_\infty. \quad (18.20)$$

It is therefore the matrix Σ_∞ that fixes the mean volatilities and correlations of the time series (for actual data, Σ_∞ can vary at time scales slower than several years). The number of parameters in this model is $N(N+1)/2+1+n$, namely N mean volatility, the matrix of $N(N-1)/2$ correlations, the single coupling constant w_∞, and the weights w_k. For a logarithmic decay, the single parameter τ_0 fixes all the weights w_k. It is therefore very parsimonious but likely too poor to accommodate the empirical data because the levels of all the lagged correlations for the volatility are essentially given by the single parameter w_∞. In such a model, the level of the lagged correlation for different time series should be very similar, and this is not supported by the empirical facts.

As shown in Chap. 12, the shape of the volatility lagged correlations are very similar across assets and asset types, but the level of the volatility lagged correlations are time series dependent. This stylized fact calls for a model of the weights where the time structure is universal but the amplitude $w_{\infty,\alpha}$ is time series dependent. A simple analytical form with these properties is

$$w_{k,\alpha} = (1 - w_{\infty,\alpha})\chi_k \tag{18.21}$$

with $\sum_k \chi_k = 1$, and χ_k have no dependency on α. The dependency on the time series α is only contained in the overall amplitude $w_{\infty,\alpha}$. With this definition, $\sum_k w_{k,\alpha} + w_{\infty,\alpha} = 1$ for all α.

With these weights, the matrix of effective covariance is taken as

$$\Sigma_{\mathrm{eff},\alpha,\beta}(t) = \sqrt{1-w_{\infty,\alpha}}\sqrt{1-w_{\infty,\beta}}\sum_{k=1}^{n}\chi_k \Sigma_{k,\alpha,\beta}(t) + \sqrt{w_{\infty,\alpha}w_{\infty,\beta}}\,\Sigma_{\infty,\alpha,\beta}. \tag{18.22}$$

The unconditional expectation $E[r_\alpha r_\beta]$ leads to

$$\begin{aligned} E[r_\alpha r_\beta] &= E[\Sigma_{\mathrm{eff},\alpha,\beta}] \\ &= \sqrt{1-w_{\infty,\alpha}}\sqrt{1-w_{\infty,\beta}}\,E[r_\alpha r_\beta] + \sqrt{w_{\infty,\alpha}w_{\infty,\beta}}\,\Sigma_{\infty,\alpha,\beta}. \end{aligned} \tag{18.23}$$

Solving the equation for given α and β leads to

$$\Sigma_{\infty,\alpha,\beta} = \frac{1-\sqrt{1-w_{\infty,\alpha}}\sqrt{1-w_{\infty,\beta}}}{\sqrt{w_{\infty,\alpha}w_{\infty,\beta}}}\,E[r_\alpha r_\beta] \tag{18.24}$$

with the diagonal element $\Sigma_{\infty,\alpha,\alpha} = E[r_\alpha^2]$. This equation relates the covariance matrix $E[\mathbf{rr}']$ to the matrix that fixes the mean level of variances and correlations Σ_∞. The relation between the mean correlation and the process parameters is

$$\rho_{\infty,\alpha,\beta} = \frac{\sqrt{w_{\infty,\alpha}w_{\infty,\beta}}}{1-\sqrt{1-w_{\infty,\alpha}}\sqrt{1-w_{\infty,\beta}}}\,\frac{\Sigma_{\infty,\alpha,\beta}}{\sqrt{\Sigma_{\infty,\alpha,\alpha}\Sigma_{\infty,\beta,\beta}}}. \tag{18.25}$$

This model has $N(N+1)/2+N+n$ parameters, with $N(N-1)/2$ for $\rho_{\alpha,\beta}$, N for $\Sigma_{\infty,\alpha,\alpha}$, N for $w_{\infty,\alpha}$, and n for the χ_k (only 1 if χ_k is given by an analytical form).

This number is minimal, with the dynamic specified by overall parameters χ_k and one parameter per time series specifying the lagged correlation $w_{\infty,\alpha}$. Then, the matrix $\Sigma_{\infty,\alpha,\beta}$ specifies the mean volatility and mean correlation for a set of time series. Notice that GARCH processes that have been proposed in a multivariate context typically have of the order of N^2 parameters (a straight generalization of GARCH(1, 1) leads to $3N^2$ parameters), and some processes have up to N^4 parameters. Even for small N, such a grow is rehibitory.

An interesting open question is whether the univariate parameter w_∞ for the time series α can be directly used in a multivariate context as $w_{\infty,\alpha}$. Such a connection would bridge the gap between the easy univariate process estimations with the more difficult multivariate estimation. The remaining multivariate parameters left to estimate is the mean covariance Σ_∞ for which a simple (long) sample mean can be used (i.e., a moment estimator). If such an approach can be followed, no maximum likelihood is needed, hence removing the difficult problem related to the inverse covariance with its small or zero eigenvalues. Another question is whether the above affine process can reproduce the multivariate empirical properties displayed in the previous chapter.

With the above affine multivariate process, the linear process is recovered when taking the limit $w_{\infty,\alpha} = w_\infty \to 0$. In this limit, the process volatility equation simplifies to

$$\Sigma_{\mathrm{eff}}(t) = \sum_{k=1}^{n} \chi_k \Sigma_k(t) \tag{18.26}$$

and the mean covariance disappears from the equation. The only parameters left are the weights $\chi_k = w_k$, and there is no dependency on the time series.

Starting again from (18.19), the effective volatility can be generalize further to allow for a different memory for different time series

$$\Sigma_{\mathrm{eff},\alpha,\beta}(t) = \sum_{l\geq 0} \sqrt{w_\alpha(l)}\sqrt{w_\beta(l)} r_\alpha(t - l\delta t) r_\beta(t - l\delta t) + w_{\infty,\alpha,\beta}\Sigma_{\infty,\alpha,\beta}. \tag{18.27}$$

The constraint on the coefficient is $\sum_l \sqrt{w_\alpha(l)}\sqrt{w_\beta(l)} + w_{\infty,\alpha,\beta} = 1$. As the lagged correlations for the volatility are very similar across most empirical financial time series, such generality seems unneeded.

18.5 More General Extensions for the Multivariate GARCH Process

Many multivariate GARCH extensions have been proposed, with various structures along the time series dimension. Depending on the hypothesis used for the structure, they typically contain of the order of N, N^2, or N^4 parameters. A fairly complete review of the existing literature can be found in [17] and [138]. In this section, only the key ideas are introduced, and the interested reader should check the reviews and original articles for more details. The different processes are discussed in the light of the empirical investigations contained in the previous chapters, and with an eye on very large systems as, for example, occurring in today's portfolio allocation.

Starting from the univariate GARCH(1, 1) equation as written in (7.2), a natural generalization is to replace the scalar variance σ_{eff}^2 by the covariance matrix Σ_{eff} and the squared return r^2 by an exterior product $r \cdot r'$. The coefficients α_0, α_1, and β_1 can be replaced by matrices, with a proper symmetrization of the equation in order to preserve the symmetry of the covariance matrix. The symmetrization is achieved by multiplying on the left and right by the square root of the coefficient matrices. Two difficulties arise with this formulation. The first one is to preserve the positivity of the covariance. This leads to restrictions of the coefficient matrices, which can be difficult to check in practice. The second issue is the large number of parameters, of the order of $3N^2$, which makes their estimation infeasible in practice. More structures should be added into the matrices of parameters in order to alleviate both issues. For example, the matrices corresponding to $\sqrt{\alpha_1}$ and $\sqrt{\beta_1}$ can be restricted to be diagonal, or even a multiple of the identity matrix. Such restrictions are introduced mainly for their mathematical convenience so as to make the model manageable.

The approach proposed earlier in this chapter follows a similar line but uses the knowledge gained about the univariate heteroscedasticity and about the multivariate covariance in order to select the leading structure. The analysis of the empirical data suggests to use the same long memory for all components in a natural cross-product form. This simple structure is based on the results of Chap. 7 (emphasizing the distinction between linear and affine processes), of Chap. 12 (showing that the heteroscedasticity has the same characteristics for all time series), and of Chap. 17 (showing that the covariance and correlation have a similar dynamics).

Another direction used to build multivariate processes is to split the covariance matrix into the correlation matrix and the diagonal variance matrix. Then, different processes can be written for the correlation matrix and for the (univariate) variances, assuming that a different dynamics occurs for each part. The set of variances can be captured by univariate processes, say a GARCH(1, 1) or a long-memory process. For the correlation matrix, a natural assumption is to take a constant matrix, leading to the CCC (constant conditional correlation) line of processes proposed by [23]. Another possibility for the correlation matrix is to postulate a GARCH(1, 1) dynamics built from the vector of innovations ϵ_α, with scalar coefficients. This approach has been introduced by [56] and is called DCC (Dynamic Conditional Correlation). Such an approach leads to $3N$ parameters for the N univariate processes and a few parameters for the correlation matrix. Moreover, the decomposition makes easier the estimation of the subproblems. Yet, the empirical exploration of Chap. 17 finds no evidence for such a split between a diagonal covariance and correlation matrices, but instead a fairly similar long-memory dynamics for the covariance and correlation matrices.

The above approach keeps the natural matrix form of the covariance, but it imposes some dependencies in the updating rules between the matrix elements. The most general update rule for a matrix is obtained by assuming complete independence for each element. This is obtained by mapping the elements of the symmetric covariance matrix into one vector, namely by picking the $N(N+1)/2$ on and below the diagonal and stacking them in a column vector. This approach called VEC-GARCH has been introduced by [24]. In a GARCH(1, 1) spirit, the update rule for the vector depends on a square transition matrix with the size of the vector, namely with $(N(N+1)/2)^2 \simeq O(N^4)$ elements. Although very general and flexible, such an approach suffers from two problems. First, the covariance matrix should be semidefinite positive. This constraint implies some relationships on the transition matrix which are given implicitly and difficult to check in practice. Second, the number of parameters is of order N^4, making the estimation infeasible even for moderately large N. Both issues are serious, and more structures should be introduced in the problem in order to reduce the complexity. Yet, there is no clear empirical evidence on the direction to follow, and many papers have been devoted to explore various mathematical restrictions of the general structure.

Another direction to model the multivariate structure is to use an eigenvalue decomposition and then to assume that the eigenvectors are fixed while setting a dynamics on the eigenvalues. A similar approach is to postulate that the observed vector of returns is obtained by a fixed linear or orthogonal transformation of a few scalar factors which have a genuine heteroscedastic dynamics. Intuitively, a few "economics" drivers are used as hidden factors, while the observed returns are a static combination of them. These models are called respectively factor-GARCH and Orthogonal-GARCH, initiated respectively by [58] and [4]. The analysis of the projectors on the dominant eigenvectors presented in Sect. 17.8 shows that such static decompositions are difficult to justify empirically.

The estimation of the parameters introduces new difficulties compared to the univariate case. For a multivariate affine GARCH process, a standard procedure consists in finding the optimal parameters with a log-likelihood maximization, assuming a given distribution for the residuals. Following the model hypothesis, the residual distribution is set to have a unit variance in this computation. After optimization, the empirical residuals will have variances close to one, in sharp distinction with the results presented in Sects. 18.3.6 and 18.3.7. Indeed, the optimized parameters are "absorbing" the problems related to the evaluation of the large inverse volatility. The result is that the covariance, with the optimal log-likelihood parameters, becomes too large with respect to the volatility of the returns. The mismatch between postulated and empirical properties of the innovations is a fundamental limitation that prevents inferences on the parameter values using the standard log-likelihood framework. This limitation is acknowledged but not identified in [60], who propose as a remedy to use a composite likelihood for subsets of size two in order to estimate parameter values.

This brief overview shows that the space of possible models is much larger in a multivariate system. The most general processes are intractable, and additional hypotheses and structures are introduced in order to reduce the complexity. Typically, the mathematical structure is split into static and dynamical parts, where the dynamical part should capture the heteroscedasticity. There is a difficult balance to strike between the flexibility required to capture the multivariate dynamics of the empirical data, the complexity of the model, and the number of parameters. Due to their complexity, the multivariate extensions of GARCH have been mostly confined to the academic literature so far. Even with strong restrictions, parsimonious multivariate GARCH processes have a number of parameters of order N. The estimation of such a large number of parameters is a very difficult task for sizes relevant to actual portfolios. As an alternative for the practical applications, the long-memory process presented earlier in this chapter gives a simple and effective way to compute covariance and covariance forecasts, essentially without free parameters. But as most applications like portfolio estimation or parameter estimations are depending on the inverse covariance matrix, the regularization of the small eigenvalues is of critical importance. Clearly, this is an area where further progresses can be expected.

Chapter 19
The Processes Compatible with the Stylized Facts

Many statistics have been presented in this book in order to extract information from financial time series and to compare with theoretical processes. These pieces of information are mostly summarized in the "mug shots" and also in other graphs and analytical derivations throughout the book. Together, these statistics allow us to build a fairly coherent picture of the stylized facts, of the processes that are compatible with them and of the processes that are unable to capture them. With these systematic statistical analyses, we are in a position to falsify some theories (in the sense of Karl Popper), namely to reject them as descriptions of financial time series. This chapter presents a summary of the models or processes that are *compatible* with the data or which can be *falsified*. Some resulting implications are discussed after each point.

- **Compatible:** Relative returns, geometric processes
  Falsified: Logarithmic returns, logarithmic processes
  With a finite time increment, the discretization matters! In order to have well-defined mathematical properties with fat-tailed innovations, a geometric process should be used, whereas the exponential of an additive process leads to diverging expectations (Chap. 6). A further advantage is that relative returns lead to a negligible skewness over a broad range of time horizons (Sect. 6.8) and that the bankruptcies for stocks appear naturally.
- **Compatible:** Student innovations
  Falsified: Normal innovations
  The volatility dynamics alone in conjunction with a normal distribution for the innovations cannot induce the observed fat-tailed distribution for the returns. Innovations should have a fat-tailed distribution, regardless of the model for the volatility. A Student distribution with a number of degrees of freedom around 5 gives a very good description of the innovation pdf (Chap. 13). Notice that it is not possible to construct a continuum limit of a process with a Student distribution over many time intervals, say, for example, δt and $2\delta t$ (because this distribution is not stable under the addition of independent random variables).

G. Zumbach, *Discrete Time Series, Processes, and Applications in Finance*, Springer Finance, DOI 10.1007/978-3-642-31742-2_19,

- **Compatible:** Multiscale ARCH volatility; GARCH(1, 1) as an approximation
  Falsified: Stochastic volatility; regime-switching volatility; FIGARCH
  The observed time-reversal asymmetry gives a powerful criterion in order to discriminate between volatility processes (Chap. 11). Stochastic volatility processes are time-reversal invariant and therefore are not able to reproduce this stylized fact. Regime-switching processes with at least three states are asymmetric, but the asymmetry is too weak. On the other side, ARCH processes have the correct amount of asymmetry (Chap. 7), depending on the detail structure of the process (the aggregated versus microscopic measures of volatilities).
  An interesting property of the multiscale ARCH processes is the fairly universal values for many of the parameters. In particular, the decay of the volatility lagged correlation is very similar across time series (Chap. 12), leading to similar parameters for a long-memory ARCH process. This is convenient, as a universal and good volatility forecast can be constructed. Such a forecast has been used to improve market risk estimation (Chap. 15), or it can be used to compute the innovations corresponding to empirical time series.
  The GARCH(1, 1) process is interesting because it is simple and captures part of the heteroscedasticity. It is therefore a very good first step beyond a constant volatility process. Yet, the multiscale processes are only slightly more complicated; they describe correctly the multiscale dynamics of the volatility, and similar parameter values can be used for most of the time series.
  The FIGARCH processes (Sect. 7.12) are very rigid due to their mathematical construction with a fractional difference operator. They describe only partly some stylized facts, and their practical implementations raise a subtle difficulty related to the finite cut-off. For all these reasons, the multiscale ARCH processes should be preferred.
  As a drawback, the ARCH processes have no continuum limits. But, the continuum limit should be abandoned anyway.
  The simple and convenient structure of the ARCH processes makes them very flexible with respect to incorporating more stylized facts in a fairly natural way. Examples are the leverage effect (Chap. 14) or the inclusion of trend/drift on the volatility (Sect. 7.13).
  The volatility dynamics induced by the multicomponent ARCH processes can be derived, leading to a new type of positive diffusion processes (Sect. 7.16). This structure is not of the stochastic volatility type.
- **Compatible:** Discrete processes
  Falsified: Continuous processes.
  All the points discussed above assume a finite time increment δt for the process, and lead to severe difficulties in the limit $\delta t \to 0$. Yet, there are no fundamental reasons to use only processes for which a genuine continuum limit can be constructed. Indeed, the abandon of the continuum limit removes a strong theoretical selection bias and liberates many avenues.
  So far, option pricing relies heavily on stochastic calculus and continuous-time processes. But as seen in Chap. 16, an option pricing scheme can be built with

a finite time increment, relative returns, and a very broad range of specifications for the volatility and drift. Similarly to the Black–Scholes option pricing, this pricing scheme is based on the construction of an equivalent martingale measure, a replicating portfolio, and a no-arbitrage argument. With finite time increment, a small δt expansion at second order plays the role of the Itô calculus. Rigorously, the small parameters for the expansions are $\sqrt{\delta t/1\ \text{year}}\,\sigma$ and $(\delta t/1\ \text{year})\,\mu$ with σ and μ annualized. Therefore, option prices can be computed (numerically) using realistic ARCH processes which incorporate the observed stylized facts and with a finite time increment.

Leaving apart continuum time processes leads to revisiting the position of the stochastic calculus. Strictly speaking, it is not applicable anymore. But analytical computations require to perform second-order Taylor expansions in the small δt parameter, essentially reproducing the proof used to derive the stochastic calculus formula. Therefore, a very similar path is followed.

- **Compatible:** Dynamical multivariate structure
  Falsified: Static multivariate relationships
  A substantial part of the financial applications concerns portfolio allocation and the risk/return trade-off. For the risk part, the Markowitz portfolio allocation assumes a covariance matrix that can be estimated using historical data and that will remain unchanged for the investment horizon. The heteroscedasticity makes the picture more complex, but also the multivariate structure shows large temporal fluctuations.
  Much remains to be done in this area. But the hypothesis of a static covariance or correlation matrix is clearly difficult to maintain.

Putting all the points together, the validated processes are using relative returns and a finite time increment, with a multiscale ARCH structure. Depending on the simplicity versus accuracy trade-off, various structures can be used for the multiscale ARCH volatility (linear versus affine, microscopic versus aggregated). Then, other stylized facts can be included (drift terms, leverage, trend/drift, etc.).

The list above presents the main points that can be validated or refuted in the light of the stylized facts. Subsequently, some tools widely used in finance should be reevaluated with respect to their applicability. The dividing lines here are more subtle and aim at the usefulness of these tools for practical applications. Contrarily to the clear dichotomy above, softer distinctions are used below.

- **Broadly applicable:** Numerical simulations
  More restricted: Analytical solutions
  Any single point in the list above creates substantial difficulties with respect to the analytical solutions of a practical financial applications. Collectively, they leave very little space for analytical tractability, if any at all. Essentially, the complexity of the processes leads mainly to using Monte Carlo simulations as a practical evaluation tool.
  This is a pity, as analytical results are much more powerful and broadly applicable than numerical evaluations. Yet, the power of today's computers is sufficient to solve realistic financial problems, with a realistic model. Therefore, one

has to choose between models, with clearly stated approximations and analytical tractability, and more complex processes, with realistic properties but only numerical tools at hand. But it is wrong today to select systematically a model or process based only on its analytical convenience.

- **Broadly applicable:** Estimation on the multiscale properties of the data
More restricted: Log-likelihood estimation

Processes have parameters that should be estimated. The standard method is a log-likelihood estimation that focuses on the properties of the innovations at the scale δt. But after the leading heteroscedasticity is discounted, the properties of the innovations are weakly depending on the details of the process. For the applications, the major goal consists in obtaining realistic statistical properties for the processes compared to the relevant stylized facts. Therefore, it is better to estimate the parameter values directly on the desired statistical properties. Roughly, this approach is similar to a generalized method of moments, with a broad panel of statistics over many time horizons used as the target to reproduce.

This approach allows us to solve naturally a common gap in the time horizons: the data are available with a given frequency (typically daily, possibly high frequency), but the problem of interest concerns longer time horizons (say portfolio allocation or option pricing). For such problems, the empirical properties should be captured correctly between the data time horizon and the problem time horizon, and the parameter estimation should reflect these constraints.

One should also have a healthy dose of skepticism with respect to the parameter values. Finance is a universe dominated by randomness, and given a possibly long but limited data sample, the accuracy of the parameter values is necessarily limited. Even with empirical samples longer than one decade, changes in the parameter values of the order of 10 % are within a reasonable range of accuracy, given the statistical errors on the corresponding empirical statistics. The slow convergence of the properties related to the volatility, as induced by the heteroscedasticity, should act as a cautionary tale, and most $1/\sqrt{N}$ error estimates are too optimistic. For this reason, it is better to have a parameterization of the processes that reflects their statistical properties, so that reality checks are intuitive.

- **Broadly applicable:** Numerical statistical tests
More restricted: Theoretical statistical tests

Rigorous statistical tests and p-values are important in order to decide if a statistical hypothesis should be accepted or rejected. The available theoretical tests are based on models, typically independence (or possibly a dependency that decays fast enough) and a normal distribution. When such a test is available and is appropriate, it should be used. But because of the heteroscedasticity, the independence hypothesis is often wrong, and a more complex null model should be selected. In this situation, only numerical statistical tests can be used. Fortunately, it is easy to conduct a rigorous hypothesis testing using Monte Carlo simulations.

Chapter 20
Further Thoughts

20.1 Multi-time Horizon Analysis

The bulk of quantitative finance uses daily data and computes statistics related to daily quantities. The long-term behavior is often analyzed with lagged correlations, emphasizing dependencies at increasing time differences for daily quantities. Although these choices are quite straightforward with daily data, the relevant methodology is less obvious with high-frequency data.

The two most distinctive characteristics of the present approach consist in *analyzing empirical data and processes at multiple time horizons* and *studying dependencies between historical and realized quantities*. Both points are much better suited to analyze financial data. First, the empirical data show no single time scale but instead a range of time scales, as it is particularly clear on the historical/realized volatility correlation plots. All processes with one characteristic time scale—ARCH or stochastic volatility—show a similar signature on this graph, and this signature differs clearly from the empirical data. Ultimately, the absence of a single characteristic time is related to the range of human time scales occurring in the objectives, constraints, and subsequent decisions. By analyzing empirical data only at a daily time horizon, this rich time structure is missed. Second, the dependencies between historical and realized quantities maximize the available information, namely better diagnostics are obtained relating quantities taken just before t and just after t. Such computations track better human behaviors that use the most recent information in order to take decisions. At a given time, a particular market agent uses its perceived or computed return or volatility, say, but he does not use much deeper past return or volatility in order to form his opinion about the future. Possibly, he might use the change of price, return or volatility, like a momentum or $\Delta\sigma$, but deep past return or volatility are of lower information content.

The power of our approach is well illustrated by the historical/realized volatility correlation. This dependency gives a discriminating two-dimensional figure that differentiates among various processes. For example, the structure is clearly symmetric for a multiscale stochastic volatility process, but it is asymmetric for multiscale ARCH processes. By contrast, the one-dimensional lagged correlation for the

G. Zumbach, *Discrete Time Series, Processes, and Applications in Finance*,

Springer Finance, DOI 10.1007/978-3-642-31742-2_20,

volatility is much less discriminating, and, for example, all multiscale processes—ARCH or stochastic volatility—can mimic the decay observed in the empirical data. Therefore, these two-dimensional correlations—functions of both the historical and realized time horizons—reveal structures that are not accessible to the more usual statistics like lagged correlations. This shows the power of financial imagery, and hence the name "finanscopics" for the related web site.

20.2 Slow Decay for the Lagged Correlation

One of the salient properties of empirical financial time series consists in the slow decay of the lagged correlation for the volatility. In Chaps. 3 and 12, a study over a very large set of time series shows that the volatility lagged correlation decays essentially as

$$\rho_\tau = \rho_1 \left(1 - \frac{\ln \tau}{\ln \tau_0} \right) \tag{20.1}$$

where τ_0 is a characteristic time of the order of a few year. The common assertion for the shape of the decay is a power law $\rho_\tau \simeq \tau^{-\nu}$, but Eq. (20.1) provides consistently for a better description of the empirical data.

However, the comparison of the historical/realized volatility correlations between empirical data and the different long-memory processes (ARCH and stochastic volatility) shows that the above description is only an approximation. The historical/realized volatility correlations show that the empirical time series have mainly a few characteristic time horizons, roughly distributed according to a geometric progression. The market process Mkt-Agg-ARCH can duplicate both the historical/realized volatility correlation and the decay for the lagged correlation. This shows that the empirical lagged correlation is probably resulting from a few characteristic times, with weights such that the form (20.1) describes the resulting lagged correlation, but this is not an exact form. Let us emphasize that the log decay analytical form describes parsimoniously the data up to a few months, but the statistical error bars grow too large to validate such an analytical description beyond one year.

The fundamental question with respect to the volatility lagged correlation is which mechanism creates this slow decay (almost) universally, with a similar characteristic time τ_0 across the financial universe. The dynamics of the market and the profile of the market participants should be such that a similar decay is created for the lagged correlation, independently of the details of the traded contracts. A hint about the potential explanation is given by a fine analysis of the parameter τ_0 for many time series. Essentially, mature markets exhibit longer characteristic times than emerging markets. A similar conclusion is reached by Di Matteo et al. [103] with a large-scale analysis of the generalized Hurst exponent on foreign exchange rates. In the context of the Mkt-Agg-ARCH, the approximate ratio for the magnitude of the market components could possibly correspond to an equilibrium between the participants at different time horizons. So far, this equilibrium mechanism should still be explained.

20.3 Definition of "the" Volatility

Volatility is a fairly elusive concept in finance. The intuition consists in to measure the price fluctuations, but the heteroscedasticity makes it difficult to translate this idea into one formula. Indeed, several definitions can be elaborated, following essentially two different purposes.

The first purpose is to measure the (historical) volatility in a given time window. In this case, a sum of squared returns in the time interval is appropriate, like (2.2). Such definitions are convenient for statistical studies as the volatility uses only information inside a given time interval. If EMA are used instead, spurious correlations appear, due to overlapping data sets.

A second purpose is to have the equivalent of an "instantaneous" volatility. The effective volatility σ_{eff} in a process is the closest approximation of an instantaneous volatility, and the justification of a particular formula and parameter set goes through the adequation of the chosen process to describe empirical time series. The finite time increment δt of the process is the best approximation of "instantaneous", and the effective volatility is indeed a one-step volatility forecast over δt. This approach captures best the dynamics of the volatility and the heteroscedasticity. The long-memory microscopic ARCH processes are a good compromise for capturing the main characteristics of the volatility dynamics, while still having simple formula like (7.37) with a small number of parameters. This approach is well suited to compute forecasts, innovations, or quantiles. If required, more accurate processes are available, albeit there are still no consensus about how to incorporate more stylized facts in a minimal way. With this approach, the long memory of the volatility requires to use a slowly decaying kernel. This makes difficult some statistical analysis because of the overlapping information sets.

The important point is that there is no unique "good" definition of volatility, but instead several definitions serving various purposes. In this sense, "the" volatility does not exists. Beside, all definitions involve time intervals, characterizing the horizon at which the volatility is measured or forecasted.

All these remarks are valid in the univariate and multivariate cases. In the multivariate case, the dynamical dependencies between assets add a new layer of complexity. The analysis of Chap. 17 shows that the dependencies follow a similar dynamics. Due to the limited information that can be extracted from the covariance matrix, much remains to be done in order to understand in detail the interplay between the dynamical structure present in the market and the random matrix theory.

20.4 Is the Heteroscedasticity that Important?

A few decades after the seminal work of Engle [59] and Bollerslev [61], and after countless papers presenting variations around heteroscedastic processes to capture this effect, large parts of finance still have not integrated this stylized fact. For example, portfolio allocations are mostly done using the Markowitz mean-variance scheme, which assumes the existence of a static covariance matrix. Another example is all the processes used to described interest rates include mean reversion effect and/or a diffusion depending on the interest rate level, but they do not include heteroscedasticity. In fact, a small fraction of researchers and practitioners would even discard heteroscedasticity because it is only a minor effect, if to be seen at all.

In order to grasp the magnitude of the volatility dynamics, the lagged correlation of the one month volatility at a lag of one month (i.e., just without overlap) is of the order of 40 to 60 %, regardless of the time series. This is not small beer! Indeed, this is the largest correlation present in financial time series, roughly an order of magnitude larger than other effects (return/return, return/volatility or leverage effect, return/return conditional on volatility, etc.). Moreover, the heteroscedasticity presents similar characteristics and magnitudes for all underlying classes: stocks, indexes, interest rates, spreads, etc. Together with fat-tailed distributions, both stylized facts are the main effects that should be added on a Bachelier process in order to obtain an accurate description of financial time series.

20.5 Fat-Tailed Distributions

The empirical probability distributions for the returns clearly have fat tails, as can be observed in the figures with a logarithmic scale for the pdf. This feature of the data has been studied quantitatively using the Hill estimator [81] and its variations [120]. Essentially, the tail exponents are in the range 3 to 6 for all time series. These values are far from a normal distribution (corresponding to an infinite tail exponent), but above two (the tail exponent of the Lévy distribution). Given this stylized fact, the questions are how to incorporate fat tails in processes and what is the origin of the fat tails.

Regarding the first question, the fat tails can be incorporated into a process either by using the innovations or by using the heteroscedasticity. Using innovations with fat-tailed distribution is straightforward, leading to return distributions with a similar exponent (slightly smaller because of the heteroscedasticity). On the other hand, Gaussian innovations can be used together with a dynamics for the volatility, so that the fat tails for the returns are resulting purely from the heteroscedasticity. For example, in a GARCH(1, 1) process, it is the feed-back loop from the return on itself that is generating a nontrivial return distribution, and the tail exponent can be computed analytically [139]. Using Gaussian innovations has important advantages on the theoretical side, as a proper continuum limit can be built with a Wiener process for the innovations. In the continuum limit, the powerful Itô calculus is available. Based on these theoretical considerations, it is tempting to use Gaussian innovations, and indeed the bulk of the literature is following this path.

All the empirical studies included in this book point clearly to fat-tailed distributions, both for the returns and the innovations, for high frequency and for daily time series. Other studies in the literature (see, e.g., [61, 67, 88, 120]) find consistently that Student distributions models better empirical data than a normal distribution. Moreover, the number of degrees of freedom for $p(\epsilon)$ is always small, of the same order as the tail exponent for the return distribution. This is also our conclusion: fat-tailed innovations capture better the empirical time series regardless of the process. For stochastic volatility processes, it is better to use fat-tailed distributions for the return innovations and for the volatility innovations. In short, all empirical evidences point toward fat-tailed innovations.

Finance is an open system, in interaction with the world at large. The quantum of external information is called typically a "piece of news". At a microscopic level, the description should include the influence of the external information, the decisions of the market participants, the trading rules, and the interaction taking place in the order queues. The simpler description of financial time series by processes for the prices is a mesoscopic description. In a mesoscopic process for the price, all the microscopic phenomena are captured in the innovations, whereas the process equations describe the dynamics of the market participants acting on particular securities.

With the availability of high-frequency data related to order queues, it is now possible to study in detail the microscopic of the price formation [65, 111, 144]. Generically, the observed distributions have fat tails for all quantities: tick-by-tick returns, volume in the queues, volumes exchanged, time intervals, etc. Indeed, at the microscopic level, there is no law of large numbers and the subsequent convergence toward a Gaussian distribution. Therefore, at least for a mesoscopic description at short time interval δt, it is consistent that fat-tailed residual distributions should be used.

The influence of external "news" is a more difficult topic because, in their vast majority, the news are not quantified but given by a string of text. The naturally "quantified" news are an economic information, like unemployment rate or GDP. These news are released in a given schedule, and the market participants have an expectation on the new numbers. They take positions accordingly, and therefore the surprise lies in the difference between the expected and unexpected part of the news. Econometric studies on this topic [6] have found that the incremental information containing in the news is very quickly incorporated in the price, within a time span of the order of 15 minutes. On the other hand, the influence of textual news is much more difficult to assess. Intuitively, their financial impacts should be measured in a logarithmic scale. The scopes of the news are clearly very different between a profit warning for one company, the outcome of a political election, or the invasion of an oil producing country. Again, this argument eliminates the normal distribution and points toward fat-tailed innovations.

All the arguments presented in the previous two paragraphs point toward fat-tailed innovations, in agreement with all empirical studies comparing Gaussian with Student distributions. Moreover, fat-tailed innovations model better the empirical time series, regardless of the analytical form used for capturing the heteroscedasticity. This indicates that the origin of the fat-tailed distribution for the return lies in the innovations and is not induced by the heteroscedasticity. Yet, using innovations differing from Gaussian creates serious difficulties on the theoretical side.

The Gaussian distribution has the special property to be closed under the addition of random variables. This property allows one to construct a Wiener process, namely a continuous Gaussian process. In the continuous-time limit, the Itô calculus is a powerful tool, which is used heavily in option pricing theory. Therefore, while replacing Gaussian innovations by fat-tailed innovations is straightforward at the level of a discretized process, this substitution is creating major difficulties when constructing a continuum limit and for pricing options. Yet, it would be wrong to restrict the innovation to a normal distribution in order to have a convenient mathematical continuum limit. Instead, the probability density for the innovations should be given by the analysis of the empirical data. Then, mathematicians and financial engineers should construct the related option pricing theory. Steps in this direction have been taken by [40, 55, 80, 116] and are presented in Chap. 16.

Another common hypothesis about the residuals consists in assuming that they are independent. At the level of the decisions made by the market participants, this is probably a good assumption. But for the external "news", this hypothesis is questionable. It is likely that an important piece of news will be followed by a stream of related information in the following days. This will create clusters of news on related topics, inducing in turn lagged correlations. Therefore, in a mesoscopic process, the innovations should have dependencies in order to capture this effect. Yet, when estimating processes on empirical data, using independent increments allows one to model correctly the time series. This indicates that the possible serial correlations of the news are too small to be noticed at this level of accuracy, and an i.i.d. distribution can be used for the residuals.

20.6 Convergence Toward a Gaussian Distribution

A sum of random variables converges to a Gaussian distribution, as the central limit theorem shows under some mild conditions. Moreover, the convergence occurs at a power law rate (namely the differences between the asymptotic Gaussian and the actual distribution shrink as a power law with increasing aggregation, typically with exponent $1/2$ and larger). Because the return at one day is the sum of a large number of transactions, we can expect the daily return to be fairly close to a Gaussian. Yet, the distribution of financial returns are far from a Gaussian law, and this can be observed clearly for returns up to a few weeks. The scaling for the moments of the distribution shows that the convergence occurs much more slowly than dictated by the central limit theorem. Indeed, the present analysis of the excess kurtosis for empirical data shows a logarithmic decay and clearly excludes a power law. Therefore the question is where does the central limit theorem fail?

The failure lies in the small characters associated with the central limit theorem: the random variables must be independent. A more general version of the theorem requires the random variables to have dependencies that decay exponentially fast (see, e.g., [66]). Yet, financial time series have dependencies in the magnitude (i.e., the volatility) that decay as a logarithm, namely much more slowly than allowed by the generalized central limit theorem. This dependency is what creates the slow convergence toward the Gaussian. Intuitively, the heteroscedasticity at multiple time scales creates the slow convergence. It is at a time scale longer than the longest characteristic time in the system that the convergence toward the central limit theorem occurs at the theoretical rate. This effect is clearly visible for processes with one time scale, with a clear cross-over at the characteristic time and a fast convergence for longer time intervals. Because the empirical data show a range of characteristic times, a cascade of cross-over is produced, leading to the observed slow convergence toward the asymptotic limit. The key point is that the heteroscedasticity is ultimately the cause of the slow convergence toward a Gaussian distribution for the returns.

Historically, the paper of Mandelbrot [95] that analyzed cotton prices was very important. The author shows that the empirical daily returns have fat tails. Because daily price changes are the sum of a large number of transactions, Mandelbrot concluded that the return distributions should be in the set of attraction of the Lévy stable distributions. This conclusion would be devastating, as the second moment of Lévy random variables does not exist, and therefore the volatilities and covariance matrices would be ill defined. Many years later, the empirical analysis of Mandelbrot turned out to be partly correct, namely price distributions have fat tails, but with a tail exponent around 4 and not below 2. Consequently, his explanation for the origin of fat-tails is wrong. There are two key conditions for the central limit theorem to hold: the independence of the variables and the existence of the second moment. It is not the later that failed, but the independence. Yet, heteroscedasticity would appear only 20 years later in finance.

20.7 Temporal Aggregation

Let us assume that some time series are well described by a particular process at a time horizon δt. An interesting question consists in finding out the equivalent process that would describe the empirical data at a time horizon $n\delta t$. For example, a good process could be available at a horizon of one day. Then, a pension fund needs to perform long-term assets and liabilities study, using long-term Monte Carlo simulations with a time increment of one year. The idea is to infer from the daily process an equivalent yearly process so that the simulations can be done efficiently. In a Gaussian random walk setting, the answer is trivial as the normal distribution is stable under the addition of random variables. In this case, the process is identical, and the parameters are modified by the usual temporal scalings.

Can this goal be achieved for a general ARCH process? For the case of the GARCH(1, 1) process, the answer is yes, but it involves some technical questions about the definition of the process (essentially, the GARCH(1, 1) process needs to be formulated in a larger space so that the aggregated process also belongs to this space). This is the content of an interesting paper by Drost and Nijman [54]. Moreover, the authors find the explicit mapping between the parameters.

In the $(\alpha_0, \alpha_1, \beta_1)$ coordinates, the mapping is fairly involved. Using instead the $(\tau, \sigma_\infty, w_\infty)$ coordinates, the mapping simplifies substantially: the mean volatility changes by the $\sqrt{n}$ scaling, the characteristic time τ is reduced by n, and only the coupling constant w_∞ involves a nonlinear equation that should be solved numerically. Indeed, a GARCH(1, 1) process has one exponential characteristic time, which is simply reduced by n when viewed at an aggregated scale $n\delta t$.

For more general ARCH processes, the aggregation question is still open.

20.8 Mean Reversion and Ornstein–Uhlenbeck Processes

The basic diffusion process used in finance is a very good description of the short- to medium-term behavior of financial time series. Yet, it fails miserably in the long term when $\sigma^2 \Delta T \gg 1$, with a variance that grows exponentially fast as seen in Sect. 6.5. Clearly, an ingredient is missing in the basic model, likely corresponding to the natural limitations present in any economic system.

A similar issue is encountered for many financial time series that naturally have a bounded domain. The simplest example is the behavior for interest rates, with a natural range from 0 to a few times 10 % per year. Similar situations occur for spreads, for example, between risk-free and risky interest rates or with credit default swaps (CDS). In order to acknowledge the limited available range for interest rates, most processes used to describe them include a mean-reversion term, similarly to an Ornstein–Uhlenbeck process. A reasonable first order-of-magnitude estimate for the mean reversion time is of the order of 5 to 10 years. Yet, the direct statistical evidences for mean reversion are quite tenuous, due to the long mean-reversion time. Interestingly, none of the canonical processes currently used to describe interest rates include heteroscedasticity (see, e.g., [33]), even though the empirical evidence for the volatility dynamics are overwhelming!

For stocks and stock indexes, a mean reversion term can also be included in the process. Instead of a fixed value, the mean-reversion value can grow (exponentially) in order to describe the time value of money and the inflation. In this way, stocks and stock indices will grow around the mean, while the variance never increases to unrealistic values. A mean-reversion time of the order of 5 to 20 years is realistic and ensures to have well-defined asymptotic properties, while the induced effects on the statistical properties of the returns up to a few months are very difficult to measure.

Very long time horizons are crucial for pension funds and life insurances. These companies have liabilities at the scales of decades, often with embedded optionalities. For example, a standard pension fund contract can include the choice at retirement between a life pension or a lump sum. Essentially, this clause is a call option on interest rates, namely the lump sum is more interesting if the interest rates are high (it is also an option on the pensioner's health). In order to measure their financing ratio and risks, life insurances need to simulate at the scale of a few decades the possible scenarios for the stock markets and the interest rates in order to check that their portfolios can match their liabilities (including the embedded optionalities). At these long time horizons, mean reversion matters. On the analytical side, the available processes at these time horizons do not have a satisfactory behavior. Moreover, it is very difficult to validate, or invalidate, a process against empirical data at the scale of decades. Clearly, much remains to be done for long horizons, including for plain stocks and stock indexes.

References

1. Acerbi, C.: Spectral measures of risk: a coherent representation of subjective risk aversion. J. Bank. Finance **26**, 1505–1518 (2002)
2. Acerbi, C., Tasche, D.: On the coherence of expected shortfall. J. Bank. Finance **26**, 1487–1503 (2002)
3. Aït-Sahalia, Y., Jacod, J.: Testing for jumps in a discretely observed process. Ann. Stat. **37**, 184–222 (2009)
4. Alexander, C.O., Chibumba, A.M.: Multivariate orthogonal factor Garch. University of Sussex. Discussion paper in Mathematics (1997)
5. Almeida, A., Goodhart, C.A.E., Payne, R.G.: The effects of macroeconomic news on high frequency exchange rate behavior. J. Financ. Quant. Anal. **33**(3), 383–408 (1998)
6. Andersen, T.G., Bollerslev, T.: Deutsche Mark–Dollar volatility: intraday activity patterns, macroeconomic announcements, and longer run dependencies. J. Finance **53**(1), 219–265 (1998)
7. Andersen, T.G., Bollerslev, T., Diebold, F.X., Labys, P.: Great realizations. Risk **13**, 105–108 (2000)
8. Andersen, T.G., Bollerslev, T., Diebold, F.X., Labys, P.: Modeling and forecasting realized volatility. Econometrica **71**, 529–626 (2003)
9. Arneodo, A., Muzy, J.-F., Sornette, D.: Causal cascade in the stock market from the "infrared" to the "ultraviolet". Eur. Phys. J. B **2**, 277–282 (1998)
10. Artzner, P., Delbaen, F., Eber, J.-M., Heath, D.: Thinking coherently. Risk Mag. **10**(11), 68–71 (1997)
11. Artzner, P., Delbaen, F., Eber, J.-M., Heath, D.: Coherent measures of risk. Math. Finance **9**(3), 203–228 (1999)
12. Bachelier, L.: Théorie de la spéculation. Ann. Sci. Éc. Norm. Super. **17** (1900)
13. Bacry, E., Delour, J., Muzy, J.-F.: Multifractal random walk. Phys. Rev. E **64**, 26103 (2001)
14. Baillie, R.T., Bollerslev, T., Mikkelsen, H.-O.: Fractionally integrated generalized autoregressive conditional heteroskedasticity. J. Econom. **74**(1), 3–30 (1996)
15. Bandi, F., Russell, J.R.: Separating market microstructure noise from volatility. J. Financ. Econ. **79**, 655–692 (2006)
16. Barndorff-Nielsen, O., Hansen, P.H., Shephard, N.: Designing realized kernels to measure the ex-post variation of equity prices in the presence of noise. Technical report, Stanford University (2006)
17. Bauwens, L., Laurent, S., Rombouts, J.K.V.: Multivariate Garch models: a survey. J. Appl. Econom. **21**, 79–109 (2006)
18. Bernstein, P.L.: Capital Ideas: The Improbable Origins of Modern Wall Street. Wiley, New York (2005)

G. Zumbach, *Discrete Time Series, Processes, and Applications in Finance*,
Springer Finance, DOI 10.1007/978-3-642-31742-2,

19. Biroli, G., Bouchaud, J.-P., Potters, M.: The student ensemble of correlation matrices: eigenvalue spectrum and Kullback–Leibler entropy. Technical report (2007). arXiv:0710.0802
20. Bisig, T., Dupuis, A., Impagliazzo, V., Olsen, R.B.: The Scale of Market Quakes (2009). See also www.olsenscale.com. arXiv:0909.1690
21. Black, F.: Studies of stock price volatility changes. In: Proceedings from the American Statistical Association, Business and Economic Statistics Section, pp. 177–181 (1976)
22. Black, F., Scholes, M.: The pricing of option and corporate liabilities. J. Polit. Econ. **81**, 637–659 (1973)
23. Bollerslev, T.: Modelling the coherence in short-run nominal exchange rates: a multivariate generalized ARCH model. Rev. Econ. Stat. **72**, 498–505 (1990)
24. Bollerslev, T., Engle, R.F., Wooldridge, J.M.: A capital asset pricing model with time-varying covariances. J. Polit. Econ. **96**, 116–131 (1988)
25. Bollerslev, T.: Generalized autoregressive conditional heteroskedasticity. J. Econom. **31**, 307–327 (1986)
26. Borland, L., Bouchaud, J.-P.: On a multi-timescale statistical feedback model for volatility fluctuations. J. Investm. Strat. **1**(1), 65–104 (2011)
27. Bormetti, G., Cisana, E., Montagna, G., Nicrosini, O.: Risk measures with non-Gaussian fluctuations. Technical report (2006). arXiv:physics/0607247
28. Bouchaud, J.-P.: The endogenous dynamics of the markets: price impact and feedback loops. Technical report (2010). arXiv:1009.2928
29. Bouchaud, J.-P.: Lessons from the 2008 crisis. In: The Endogenous Dynamics of Markets: Price Impact, Feedback Loops and Instabilities. Risk Publication (2011)
30. Bouchaud, J.-P., Gefen, Y., Potters, M., Wyart, M.: Fluctuations and response in financial markets: the subtle nature of 'random' price changes. Quant. Finance **4**(2), 176–190 (2004)
31. Bouchaud, J.-P., Matacz, A., Potters, M.: Leverage effect in financial markets: the retarded volatility model. Phys. Rev. Lett. **87**(22), 228701 (2001)
32. Breymann, W., Zumbach, G., Dacorogna, M.M., Müller, U.A.: Dynamical deseasonalization in OTC and localized exchange-traded markets. Internal document WAB.2000-01-31, Olsen & Associates, Seefeldstrasse 233, 8008 Zürich, Switzerland (31 January 2000)
33. Brigo, D., Mercurio, F.: Interest Rate Models—Theory and Practice. Springer, Berlin (2006)
34. Briner, B.G., Connor, G.: How much structure is best? A comparison of market model, factor model and unstructured equity covariance matrices. J. Risk **10**, 3–30 (2008)
35. Campbell, S.D.: A review of backtesting and backtesting procedures. J. Risk **9**(2), 1–17 (2006)
36. Carr, P., Geman, H., Madan, D., Yor, M.: The fine structure of asset returns: an empirical investigation. J. Bus. **75**, 305–332 (2002)
37. Chen, Y., Chou, R., Kuan, C.: Testing time irreversibility without moment restrictions. J. Econom. **95**, 199–218 (2000)
38. Chicheportiche, R., Bouchaud, J.-P.: The fine-structure of volatility feedback. Technical report (2012). Available at http://ssrn.com/abstract=2081675
39. Christie, A.A.: The stochastic behavior of common stock variances: value, leverage and interest rate effects. J. Financ. Econ. **10**(4), 407–432 (1982)
40. Christoffersen, P., Elkamhi, R., Feunou, B., Jacobs, K.: Option valuation with conditional heteroskedasticity and non-normality. Rev. Financ. Stud. **23**(5), 2139–2183 (2010)
41. Chung, C.F.: Estimating the fractionally integrated GARCH model. Unpublished working paper, National Taiwan University, Taipei, TW (1999)
42. Clark, P.K.: A subordinated stochastic process model with finite variance for speculative prices. Econometrica **41**(1), 135–155 (1973)
43. Cont, R.: Empirical properties of asset returns: stylized facts and statistical issues. Quant. Finance **1**(2), 223–236 (2001)
44. Cont, R., Tankov, P.: Financial Modelling with Jump Processes. Chapman & Hall/CRC, London/Boca Raton (2004)
45. Corradi, V.: Reconsidering the continuous time limit of the GARCH(1, 1) process. J. Econom. **96**, 145–153 (2000)

46. Corsi, F., Zumbach, G., Müller, U.A., Dacorogna, M.M.: Consistent high-precision volatility from high-frequency data. Econ. Notes, Rev. Bank. Finance Monet. Econ. **30**(2), 183–204 (2001)
47. Curci, G., Corsi, F.: Discrete sine transform for multiscale realized volatility measures. Quant. Finance (2011)
48. Dacorogna, M.M., Gençay, R., Müller, U.A., Olsen, R.B., Pictet, O.V.: An Introduction to High-Frequency Finance. Academic Press, San Diego (2001)
49. Dacorogna, M.M., Müller, U.A., Nagler, R.J., Olsen, R.B., Pictet, O.V.: A geographical model for the daily and weekly seasonal volatility in the FX market. J. Int. Money Finance **12** (1993)
50. Dacorogna, M.M., Müller, U.A., Olsen, R.B., Pictet, O.V.: Modelling short-term volatility with GARCH and HARCH models. In: Dunis, C., Zhou, B. (eds.) Nonlinear Modelling of High Frequency Financial Time Series, pp. 161–176. Wiley, Chichester (1998)
51. Dacorogna, M.M., Müller, U.A., Pictet, O.V., de Vries, C.G.: Extremal Forex returns in extremely large data sets. Extremes **4**(2), 105–127 (2001)
52. Ding, Z., Granger, C.: Modeling volatility persistence of speculative returns: a new approach. J. Econom. **73**, 185–215 (1996)
53. Ding, Z., Granger, C.W.J., Engle, R.F.: A long memory property of stock market returns and a new model. J. Empir. Finance **1**, 83–106 (1993)
54. Drost, F.C., Nijman, T.E.: Temporal aggregation of GARCH processes. Econometrica **61**, 909–927 (1993)
55. Duan, J.-C.: The GARCH option pricing model. Math. Finance **5**, 13–32 (1995)
56. Engle, R.: Dynamic conditional correlation: a simple class of multivariate generalized conditional heteroskedasticity models. J. Bus. Econ. Stat. **20**, 339–350 (2002)
57. Engle, R., Ng, V.: Measuring and testing the impact of news and volatility. J. Finance **48**, 1749–1778 (1993)
58. Engle, R., Ng, V., Rothschild, M.: Asset pricing with a factor ARCH covariance structure: empirical estimates for treasury bills. J. Econom. **45**, 213–238 (1990)
59. Engle, R.F.: Autoregressive conditional heteroskedasticity with estimates of the variance of U.K. inflation. Econometrica **50**, 987–1008 (1982)
60. Engle, R., Shephard, N., Sheppard, K.: Fitting vast dimensional time-varying covariance models. Preprint (2008)
61. Engle, R.F., Bollerslev, T.: Modelling the persistence of conditional variances. Econom. Rev. **5**, 1–50 (1986)
62. Engle, R.F., Ito, T., Lin, W.-L.: Meteor showers or heat waves? Heteroskedastic intra-daily volatility in the foreign exchange market. Econometrica **58**, 525–542 (1990)
63. Fama, E.F., French, K.R.: Common risk factors in the returns of stocks and bonds. J. Financ. Econ. **33**, 3–56 (1993)
64. Fama, E.F.: Mandelbrot and the stable Paretian distribution. J. Bus. **36**, 420–429 (1965)
65. Farmer, J.D., Gerig, A., Lillo, F., Mike, S.: Market efficiency and the long-memory of supply and demand: is price impact variable and permanent or fixed and temporary. Quant. Finance **2**(6), 107–112 (2006)
66. Feller, W.: An Introduction to Probability Theory and Its Applications, 2nd edn. Wiley Series in Probability and Mathematical Statistics, vol. II. Wiley, New York (1971)
67. Fergusson, K., Platen, E.: On the distributional characterization of daily log-returns of a world stock index. Appl. Math. Finance **13**(1), 19–38 (2006)
68. Fillol, J.: Multifractality: theory and evidence an application to the French Stock Market. Econ. Bull. **3**(31), 1–12 (2003)
69. Fisher, A.J., Calvet, L.E., Mandelbrot, B.B.: Multifractality of Deutschemark US dollar exchangerates. Cowles Found. Discuss. Pap. **1166**, 1–40 (1997)
70. Fong, W.M.: Time irreversibility tests of volume-volatility dynamics for stock returns. Econ. Lett. **81**, 39–45 (2003)
71. Frisch, U.: Turbulence, the Legacy of A.N. Kolmogorov. Cambridge University Press, Cambridge (1995)

72. Gabaix, X.: A theory of power law distributions in financial market fluctuations. Nature **423**, 267–270 (2003)
73. Gatheral, J.: Developments in volatility derivatives pricing. Presentation at "Global derivative", Paris, May 23 (2007)
74. Ghashghaie, S., Breymann, W., Peinke, J., Talkner, P., Dodge, Y.: Turbulent cascades in foreign exchange markets. Nature **381**, 767 (1996)
75. Glosten, L.R., Jagannathan, R., Runkle, D.E.: On the relation between the expected value and the volatility of the nominal excess return on stocks. J. Finance **48**, 1779–1801 (1993)
76. Gnanadeskian, R., Kettenring, J.R.: Robust estimates, residuals, and outlier detection with multiresponse data. Biometrika **28**, 81–124 (1972)
77. Granger, C., Ding, Z.: Varieties of long memory models. J. Econom. **73**, 61–77 (1996)
78. Hansen, P.R., Lunde, A.: Realized variance and market microstructure noise. J. Bus. Econ. Stat. **24**, 127–161 (2006)
79. Hauksson, H., Dacorogna, M.M., Domenig, T., Müller, U.A.: Multivariate extremes, aggregation and risk estimation. Quant. Finance **1**(1), 79–95 (2001)
80. Heston, S., Nandi, S.: A closed-form GARCH option pricing model. Rev. Financ. Stud. **13**, 585–626 (2000)
81. Hill, B.M.: A simple general approach to inference about the tail of a distribution. Ann. Stat. **3**(5), 1163–1173 (1975)
82. Hosking, J.R.M.: L-moments: analysis and estimation of distributions using linear combinations of order statistics. J. R. Stat. Soc. B **52**(1), 105–124 (1990)
83. Hull, J.C.: Options, Futures and Other Derivative Securities. Prentice-Hall International, Englewood Cliffs (1993)
84. Kim, Y.S., Rachev, S.T., Bianchi, M.L., Fabozzi, F.J.: Financial market models with levy processes and time-varying volatility. J. Bank. Finance **32**, 1363–1378 (2008)
85. Laloux, L., Cizeau, P., Bouchaud, J.-P., Potter, M.: Noise dressing of financial correlation matrices. Phys. Rev. Lett. **83**, 1467 (1999)
86. Ledoit, O., Wolf, M.: Honey, I shrunk the sample covariance matrix. J. Portf. Manag. **4**(30), 110–119 (2004)
87. Ledoit, O., Wolf, M.: Improved estimation of the covariance matrix of stock returns with an application to portfolio selection. J. Empir. Finance **10**, 603–621 (2004)
88. Linden, M.: A model for stock return distribution. Int. J. Finance Econ. **6**(2), 159–169 (2001)
89. Lindskog, F.: Linear correlation estimation. Technical report RiskLab, ETH, Zürich, Switzerland (2000)
90. Liu, Y., Cizeau, P., Meyer, M., Peng, C.-K., Stanley, H.E.: Correlation in economic time series. Physica A **245**, 437 (1997)
91. Lux, T.: Detecting multi-fractal properties in asset returns: the failure of the scaling estimator. Universitat Kiel (2003)
92. Lynch, P., Zumbach, G.: Market heterogeneities and the causal structure of the volatility. Quant. Finance **3**, 320–331 (2003)
93. Maillet, B., Michel, T.: An index of market shocks based on multiscale analysis. Quant. Finance **3**(2), 88–97 (2003)
94. Mandelbrot, B.B.: Fractals and Scaling in Finance. Springer, Berlin (1997)
95. Mandelbrot, B.B.: The variation of certain speculative prices. J. Bus. **36**, 394 (1963)
96. Mandelbrot, B.B., Fisher, A.J., Calvet, L.E.: A multifractal model of asset returns. Cowles Found. Discuss. Pap. **1164**, 1–44 (1997)
97. Mantegna, R.N., Stanley, H.E.: An Introduction to Econophysics—Correlations and Complexity in Finance. Cambridge University Press, Cambridge (2000)
98. Mantegna, R.N., Stanley, H.E.: Turbulence and financial markets. Nature **383**, 587–588 (1996)
99. Mantegna, R.N., Stanley, H.E.: Stock market dynamics and turbulence: parallel analysis of fluctuation phenomena. Physica A **239**, 255–266 (1997)
100. Marchenko, V.A., Pastur, L.A.: Distribution of eigenvalues for some sets of random matrices. Math. USSR Sb. **1**, 457–483 (1967)

101. Matia, K., Ashkenazy, Y., Stanley, H.E.: Multifractal properties of price fluctuations of stocks and commodities. Europhys. Lett. **61**, 422 (2003)
102. Di Matteo, T.: Multi-scaling in finance. Quant. Finance **7**(1), 21–36 (2007)
103. Di Matteo, T., Aste, T., Dacorogna, M.M.: Long-term memories of developed and emerging markets: using the scaling analysis to characterize their stage of development. J. Bank. Finance **29**, 827–851 (2005)
104. McAleer, M., Medeiros, M.C.: Realized volatility: a review. Econom. Rev. **27**, 10–45 (2008)
105. Menn, C., Rachev, S.T.: A GARCH option pricing model with α-stable innovations. Eur. J. Oper. Res. **163**, 201–209 (2005)
106. Merton, R.C.: Theory of rational option pricing. Bell J. Econ. Manag. Sci. **4**, 141–183 (1973)
107. Merton, R.C.: On estimating the expected return on the market: an exploraty investigation. J. Financ. Econ. **8**, 323–361 (1980)
108. Meucci, A.: Risk and Asset Allocation. Springer Finance (2007)
109. Mina, J., Xiao, J.: Return to riskmetrics: the evolution of a standard. Technical report, Risk Metrics Group (2001). Available at http://www.riskmetrics.com/pdf/rrmfinal.pdf
110. Morgan, J.P.: RiskMetrics technical document. Technical report, J.P. Morgan and International marketing (1996)
111. Moro, E., Vicente, J., Mayano, L.G., Gerig, A., Farmer, J.D., Vaglica, G., Lillo, F., Mantegna, R.N.: Market impact and trading profile of hidden orders in stock markets. Phys. Rev. E **6**(80), 1–8 (2009)
112. Muzy, J.-F., Delour, J., Bacry, E.: Modelling fluctuations of financial time series: from cascade process to stochastic volatility model. Eur. Phys. J. B **17**, 537–548 (2000)
113. Muzy, J.F., Bacry, E.: Multifractal stationary random measures and multifractal random walks with log-infinitely divisible scaling laws. Phys. Rev. E **66**, 56121 (2002)
114. Nelson, D.B.: Conditional heteroskedasticity in asset returns: a new approach. Econometrica **59**, 347–370 (1991)
115. Nelson, D.B.: Stationarity and persistence in the GARCH(1, 1) model. Econom. Theory **6**, 318–334 (1990)
116. O'Neil, C., Zumbach, G.: Using relative returns to accommodate fat tailed innovations in processes and option pricing. Technical report, Riskmetrics group (2009). Quant. Finance (to appear). Available at www.ssrn.com
117. Pafka, S., Kondor, I.: Evaluating the riskmetrics methodology in measuring volatility and value-at-risk in financial markets. Physica A **299**, 305–310 (2001)
118. Pafka, S., Kondor, I., Potters, M.: Exponential weighting and random matrix theory based filtering of financial covariance matrices. Technical report (2004)
119. Phillips, P.C.B., Yu, J.: Discussion of Hansen & Lunde. J. Bus. Econ. Stat. **24**, 202–208 (2006)
120. Pictet, O.V., Dacorogna, M.M., Müller, U.A.: Hill, bootstrap and jackknife estimators for heavy tails. In: Taqqu, M. (ed.) A Practical Guide to Heavy Tails: Statistical Techniques for Analysing Heavy Tailed Distributions, pp. 283–310. Birkhäuser, Berlin (1998)
121. Plerou, V., Gopikrishnan, P., Rosenow, B., Amaral, L.A.N., Guhr, T., Stanley, H.E.: A random matrix approach to cross-correlations in financial data. Phys. Rev. E **65**, 066126 (2002)
122. Plerou, V., Gopikrishnan, P., Rosenow, B., Amaral, L.A.N., Stanley, H.E.: Universal and non-universal properties of cross-correlations in financial time series. Phys. Rev. Lett. **83**, 1471 (1999)
123. Pomeau, Y.: Symétrie des fluctuations dans le renversement du temps. J. Phys. **43**, 859 (1982)
124. Poon, S.-H.: A Practical Guide to Forecasting Financial Market Volatility. Wiley, London (2005)
125. Poon, S.-H., Granger, C.W.J.: Forecasting volatility in financial markets. J. Econ. Lit. **XLI**, 478–539 (2003)
126. Potters, M., Cont, R., Bouchaud, J.-P.: Financial markets as adaptative ecosystems. Europhys. Lett. **41**, 239 (1998)
127. Potters, M., Bouchaud, J.-P.: Theory of Financial Risk and Derivative Pricing. Cambridge University Press, Cambridge (2003)

128. Potters, M., Bouchaud, J.-P., Laloux, L.: Financial applications of random matrix theory: old laces and new pieces. Acta Phys. Pol. B **36**, 2767 (2005)
129. Press, W., Teukolsky, S., Vetterling, W., Flannery, B.: Numerical Recipes in C++. Cambridge University Press, Cambridge (2002)
130. Ramsey, J.B., Rothman, P.: Characterization of the time irreversibility of economic time series: estimators and test statistics. Working paper #88-39, C.V. Starr Center for Applied Economics, NYU (1988)
131. Ramsey, J.B., Rothman, P.: Time irreversibility and business cycle asymmetry. J. Money Credit Bank. **28**(1), 1–21 (1996)
132. Richter, C.F.: Elementary Seismology vol. 1. Freeman, San Francisco (1958)
133. Roll, R.: A simple implicit measure of the effective bid-ask spread in an efficient market. J. Finance **39**(4), 1127–1139 (1984)
134. Rosenbaum, M.: A new microstructure noise index. Quant. Finance **11**(6), 883–899 (2011)
135. Rosenblatt, M.: Remarks on multivariate transformation. Ann. Math. Stat. **23**, 1052–1057 (1952)
136. Sentana, E.: Quadratic ARCH models. Rev. Econ. Stud. **62**(4), 639 (1995)
137. Shephard, N.: Stochastic Volatility. Oxford University Press, Oxford (2003)
138. Silvennoinen, A., Teräsvirta, T.: Multivariate GARCH models. In: Andersen, T.G., Davies, R.A., Kreiss, J.-P., Mikosch, T. (eds.) Handbook of Financial Time Series. Springer, Berlin (2008)
139. Starica, C.: Estimation of the extreme value distribution for constant conditional correlation models. Proceedings of the HFDF-II Conference in Zurich **2**, 1–30 (1998)
140. Taylor, S.J.: Modelling Financial Time Series, 2nd edn. World Scientific, Singapore (2008)
141. Teyssière, G.: Double long-memory financial time series. QMW Working paper 348, University of London, UK (1996)
142. Tsay, R.S.: Analysis of Financial Time Series, 2nd edn. Wiley, New York (2005)
143. Wishart, J.: The generalized product moment distribution in samples from a normal multivariate population. Biometrika A **20**, 32–43 (1928)
144. Wyart, M., Bouchaud, J.-P., Kockelkoren, J., Potters, M., Vettorazzo, M.: Relation between bid-ask spread, impact and volatility in order driven markets. Quant. Finance **8**(1), 41–57 (2008)
145. Zangari, P.: Equity risk factor models. In: Litterman, B. (ed.) Modern Investment Management. Wiley, New York (2003)
146. Zhang, L.: Efficient estimation of stochastic volatility using noisy observations: a multi-scales approach. Bernoulli **12**, 1019–1043 (2006)
147. Zhang, L., Mykland, P.A., Aït-Sahalia, Y.: A tale of two time scales: determining integrated volatility with noisy high frequency data. J. Am. Stat. Assoc. **100**, 1394–1411 (2005)
148. Jianga, Z.-Q., Zhou, W.-X.: Multifractality in stock indexes: fact or fiction? Physica A **387**(14), 3605–3614 (2008)
149. Zhou, B.: High-frequency data and volatility in foreign-exchange rates. J. Bus. Econ. Stat. **14**(1), 45–52 (1996)
150. Zumbach, G.: The Pitfalls in Fitting GARCH Processes. Advances in Quantitative Asset Management. Kluwer Academic, Dordrecht (2000)
151. Zumbach, G.: How the trading activity scales with the company sizes in the FTSE 100. Quant. Finance **4**, 441–456 (2004)
152. Zumbach, G.: Volatility processes and volatility forecast with long memory. Quant. Finance **4**, 70–86 (2004)
153. Zumbach, G.: A gentle introduction to the RM2006 methodology. Technical report, Riskmetrics Group (2006). Available at http://www.riskmetrics.com and www.ssrn.com
154. Zumbach, G.: The riskmetrics 2006 methodology. Technical report, RiskMetrics Group (2006). Available at: www.riskmetrics.com and www.ssrn.com
155. Zumbach, G.: Back testing risk methodologies from one day to one year. J. Risk **9**(2), 55–91 (2007)
156. Zumbach, G.: Time reversal invariance in finance. Quant. Finance **9**, 505–515 (2009)

157. Zumbach, G.: Volatility conditional on price trends. Quant. Finance **10**, 431–442 (2010)
158. Zumbach, G.: Volatility forecasts and the at-the-money implied volatility: a multi-components ARCH approach and its relation with market models. Quant. Finance **11**, 101–113 (2010)
159. Zumbach, G.: Characterizing heteroskedasticity. Quantitative Finance (2011)
160. Zumbach, G.: The empirical properties of large covariance matrices. Quant. Finance **11**, 1091–1102 (2011)
161. Zumbach, G.: The statistical properties of the innovations in multivariate ARCH processes in high dimensions. Quantitative Finance (2011)
162. Zumbach, G., Corsi, F., Trapletti, A.: Efficient estimation of volatility using high frequency data. Technical report (2002). Available at http://papers.ssrn.com/sol3/papers.cfm?abstract_id=306002
163. Zumbach, G., Dacorogna, M.M., Olsen, J.L., Olsen, R.B.: Market crises, shock of the new. Risk Mag. **13**(3), 110–114 (2000)
164. Zumbach, G., Dacorogna, M.M., Olsen, J.L., Olsen, R.B.: Measuring shock in financial markets. Int. J. Theor. Appl. Finance **3**(3), 347–355 (2000)
165. Zumbach, G., Fernández, L.: Fast and realistic European ARCH option pricing and hedging. Technical report, Swissquote Bank (2011, submitted). Available at www.ssrn.com
166. Zumbach, G., Fernández, L.: Option pricing with realistic ARCH processes. Technical report, Swissquote Bank (2011, submitted). Available at www.ssrn.com
167. Zumbach, G., Fernández, L., Weber, C.: Realistic processes for stocks from one day to one year. Technical report, Swissquote Bank (2010, submitted). Available at www.ssrn.com
168. Zumbach, G., Finger, C.: An historical perspective on market risks using the DJIA index over one century. Wilmott Magazine (2010)
169. Zumbach, G., Lynch, P.: Heterogeneous volatility cascade in financial markets. Physica A **298**(3–4), 521–529 (2001)
170. Zumbach, G., Müller, U.A.: Operators on inhomogeneous time series. Int. J. Theor. Appl. Finance **4**(1), 147–178 (2001)
171. Zumbach, G., Pictet, O., Masutti, O.: Genetic Programming with Syntactic Restrictions Applied to Financial Volatility Forecasting. Computational Methods in Decision-Making, Economics and Finances. Kluwer Academic, Dordrecht (2002)

Index

G. Zumbach, *Discrete Time Series, Processes, and Applications in Finance*,
Springer Finance, DOI 10.1007/978-3-642-31742-2,

MIX
Papier aus verantwortungsvollen Quellen
Paper from responsible sources
FSC® C105338

If you have any concerns about our products,
you can contact us on
ProductSafety@springernature.com

In case Publisher is established outside the EU,
the EU authorized representative is:
Springer Nature Customer Service Center GmbH
Europaplatz 3, 69115 Heidelberg, Germany

Printed by Libri Plureos GmbH
in Hamburg, Germany